BEITRÄGE ZUR REGIONALEN GEOLOGIE DER ERDE BAND 17

BEITRÄGE ZUR REGIONALEN GEOLOGIE DER ERDE

Herausgegeben von Prof. Dr. F. Bender, Hannover,
Prof. Dr. V. Jacobshagen, Berlin, Prof. Dr. J. D. de Jong,
Bennekom/Niederlande, und Prof. Dr. G. Lüttig, Erlangen
Band 17

Victor E. Khain

GEOLOGY OF THE USSR

First part

Old cratons and Paleozoic fold belts

1985

GEBRÜDER BORNTRAEGER · BERLIN · STUTTGART

GEOLOGY OF THE USSR

First part

Old cratons and Paleozoic fold belts

by

Victor E. Khain

With 57 figures in the text

1985

GEBRÜDER BORNTRAEGER · BERLIN · STUTTGART

Address of the author:

Professor Dr. V. E. Khain
Geological Institute
USSR Academy of Sciences
Pyjevski per., 7
USSR-109017 Moscow 17

Preface

Having received an offer from Gebrüder Borntraeger Publishing House Berlin–Stuttgart to write a book on the geology of the USSR for the series "Beiträge zur regionalen Geologie der Erde", I accepted this offer willingly, being aware of the fact that there is no recent work on the geology of the Soviet Union published outside that country, and that many scientists are unable to use the extensive Soviet literature for a lack of knowledge of the Russian language. The classical work by NALIVKIN, "Geologie of the USSR", which was first published in 1960, was the last significant review of the geology of the USSR that was translated into English. Since then, new data, which change the old concepts in many respects, were obtained for practically all regions of the Soviet Union. Another circumstance is of no less importance: the structure of the fold system regions of the Soviet Union is now viewed in the light of fundamentally new ideas of theoretical geology; the major role of overthrust nappes has been recognized; ophiolitic complexes, which are regarded as relics of the oceanic crust, have been studied in detail; paleomagnetic determinations furnished evidence of appreciable relative horizontal movements of ancient cratons.

The author, discussing the geology of the vast territory of the USSR, places the main emphasis on its tectonic structure, paleotectonics, and the sequence of the main lithological formations. Stratigraphic material and, in particular, paleontological proof of the age of stratigraphic subdivision, is not given in the book, lest it would make the book unwieldy. The reader is, therefore, referred to the extensive editions of the "Geology of the USSR" and especially "Stratigraphy of the USSR", where exhaustive information on this subject can be found. As for paleogeography, we advise the reader to refer to the four-volume "Atlas of Lithologic-Paleogeographic Maps of the USSR" and to the explanatory notes for the Atlas, "Paleogeography of the USSR", also published in four issues. The reader will find a list of the principal general works on USSR geology, published during the last 15 to 20 years.

In this description of fold systems, situated near Soviet borders with other countries, the author found it appropriate, for the completeness of the discussion, to follow their continuations (or intermediate links) on the territories of adjacent countries.

I would like to conclude this preface by expressing my appreciation of the initiative in publication of the book to Professors H. R. von GAERTNER and G. LÜTTIG, as well as to Dr. E. NÄGELE. I hope that this book will contribute to familiarizing a wide circle of non-Soviet geologists with the principal structural features of the vast and varied territory of the Soviet Union, and will serve them as a basis for a more profound appraisal of the Soviet geological literature.

Moscow, March 1984   V. E. KHAIN

Contents

Preface V

1. Principal structural elements of the USSR territory 1
2. The ancient Eastern European Craton 6
   2.1 Boundaries and main structural elements 6
   2.2 The Baltic Shield 10
   2.3 The Ukrainian Shield and the Voronezh Massif 27
   2.4 The Russian Platform 35
       2.4.1 Inner structure of the basement 35
       2.4.2 The structure of the sedimentary cover 38
   2.5 Main stages of evolution 48
       2.5.1 The Archaean stage 48
       2.5.2 The Early Proterozoic (Karelian) stage 49
       2.5.3 The Middle Proterozoic (Svecofennian) stage 50
       2.5.4 The beginning of the Late Proterozoic (Gothian) stage 50
       2.5.5 The middle and the end of the Late Proterozoic (Dalslandian and Early
             Baikalian stages) 51
       2.5.6 The Late Baikalian−Caledonian stage (Vendian−Early Devonian) 53
       2.5.7 The Hercynian stage (Middle Devonian−Middle Triassic) 55
       2.5.8 The Cimmerian and Alpine stages (Late Triassic−Quaternary period) 60
   2.6 Conclusions on the history and structure of the Eastern European Craton 62
3. The Timan−Pechora epi-Baikalian Platform 67
   3.1 Boundaries and principal structural subdivisions 67
   3.2 The Kanin−Timan Ridge and the nature of the basement of the Timan−Pechora
       Platform 69
   3.3 The Pechora Syneclise 72
   3.4 Main stages of evolution 73
       3.4.1 Dalslandian−Baikalian stage 74
       3.4.2 Caledonian stage (Ordovician−Late Devonian) 74
       3.4.3 The Middle Devonian and early Late Devonian 75
       3.4.4 Hercynian stage (Late Devonian−Early Permian) 75
       3.4.5 Late Hercynian−Early Cimmerian stage (Middle Permian−Triassic) 76
       3.4.6 Late Cimmerian−Alpine stage (Jurassic−Paleogene) 76
       3.4.7 Late Alpine stage (Neogene−Quaternary) 77
4. The ancient Siberian Craton 79
   4.1 Boundaries and principal subdivisions of the Siberian Craton 79
   4.2 Aldan Shield, Anabar Massif, and general structure of craton basement 82
   4.3 Structure of the sedimentary cover of the Mid-Siberian Platform 94
   4.4 Main stages of craton evolution 102
       4.4.1 Archaean stage 102
       4.4.2 Early Proterozoic stage 103

      4.4.3 Middle Proterozoic stage 104
      4.4.4 Late Proterozoic stage including Baikalian 104
      4.4.5 Caledonian stage (Cambrian to Middle Devonian) 107
      4.4.6 Hercynian stage (Late Devonian to Middle Triassic) 112
      4.4.7 Late Cimmerian stage (Late Triassic to Early Cretaceous) 113
   4.5 Some conclusions on the history and structure of the Siberian Craton 115
5. Taimyr–Severnaya Zemlya fold region 118
   5.1 General features of the relief and main views on tectonics of the region 118
   5.2 The Kara Massif 120
   5.3 The Northern Severnaya Zemlya zone 123
   5.4 The Central Taimyr zone 124
   5.5 The Southern Taimyr zone 125
   5.6 The Yenisei–Khatanga Trough 125
   5.7 Main stages of evolution 126
      5.7.1 Early Precambrian stage 126
      5.7.2 Late Precambrian stage 127
      5.7.3 Caledonian and Early Hercynian stage (Cambrian to Early Carboniferous). 127
      5.7.4 Late Hercynian–Early Cimmerian stage (Middle Carboniferous to Late
             Triassic) 128
      5.7.5 Alpine stage (Jurassic to Quaternary) 129
   5.8 Some deductions 130
6. Hercynian geosynclinal fold system of the Urals 132
   6.1 Principal features of the relief, state of geological knowledge and general structural
      zonality 132
      6.1.1 Zone of Uralian foredeeps 135
      6.1.2 The Frontal Bashkirian Uplift 136
      6.1.3 The Western Uralian zone 136
      6.1.4 The Central Uralian Uplift zone (Anticlinorium or Meganticlinorium) . . . 137
      6.1.5 The Tagil–Magnitogorsk zone (Synclinorium, Megasynclinorium) 138
      6.1.6 The Eastern Uralian Uplift zone (Anticlinorium or Meganticlinorium) . . . 139
      6.1.7 The Eastern Uralian Synclinorium zone 139
      6.1.8 The Transuralian Uplift zone 140
      6.1.9 The Tyumen'–Kustanai zone (Synclinorium) 141
   6.2 The Southern Urals 141
   6.3 The Middle Urals 146
   6.4 The Northern Urals 150
   6.5 The Polar Urals 152
   6.6 Pai Khoi and Novaya Zemlya, structure of the Barents Sea 155
   6.7 Main stages of evolution 160
      6.7.1 Pre-Middle Riphean history 160
      6.7.2 The Grenville and Baikalian stages (Middle and Late Riphean, Vendian, and
             Early Cambrian) 161
      6.7.3 The Caledonian stage (Late Cambrian–Ordovician–Early Devonian) . . . 164
      6.7.4 The Hercynian geosynclinal stage (Middle Devonian–Early
             Carboniferous) 167
      6.7.5 The Hercynian orogenic stage (Middle Carboniferous–Early Triassic) . . . 168
      6.7.6 The Platform stage (Jurassic–Eocene) 171
      6.7.7 The Neotectonic stage (Oligocene–Quaternary) 171
   6.8 Some conclusions and problems 172
7. Paleozoic geosynclinal fold system of Central Kazakhstan and Northern Tien Shan 174
   7.1 Boundaries and main structural subdivisions, state of geological knowledge 174

7.2   The Kokchetav–Muyunkum Median Massif   178
7.3   The Ishim–Talas zone of Early Caledonides   184
7.4   The Karatau–Naryn Caledonian–Hercynian fold zone   185
7.5   The Turgai–Syr Darya Median Massif   187
7.6   The Yerementau–Chuili Caledonian eugeosynclinal fold system.   188
7.7   The Chinghiz–Tarbagatai Late Caledonian and Hercynian fold system   191
7.8   The Dzhungaro–Balkhash Hercynian fold system   191
7.9   The platform structure of Central Kazakhstan and the orogenic structure of
       Northern Tien Shan   198
7.10   Main stages of evolution   199
       7.10.1   Pre-Baikalian history   199
       7.10.2   The Late Archaean stage   201
       7.10.3   The Early Proterozoic (Karelian) stage.   201
       7.10.4   The Middle Proterozoic–Early Riphean stage   202
       7.10.5   The Middle Riphean (Grenvillian) stage   202
       7.10.6   The Baikalian stage (Late Riphean–Vendian)   203
       7.10.7   The Caledonian stage (Vendian–Middle Devonian)   203
       7.10.8   The Hercynian geosynclinal stage (second half of the Devonian–first half
                of the Early Carboniferous)   212
       7.10.9   The Hercynian early orogenic stage (end of Early Carboniferous–begin-
                ning of the Permian)   213
       7.10.10 The Hercynian late orogenic stage (Permian period)   214
       7.10.11 The Cimmerian–Early Alpine platform stage (Triassic–Paleogene)   215
       7.10.12 The Late Alpine stage of orogenic reactivation (Neogene–Quaternary) . .   216
7.11   Structural and evolutional features of the region   217
8. Southern Tien Shan Hercynian geosynclinal fold system   221
8.1   Boundaries, principal structural subdivisions, geological ooverage.   221
8.2   The Central (Gissar–Alai) segment of Southern Tien Shan   224
8.3   The western (Kyzyl Kum) segment of Southern Tien Shan and its relationship with
       the Urals   228
8.4   The eastern segment of Southern Tien Shan   231
8.5   Recent structure of Southern Tien Shan. The Ferghana Intermontane Basin   234
8.6   Main stages of evolution   237
       8.6.1   Pre-geosynclinal and early geosynclinal (Late Riphean-Early Carbonifer-
               ous) history   237
       8.6.2   The early orogenic stage (end of the Early to Middle Carboniferous)   239
       8.6.3   The late orogenic stage (Late Carboniferous–Middle Triassic).   240
       8.6.4   The platform stage (Late Triassic–Early Paleogene)   241
       8.6.5   Stage of orogenic reactivation (Oligocene–Quaternary)   242
8.7   Conclusions.   243
References   245
Author Index   261
Subject Index   264

1. Principal structural elements of the USSR territory

The present continental structure of the Soviet Union was shaped around two centers, fragments of the Middle Precambrian Megagea: the Eastern European and Siberian cratons. The Siberian Craton is situated entirely within Soviet territory. The Eastern European Craton continues in the west to Finland, Sweden, Norway, Denmark and Poland. Both cratons have basements of Early Precambrian, mainly Archaean rocks, more than 2.4 billion years old in the Siberian Craton, and partly also Lower Proterozoic rocks, up to 1.7 billion years old, in the Eastern European Craton. The two cratons took shape in the Late Precambrian (Middle and Late Riphean), when the Uralo–Okhotsk (Uralo–Mongolian, according to MURATOV, 1976), Mediterranean, Pacific, and Arctic geosynclinal belts evolved. The Uralo–Okhotsk Belt separated the Eastern European from the Siberian Craton; its latitudinal eastern part, sometimes identified as an independent Central Asian belt, separated the Siberian and Sinian (Sino–Korean) cratons. The Mediterranean Belt separated the Eastern European Craton from the African (African–Arabian) Craton, more specifically, from the Gondwana Supercraton. The Pacific Belt defined the eastern edge of the Siberian Craton; the northern edge of this craton was marked by formation of the Arctic Belt which separated the Siberian Craton from the Hyperborean Craton (the region of the De Long Islands and the Chukchi oceanic plateau northeast of the Novosibirsk Archipelago are relics of the latter).

These geosynclinal belts (paleooceans) in places include fragments of the ancient continental crust whose breakdown and spreading apart produced them. These fragments are median masses (microcontinents).

These geosynclinal belts later underwent closure and consolidation, usually starting from the periphery and, in some cases, from the framing of major median massifs such as the Kokchetav–Muyunkum in Central Kazakhstan. This resulted in the accretion of increasingly young fold systems, from Baikalides to Alpides, to the Early Precambrian cratons. This accretion was not, however, uniform, since the direct or almost direct neighbours of the old cratons are fold systems of various, down to Alpine, age (the Carpathians).

The Siberian Craton in the northwest, west, southwest, and south, and the Eastern European Craton in the northeast are surrounded by zones which consolidated at the very end of the Precambrian, during the Baikalian epoch of tectogenesis. These are, respectively, the Taimyr–Severnaya Zemlya region, the Yenisei Range, the Eastern Sayan, and the Baikal–Patom Upland in Siberia, and the Timan–Pechora region in the European USSR (Fig. 1).

The Baikalian framing of the Siberian Craton is superseded by the zone of Salairian, i. e., Late Cambrian consolidation, as one proceeds into the Uralo–Okhotsk Belt. This zone includes (within the USSR) the Kuznetsk Alatau, the basement of the Minusa Basin, the southwestern slope of the Eastern Sayan, the northern periphery of the Tuva-

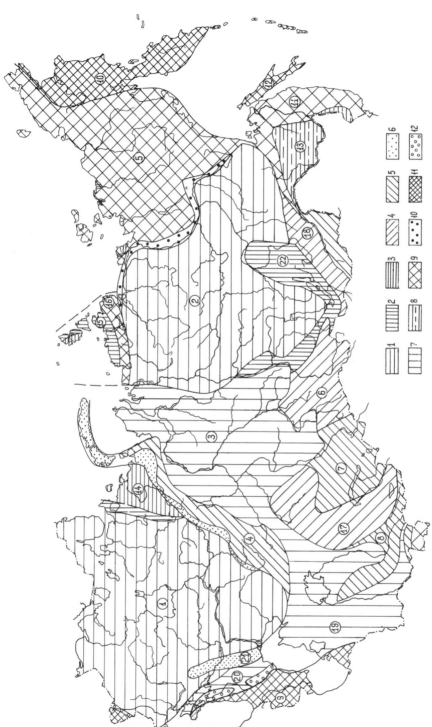

Fig. 1. General structure of the USSR territory. 1 = old cratons; 2 = Baikalian fold systems; 3 = Epibaikalian platforms; 4 = Salairian and Caledonian fold systems; 5 = Hercynian fold systems; 6 = Hercynian foredeeps; 7 = Epihercynian Platforms (including segments with older or younger basement); 8 = Precambrian basement reworked in Paleozoic and Mesozoic times; 9 = Mesozoic (Cimmerian) fold systems; 10 = Cimmerian foredeeps; 11 = Alpine fold systems; 12 = Alpine foredeeps. Encircled numbers: 1 = East European Craton; 2 = Siberian Craton; 3 = West Siberian Platform; 4 = Uralian system; 5 = Verkhoyansk–Chukotka region; 6 = Altai–Sayan region; 7 = Dzhungaria–Balkhash region; 8 = Tien Shan system; 9 = Caucasus; 10 = Koryakia–Kamchatka system; 11 = Sikhote–Alin system; 12 = Sakhalin; 13 = Bureya Massif; 14 = Timan–Pechora Platform; 15 = N. Taimyr; 16 = S. Taimyr; 17 = Central Kazakhstan region; 18 = Transbaikalian region; 19 = Turanian Platform; 20 = Scythian Platform; 21 = Donets–Caspian fold system; 22 = East Sayan–Peribaikalian region.

Mongolian Median Massif, the Dzhida- and Uda-Vitim zones in Western Transbaikalia. Further west, the Salairides in the same Altai–Sayan region are surrounded by the Caledonides of the Western Sayan and Altai mountains, in Transbaikalia directly by the Daurian Hercynides.

Central Kazakhstan and the Northern Tien Shan are another region in the Uralo–Okhotsk Belt, where the Caledonides are widely developed. They occur in the framing of the Precambrian median Kokchetav–Muyunkum Massif.

The Hercynides play a major role in the structure of the Uralo–Okhotsk Belt. They primarily include the Uralian system whose eastern part and southern part are buried beneath a cover of Meso-Cenozoic platform sediments. The Southern Tien Shan, the Dzhungaro–Balkhash zone in the southern part of Central Kazakhstan, and the Rudny Altai are of the same age. The Ob'–Zaisan system, which occupies an axial position in the structure of the western Uralo–Siberian segment of the Uralo–Okhotsk Belt, belongs to the Late Hercynides. Its southern continuation runs into Mongolia, and the northern continuation plunges under the cover of the Western Siberian Platform where, as some investigators believe, it reaches the Yamal Peninsula. The fold zone of the Byrranga Range in the south of Taimyr is of Late Hercynian or, more likely, Early Cimmerian age.

The Hercynian tectogenesis manifested itself vigorously also in the northern half of the Mediterranean Belt, practically everywhere in its Soviet part except southern Trans-caucasia. The region of epi-Hercynian consolidation encompasses a broad band which extends from Southern Moldavia, across the northwestern part of the Black Sea, the Crimean Plain and the Sea of Azov to cis-Caucasia, spreading to the Donets Ridge and its buried eastern continuation, and further on, across the Caspian Sea to Central Asia (the Turanian Lowland); it also includes the Front Range of the Central Caucasus and the Northern Pamir. Although the Hercynian deformations were decisive for this whole region, there are several major massifs in it (Northern Ustyurt, Karabogaz, Karakum, etc.) which consolidated already in the Baikalian epoch, and a system of relatively narrow fold zones of Early Cimmerian age, which correspond to Triassic tafrogeosynclines, i. e., products of partial regeneration of geosynclinal conditions in the Early or Middle Triassic. One such zone, the Gorny Mangyshlak one, is exposed directly on the surface; the rest are buried under platform Meso-Cenozoic sediments.

As in the south of the European USSR, the Early Mesozoic regeneration manifested itself also within the Mongolo–Okhotsk segment of the Uralo–Okhotsk Belt, in Eastern Transbaikalia and the Amur region. This regeneration became more pronounced towards the east, as increasingly wider and deeper troughs were formed. These were filled with terrigenous formations of increasingly typical geosynclinal appearance; they overlie, with sharp unconformity, the Hercynian and older, up to Baikalian, folded basement. These troughs began to close in the west as early as the Middle Jurassic. In the east, this process continued to the Early Cretaceous, being accompanied by the intrusion of numerous granitoid plutons which are quite common also in the older framing. Thus, complete consolidation of the Mongolo–Okhotsk segment took place in the Late Cimmerian epoch of tectogenesis, although the earlier, Hercynian tectogenesis was of major significance in shaping its structure (from the metallogenetic aspect, however, the Late Cimmerian epoch played a more significant role).

The much more spacious Verkhoyano–Chukotka region, where the Late Cimmerian tectogenesis was pronounced, is situated east and northeast of the Siberian Craton. It is superposed on its eastern continuation; the period of its vigorous development started in the Viséan age of the Early Carboniferous, as a very thick stratum of terrigenous sediments of the so-called Verkhoyansk complex began to accumulate in the interval to the Middle or Late Jurassic. The Verkhoyano–Chukotka region includes several older massifs (Okhotsk, Kolyma, Omolon, Taigonos, and Chukotka) which are salients of the Early Precambrian basement.

In the east and southeast, the Verkhoyano–Chukotka region borders on a region of younger fold systems which evolved mainly on the oceanic crust of the Pacific northwestern periphery. There may, however, be fragments of the older continental curst within it, in Western Kamchatka and in the central part of the Sea of Okhotsk. This region includes the Koryak Upland, Kamchatka, Sakhalin, and the Primorye territory (Sikhota–Alin'), as well as the areas of the Bering, Okhotsk, and Japan seas. In the direction of the Pacific Ocean, it forms a sequence of zones of Middle Cretaceous (Sikhota–Alin', the northern shore of the Sea of Okhotsk, and the northwestern part of the Koryak Upland), Late Cretaceous–Early Paleogene (Sakhalin, the Lesser Kurile Range, Western Kamchatka, and the central part of the Koryak Upland), and younger (the southern part of the Koryak Upland, Eastern Kamchatka, the Kurile and Commander Islands) fold systems. This Far Eastern region is separated from the Verkhoyano–Chukotka region by the wide Okhotsk–Chukotka volcanic-plutonic belt which developed from the Aptian to the beginning of the Paleogene. Offshoots of this belt in the south follow the northern periphery of the Mongolo–Okhotsk segment and the eastern periphery of the Sikhota–Alin' system.

The Sikhota–Alin' system links with the Mongolo–Okhotsk system in the north, along the lower reaches of the Amur; elsewhere they are separated by a major salient of the Early Precambrian basement. The latter is represented on Soviet territory by the Bureya (in the Middle Amur region) and Khanka (in the Primorye region) massifs, and initially formed the northern promontory of the Sino-Korean Craton.

The southern part of the Mediterranean Belt continued its active evolution, undergoing deep subsidence in the Mesozoic and Paleogene; in the Oligocene and beginning of the Miocene it entered the orogenic stage of evolution, during which the fold mountain structures of the Carpathians, Greater and Lesser Caucasus, and Kopet Dagh were formed. The Late Cimmerian features of the Crimean Mountains, the Greater Balkhan, the Central and Southeastern Pamirs, as well as Hercynian structures of the Greater Caucasus Front Range and Northern Pamir were directly involved in this process. Moreover, a vigorous process of recent orogenic reactivation spread to the Tien Shan, the southeastern part of Central Kazakhstan (Karatau, the Dzhungarian Alatau, and Chinghiz–Tarbagatai), Altai, Sayan, the Baikal region and Transbaikalia, the southern part of the Aldan Shield of the Siberian Craton (including the Stanovoi Range zone), and the Verkhoyano–Chukotka region. Zones of various consolidation periods, from Early Precambrian to Cimmerian, were affected. These recent movements also revived the mountainous relief of a large part of the Uralian system, the Baltic Shield (Kola Peninsula), the Timan Range, the Byrranga Range in Southern Taimyr, and formed the Putorana Upland in the northwestern part of the Siberian Craton.

Foredeeps and intermontane troughs filled with thick molasse strata formed simultaneously and in conjunction with Alpine mountain structures and mountain structures of recent reactivation. These are the cis-Carpathian, Indol-Kuban, Terek–Caspian, cis-

Kopet Dagh, Rion, Kura, Western Turkmenian, Tadzhik, Ferghana, Naryn, Issykkul', Chu, and some other troughs. The Black Sea and Southern Caspian depressions resemble this category of structures. They occur in the same band as the intermontane troughs of Transcaucasia and Western Turkmenia, but differ in the type of crust, which is similar to the oceanic.

However, not the entire territory of the Mediterranean and Uralo–Okhotsk belts underwent recent orogenic activation. Vast territories, where the sedimentary cover had accumulated since Jurassic time, experienced no significant deformations afterwards. These are the Moldavian, Scythian (northwestern Black Sea region, the Crimean Plain, the Sea of Azov, and cis-Caucasia), Turanian (the plains of Central Asia), and Western Siberian platforms; the two latter are connected by the Turgai Trough which separates the Urals from Central Kazakhstan. The Zeya–Bureya Depression was superposed on the Bureya Massif. The platform cover underlies also the Barents Sea area, overlapping various structural elements, including the northern continuation of the Timan and Pechora depression Baikalides, and the ancient Barents Sea Massif which is a fragment of the Eastern European Craton. This cover is extensive also in the Kara Sea, on the continuation of the Western Siberian Platform, and in the Arctic further east (Laptev, New-Siberian, and Chuckchi seas), where the folded basement is younger (down to Late Cimmerian).

2. The ancient Eastern European Craton

2.1 Boundaries and main structural elements

The Eastern European Craton (Fig. 2) strictly speaking, is a continental block with crystalline basement, which had consolidated by 1,750–1,650 m.y. B.P., i.e., at the end of the Middle Proterozoic, during the Late Karelian or Svecofennian tectonic epoch, but was partly reworked in the Late Proterozoic, up to 1,000 m.y. B.P. The boundaries of this block took shape at different times, but mainly between 1,350–1,400 and 250–200 m.y. B.P.

The northeastern boundary of the platform is very clearly traceable from the Varanger Fjord, via the isthmus between the Rybachy and Sredny peninsulas, the straits between Kildin Island and the Kola Peninsula, and further along the coast of the Kola Peninsula, across the base of the Kanin Peninsula, and along the southwestern slope of the Timan Range to the Polyudov Range at the junction between the Timan and the Urals. Over this length, the boundary manifests itself in an overthrust of the Baikalian miogeosynclinal fold system on the ancient basement of the epi-Karelian Platform. The overthrust nature of this dislocation is most pronounced near the Varanger Fjord and Rybachy and Sredny peninsulas, as well as along the Timan where it was traversed by a seismic profile. It may be assumed that considerable overlapping of the platform basement by a Baikalian complex took place along the overthrust. At several points of the Kanin–Timan section, the overthrust line is displaced by transverse faults of obviously strike-slip nature (a similar phenomenon will be shown later to be characteristic of the southwestern and southern boundaries of the platform).

The present eastern boundary of the platform runs in meridional direction from the Polyudov Range to the upper reaches of the Emba River along the western edge of the cis-Uralian Foredeep. It is, however, now known that not only the cis-Uralian Foredeep but also the entire western, miogeosynclinal zone of the Urals is underlain by a basement similar to that of the Eastern European Craton, and that the region containing this type of basement is bounded only by a fault running along the eastern slope of the Uraltau Uplift. This is the Major Uralian Fault.

Between the upper reaches of the Emba and the Caspian Sea, the boundary of the Eastern European Craton has a southwestern direction over a relatively small length, coinciding with the so-called Southern Emba Fault zone. This zone separates the southeastern margin of the North Caspian Syneclise from the narrow Early Hercynian fold zone. The latter is a continuation of the Uralian Zilair Trough and tapers off in the vicinity of the Caspian shore. However, seismic profiles permit tracing the deeply buried continuation of the Early Precambrian platform basement through this zone to the borders of Northern Ustyurt. The latter thus appears to be a fragment of the Eastern European Craton.

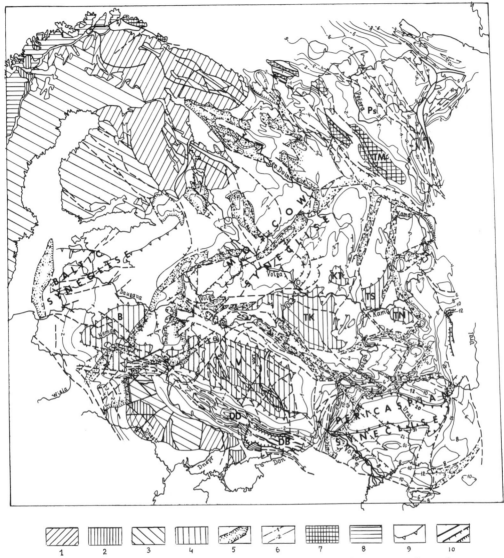

Fig. 2.   Structure of the East European Craton. 1–3 = Precambrian basement on the surface; 1 = predominantly Archaean, 2 = Archaean, reworked in Early Proterozoic, 3 = Lower Proterozoic; 4 = salients of Precambrian basement near the surface; 5 = Riphean aulacogens; 6 = contours of the basement surface; 7 = outcrops of Riphean metamorphic complex (Baikalian); 8 = Scandinavian Caledonides; 9 = contours of anteclises and swells; 10 = faults: a) on the surface, b) in the basement, c) bordering aulacogens. Positive and some negative structures of the Russian Platform: B = Belorussian Anteclise; V = Voronezh Anteclise; Tk = Tokmov Swell; Kt = Kotel'nich Swell; TN = North Tatarian Swell; TS = South Tatarian Swell; Tm = Timan Range; DD = Dnieper –Donets Aulacogen; Db = Donets Basin; Ps = Pechora Syneclise.

Further west, the boundary of the platform has a generally latitudinal direction over a long distance, from the northeastern Caspian shore to the northwestern Black Sea shore.

Along this southern boundary, the ancient platform borders the Hercynides, mainly
buried under a Meso-Cenozoic sedimentary cover and exposed at the surface only in the
Donets Range. The Donets fold system structure or, in other words, the Donets Basin, is
a bay-like protrusion of the Hercynian geosyncline (or of its foredeep) in the body of the
ancient craton, and disrupts the generally smooth course of its southern boundary. The
Donets-Caspian zone of Hercynides is obviously a Middle Paleozoic new formation; the
original platform boundary must have passed along the southern edge of the Rostov
Promontory of the Ukrainian Shield and thence along the Manych and Central Ustyurt
fault zones south of the Northern Ustyurt Block.

From the northwestern shore of the Black Sea in the southern Ukraine (Ismail Region)
and Moldavia, to the southern Baltic shore in Polish Pomerania, the boundary of the
Eastern European Craton runs fairly straight to the northwest, it follows an obviously
major fault zone which H. STILLE (1947) called the Balto-Podolian Lineament, but it is
also known as the Teysseyre–Tornquist line, after a Polish and a German scientist who
were the first to notice it. The Alpine foldnappe Carpathian structure, more specifically,
the Neogene cis-Carpathian Foredeep directly borders this southwestern boundary of
the platform over a considerable distance. However, where the Carpathian arc turns
away from the edge of the ancient platform (in the north in Southern Poland, in the south
in Romania and Moldavia), older formations emerge from under the Alpine complex; the
Baikalian greenschist complex forms the folded basement here. Consequently, the Galician (after SEMENENKO, 1972) Baikalian system originally bordered the ancient platform,
as the Timanide system of the same age borders the northeastern edge. Hercynides and
Early Cimmerides intruded between the Baikalides and the ancient platform margin in
the Northern Dobrogea. Caledonides intruded in Poland, being continued to Pomerania
and Rügen Island.

Least defined is the position of the Eastern European Craton boundary between Rügen
and Southern Norway. This problem has been discussed for long; different answers were
given. The craton boundary most probably runs from the northern end of Rügen Island
to the base of Jütland and thence in the same latitudinal direction to the North Sea, where
it crosses the probable continuation of the Scandinavian Caledonide front. This is in
maximum agreement with both geological and geophysical (GAFAROV, 1971) data, and
was suggested by N. S. SHATSKY as far back as 1946.

Starting from there, the northwestern boundary of the craton more certainly runs (on
land) along the front of the Scandinavian Caledonides; it crosses at almost a right angle the
older Balto-Podolian Lineament which later underwent repeated rejuvenation. Tectonic
windows in the Caledonides reveal basement formations identical with the Eastern
European Craton basement; in places they are overlapped by a thin Lower Paleozoic
cover. This shows that this basement continues for more than 200 km beneath the
Caledonides and reaches the Norwegian coast. Moreover, similar ancient rocks are
exposed on the Lofoten Islands whose tectonic position is, however, unclear.

The contour of the Eastern European Craton closes in the Varanger area, where the
Timan Baikalides emerge from beneath the Caledonides which they join at the edge. The
platform outlines are readily seen to have sharp angles (polygonal); they can be inscribed
in an irregular heptagon. Neglecting the effect of Hercynian reworking of the southeastern platform corner results in a hexagon, whose southeastern vertex reaches the Aral Sea.
The northeastern and southwestern boundaries of the platform are parallel, having a

northeasterly direction, and are of the same age; the pre-Baikalian, mainly Early Pre-cambrian Karelian basement is along them replaced by the older Baikalian basement. The southern and eastern boundaries were initially also Baikalian (part of the eastern bound-ary is older), but later underwent Hercynian reworking. Finally, the northwestern boundary is in fact of Baikalian rather than Caledonian age, since the Scandinavian (Grampian) Geosyncline originated at the end of the Proterozoic. To put it differently, the general shape of the Eastern European Craton was outlined in Late Precambrian time.

The Baltic Shield (Fennoscandia, as called by the Finnish geologist RAMSAY 1907) and the Russian Platform are two major elements of the internal structure of the Eastern European Craton. The Baltic Shield, which practically (see above) includes the entire Scandinavian Peninsula, the Kola Peninsula, Finland, and Karelia, is the most stable positive element of the craton; it underwent uplift almost throughout the last billion years or, if the Southern Scandinavian Block is disregarded, even a billion and a half years. Only in the Cambrian and Silurian did the western and southern margins of the shield undergo slight downwarping (or eustatic submergence?), with the accumulation of dozens or a few hundred meters of shallow-water marine sediments. The prolonged uplift caused deep erosion of the crystalline basement; these exposed bed rocks metamorphosed in the amphibolite or even granulite facies, i.e., bed rocks which originally had lain at a depth of up to 15 km or more.

The Baltic Shield occupies the northwestern corner of the platform; the Precambrian formations constituting it in the east and south plunge under a sedimentary cover; the region where the latter occurs more or less continuously was called Russian Platform (Russische Tafel) by E. SUESS (1885). It remains unclear whether or not the area of outcrops of the Early Precambrian basement in the Ukraine should be considered as part of the platform. This area is called either Ukrainian Crystalline Massif or Ukrainian Shield, depending on the particular answer to this question. The Ukrainian Shield (or Massif) is also a quite large zone of prolonged uplift which mainly began at the transition from Middle to Late Proterozoic, i.e., not later than the uplift of the Baltic Shield. The Ukrainian Shield underwent short-term submergence only in the Early Carboniferous (Viséan) and Late Cretaceous; even this did not involve the whole area. There are, therefore, sufficient grounds for calling it shield rather than massif, though separation of this shield from the Russian Platform breaks its geometric continuity. The Baltic and Ukrainian shields, as in some other platforms, the Siberian in particular, occupy peripheral rather than central positions in the Eastern European Craton.

Within the Russian Platform in a narrow sense, i.e., without the Ukrainian Shield, the basement occurs at depths ranging from zero (the Belorussian and Voronezh massifs) to five kilometers or more. Quite conspicuous against this background are structures like the Dnieper-Donets Aulacogen, where the basement lies at a depth of up to 8–10 km, and especially the North Caspian Syneclise in whose central part the sedimentary cover is thicker than 20 km, according to recent data (FOMENKO, 1972). The North Caspian Syneclise might possibly be regarded as a separate structural element equivalent to the Russian Platform. Also identified as similar deeply buried elements of the platform periphery are the Dniester and Black Sea pericratonic troughs which in fact are slopes of the Ukrainian Shield (the basement there lies at a depth of up to 5–6 km), as well as the Peri-Timan Trough (more than 5 km) in the northeast.

2.2 The Baltic Shield

The Baltic Shield or Fennoscandia is the largest and the most stable positive structural element of the Eastern European Craton. It encompasses the entire territory of Finland, the Murmansk region, and the Karelian ASSR, the greater part of Sweden, and the south and extreme north of Norway (Fig. 3). In the northwest, the Early Precambrian rocks of the Baltic Shield are hidden beneath the tectonic nappes of the Scandinavian Caledonides, but appear again as a whole series of tectonic windows situated approximately along the main water divide of the peninsula and reach the Norwegian Sea shore. Formations of the same age occur also on the Lofoten Islands; however, their connection with the shield basement is much less obvious. Nevertheless, the windows in the Caledonides indicate that the Baltic Shield undoubtedly continues northwest under the Caledonides over a distance exceeding 200 km. In the northeast, along the Murmansk coast, Late Precambrian geosynclinal folded formations of the northwestern continuation of the Timan system are thrust over the Baltic Shield. In the east and south, the shield basement plunges quite smoothly under the sedimentary cover of the Russian Platform. In the southwest, subsidence is much sharper, because here it is due to the flexures and faults of the "Teysseyre–Tornquist line".

Despite the quite long history of geological investigation, which includes classic works in Precambrian geology, such as by W. RAMSAY (1902), J. SEDERHOLM (1932), P. ESKOLA (1925), O. BACKLUND, A.A. POLKANOV (1947), N.G. SUDOVIKOV (1935), and others, and the quite extensive radiogeochronological data, many fundamental problems of stratigraphy and tectonics of the Precambrian basement of the shield remain unsolved. Quite controversial, in many cases directly opposed views exist. Under these conditions, the most objective analysis of the shield structure seems to be, first, its subdivision into megablocks approximately equivalent to the Canadian Shield "provinces". These differ both geochronologically and in the deep structure, in particular, the crustal thickness (KRATS & LOBACH-ZHUCHENKO, 1970), and are in most cases separated by deep faults (TSIRYUL'NIKOVA et al., 1968); this, by the way, has now been confirmed also for the Canadian Shield (the Grenville front). The next step must include separate analysis of the structure of each megablock, in some cases with its subdivision into individual blocks; the age of the rocks constituting some of the blocks will have to remain undetermined. We note in advance that the conclusion of POLKANOV & GERLING (1961), that the age of the Baltic Shield basement becomes increasingly younger in the direction from northeast to southwest, i.e., from the Kola Peninsula to the south of the Scandinavian Peninsula, has been confirmed. It is therefore worthwhile to discuss the megablock structure in this particular sequence.

The boundaries between the megablocks run in the north of the shield in the NW–SE direction and in the south in the NNW–SSE or even meridional direction. This is in conformity with the orientation of their inner structure, with the exception of the Svecofennian Megablock.

The **Kola Megablock** occupies nearly the entire peninsula, except the southern coast which belongs to the White Sea Megablock (see below). The greatest number of ancient radiometric datings, including those indicating more than 3,000 m.y., were obtained within this megablock. The oldest rocks of the megablock include those of the Kola series which generally make up the Central Kola Block; they are represented by biotite and

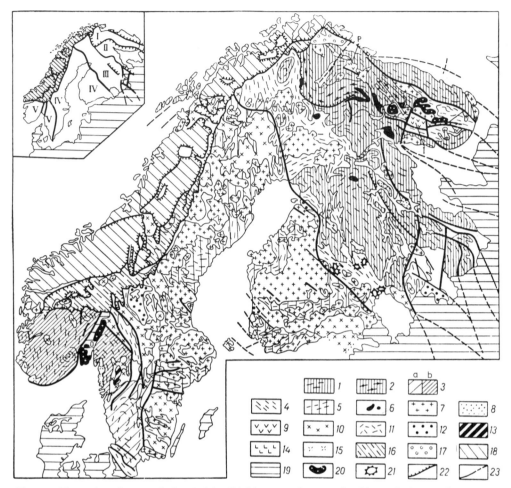

Fig. 3. Tectonic scheme of the Baltic Shield (according to the Tectonic Map of Europe, 1:10.000.000, 1977). 1 = gneisses and granite-gneisses; Archaean, amphibolite facies; 2 = ibid., granulite facies; 3 = reworking of the ancient basement: a) Gothian, b) Gothian–Dalslandian; 4 = metavolcanic and metasedimentary Lower Proterozoic; 5 = Svecofennian geosynclinal complex; 6 = ultramafics and gabbroid rocks of the Lower and Middle Proterozoic; 7 = Early Karelian and Svecofennian synkinematic granites; 8 = Upper Karelian (Karelian s.str.) – Sariolian, Jatulian, Suisaarian; 9 = volcano-sedimentary filling of the Pechenga and Imandra–Varzuga troughs; 10 = Late Svecofennian and Jotnian granites, rapakivi inclusive; 11 = Sub-Jotnian, Hoglandian and their equivalents; 12 = Jotnian; 13 = Dalslandian; 14 = anorthosites of Nordland; 15 = granites of the Bohus–Arendahl type; 16 = sparagmite in the Caledonian nappes; 17 = ibid., outside the nappes; 18 = Caledonian nappes, exclusive of sparagmite; 19 = Phanerozoic sedimentary cover; 20 = Caledono–Hercynian alkaline plutons; 21 = gneiss domes; 22 = Caledonian overthrusts; 23 = deep faults. P = Pechenga Trough; I = Imandra–Varzuga Trough; O = Oslo Graben. Inset: I = Kola Megablock; II = Belomorian Megablock; III = Karelian Megablock; IV = Svecofennian Megablock; V = South Scandinavian Megablock.

garnet-biotite gneisses, amphibolites, and amphibolic gneisses. Banded magnetite quartzites occur in the middle part of the series; high-alumina crystalline schists predominate in the upper part. The metamorphism corresponds to the amphibolite, partly granulite facies. Datings giving an age of 3,300 ± 100 m.y. were obtained for crystalline schists of the granulite facies by the K/Ar method on pyroxenes. This is regarded as the age of the earliest metamorphism of the Kola series. The gneiss-granites and plagiomicrocline granites, coeval with diaphthoresis oft the Kola series rocks in the amphibolite facies, have an age of 2,800–2,700 m.y. ("Geochronological Boundaries ...", 1972); somewhat younger (2,500 m.y.) granites were found in the west. Younger radiometric datings (about 1,800–1,700 m.y.) correspond to the Late Karelian tectonomagmatic epoch with which the formation of pegmatites in the Kola Megablock is associated.

The structure of the Kola series is extremely complex; it consists of folds of many orders, including isoclinal, of various strikes, complicated by faults; migmatization, boudinage, and blastomilonitization manifest themselves very extensively. Alongside granitoids, intrusive bodies of gabbro-amphibolites, hypersthene and bipyroxene diorites, and gabbro-diorites, as well as other basic and ultrabasic rocks are associated with the Kola series.

The oldest granite-gneiss basement of an average tonalitic composition, usually consisting of gray rocks (similar to the "gray gneisses" of other ancient crystalline shields of the Earth), has been identified with increasingly higher reliability in the Central Kola Block. This basement contains diorites, granodiorites, trondhjemites, tonalites, plagiogranites, enderbites and acid metavolcanics. There are no reliable radiometric datings of this complex, U/Pb ages on zircons give 2,800–2,680 m.y. The salients of the oldest basement are structurally shaped like domes or isometric blocks with rounded angles. A meridional or northeastern orientation predominates in its inner structure.

Some researchers (MASLENNIKOV, 1968, and others) considered the gabbro-norite-peridotites and pyroxenites of the Monchegorsk nickel-bearing pluton to be the oldest rocks of the Kola Megablock, or even representatives of the initial oceanic crust. It was indicated that they are separated from the Kola series blocks by conglomerates which are basal for this series. The very old age of the Monche-tundra rocks seemed to have been confirmed by dating them with the U/Pb method (3,800 ± 300 m.y.) (TUGARINOV & BIBIKOVA, 1975). However, most investigators of the Kola Peninsula believe now that the Monchegorsk Pluton is younger than the Kola series and that it formed when the cratonic regime became established. The pluton contacts with the enclosing rocks are regarded as tectonic, and the conglomerate mentioned above as a breccia. The latter was found to contain fragments of Kola series rocks.

In the east, the Kola series plunges under a younger, probably Lower Proterozoic Keyvy series which makes up the Keyvy-tundra synclinorium. It consists of various gneisses, amphibolites, and high-alumina crystalline schists. According to their initial composition, these are volcanics, from basic to acid, and also clastic rocks (graywackes, arkoses, etc.). The content of acid volcanics and coarse fragmental rocks increases toward the top of the sequence. The uppermost, relatively weakly metamorphosed dolomite-sandstone part of the series, containing stromatolites, may belong to the Middle Proterozoic. The age of the synkinematic alkaline Keyvy granites intruding the Keyvy series is about 1,900 m.y.

In the northeast, the Central Kola Block borders on the Murman Block, overthrust on it. The Murman Block is composed of various types of granite, mainly microclinic and plagiomicroclinic with biotite, presumably of the same age as the Archaean granites of the

Central Kola Block. The oldest oligoclase gneiss-granites occur in the form of xenolithic blocks. Alongside these two groups of granites, there exists a younger granitoid group aged 2,200–2,000 m.y., which corresponds to the Karelian reactivation.

A zone of the Polmos–Poros series having a synclinal structure (the Voronye–Kolmo-zero zone) has been identified at the boundary between the Central Kola Block and the Murman Block. The Polmos–Poros series starts with granitic conglomerates and is composed of gneisses and amphibolites (in the lower part) and aluminous greenschists (in the upper part). The age of the rock metamorphism is about 2,500–2,400 m.y. (K/Ar method, on amphiboles) and that of the granites and pegmatites intruding them even higher – 2,870–2,670 m.y. (Rb/Sr method, whole rocks); the series must therefore belong to the Upper Archaean and probably represents a greenstone belt.

The Tundra series of very variable composition and variable grades of metamorphism is customarily identified along the southern and eastern periphery of the Kola Mega-block. There are serious doubts about the independent position of these rocks which are usually associated with the Lower Proterozoic; they most likely include elements of various Lower Precambrian series which underwent pronounced reworking including diaphthoresis.

Two quite peculiar troughs filled with thick terrigenous-volcanic strata of the upper-Lower to Middle Proterozoic are situated along the tectonic boundary between the Kola Megablock and the White Sea Megablock adjacent on the south. These are the Pechenga and Imandra–Varzuga troughs; the latter is several times longer (about 350 km) than the former; both are up to 30–40 km wide. Structurally, both troughs are one-sided grabens (half-grabens) with a monoclinal southwestern dip at an angle of 10–15 to 30–50° of the fill (Fig. 4). The monocline is broken into longitudinal blocks by additional faults. The sedimentary-volcanic fill of the grabens is 8–10 km thick in the Imandra–Varzuga series and 4.5–5 km in the Pechenga one. Both series, especially the Pechenga one, are characterized by a macrocyclic structure: the lower part of the cycles consists of sedimentary-tuffogene rocks; the upper consists of volcanic rocks, mainly basic to picrites, partly also intermediate and acid. The sedimentary members contain dolomite interlayers with stromatolites and oncolites. The sequence of the Imandra–Varzuga Trough is much fuller, because of the lowermost strata, than the sequence of the Pechenga Trough. Present here is a series also representing a cyclic sequence of clastic sediments and

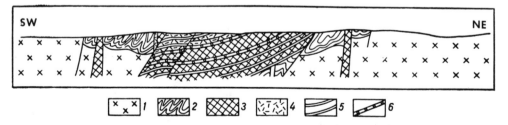

Fig. 4.   Schematic profile of the Imandra–Varzuga Trough (after SIMON, 1973). 1 = gneiss-granites and "granitic" layer of the Earth's crust; 2 = rocks of the lower structural stage of the Karelides; 3 = basic lavas and sills of the lower series; 4 = volcanics of the upper series; 5 = layers of sedimentary rocks; 6 = volcanic breccias.

volcanics, from picrites to dacites, with a total thickness of up to 5–7 km. In composition and stratigraphic position this series corresponds to the Sumian complex of the Karelian Megablock and, consequently, to the Lower Proterozoic (see below). An unconformity separates it from the Imandra–Varzuga series proper. Metamorphism of both the Pechenga and Imandra-Varzuga series is mainly greenschist, in places along the major southern boundary faults, up to amphibolite. Both series enclose basic and ultrabasic intrusions (copper- and nickel-bearing in the Pechenga Trough), and are penetrated by granitoids. The metamorphism here is of Late Karelian age: about 1,800–1,700 m.y. The basites and hyperbasites are slightly older (2,000 to 1,800 m.y.). In "Geochronological Boundaries…" (1972) the probable age of the Pechenga-Varzuga complex is estimated at 1,900–1,680 m.y. This, however, can refer only to the Pechenga series and the upper part of the Imandra–Varzuga series.

The tectonic nature of the Pechenga and Imandra–Varzuga troughs is still unclear. Some investigators consider them to be final orogenic structures of the Karelides, others assume a subcratonic character. This author believes that these troughs are protoaulacogens (in analogy with the similar Aldan Shield structures of the Siberian Platform – see Chapter 3) or, in other words, paleorifts which correspond to the beginning of the Kola Megablock cratonic stage of evolution[1].

In fact, by the beginning of accumulation of the Pechenga–Varzuga complex, both the Kola and the White Sea (Belomorian) megablocks must have been sufficiently consolidated. Formation of typical cratonic intrusions took place at the end of that time. These are alkaline plutons of the Keyvy tundras, with the multiphase (from ultrabasic rocks to nepheline syenites and alkaline granites) Gremyakha–Vyrmes Pluton situated near the southern boundary of the Kola Megablock, and porphyroid "aragranites" in the west of the block.

In its subsequent history, the Kola Megablock underwent appreciable reactivation during the final periods of the Salairian (Late Baikalian), Caledonian, and Hercynian tectonic stages. Minor annular plutons of ultrabasic-alkaline rocks and carbonatites formed during the first two periods; the unusually large Khibiny and Lovozero subvolcanic plutons of nepheline syenites formed during the third period. These plutons are associated with a northeast running fault at the boundary between the Central Kola and the Keyvy blocks.

The **White Sea (Belomorian) Megablock** underlies the greater part of the White Sea; the Archaean rocks forming it crop out along its northern, southern, and western coasts, and continue in the direction of Lapland in the northwest. The White Sea series consists of various gneisses, mainly biotitic and garnet-biotitic; the upper part of the series partly contains aluminous gneisses and amphibolites which predominate in the middle part. The primary composition of the rocks must have been terrigenous-volcanic. Their total thickness was determined as 8–10 km or more, though this may be an overestimation. The metamorphic facies is amphibolite; relics of granulite facies are also encountered. The structure is extremely complex, resulting from several phases of deformation which

---

[1] Protoaulacogens, as understood by this author, differ from the paleoaulacogens described by LEITES, MURATOV & FEDOROVSKY (1970). The paleoaulacogens of LEITES and his coworkers correspond, in the opinion of the present author, to Early Precambrian greenstone belts.

differed in the stress orientation. These phases were accompanied by regional metamorphism and intrusions, and can be dated radiometrically. The first phase manifested itself in the formation of folds of not yet ascertained trends, metamorphism of the granulite facies, basic and ultrabasic intrusions as well as in migmatization and metasomatosis. It was followed by northwesterly folding, metamorphism (regressive) of the amphibolite grade, and the formation of plagiogranites. The second phase led to the formation of sublatitudinal folds and was accompanied by lherzolite, gabbro-norite, granite, and pegmatite intrusions. This is the principal "White Sea fold system", which is now dated at 2,790 (Rb/Sr) – 2,650 (U/Pb) m.y., i.e., coinciding with the Archaean/ Proterozoic boundary and with the principal epoch of Kola series granitization[2].

The third phase, more specifically, epoch, which occurred at the end of the Early and Middle Proterozoic, was approximately coeval with the major deformations of the Karelides. Associated with this epoch was repeated folding, first of northeasterly, then meridional, and, finally northwesterly trend, concordant with the Karelides and with the southwestern boundary of the megablock. It was also accompanied by migmatization, granite and pegmatite intrusions, and finally intrusions of diabase-porphyrite dikes (M. M. STENAR', see SOKOLOV et al., 1973). This epoch of tectono-magmatic reactivation has been dated at 1,800–1,700 m.y. on pegmatites.

Particular formations of the so-called granulite complex have been identified in the northwestern continuation of the White Sea complex, mainly in Finland and Norway. According to new data, it overthrusts the White Sea complex and then covers it completely. The complex includes both acid rocks close to gneisses, and basic rocks including amphibolites and charnockites. Some investigators regarded this complex as extremely ancient, at any rate pre-White Sea or even "pre-Archaean" (ZHDANOV, 1965). More recent data, however, both geological and radiometric, rather point to the synchronous nature of the granulite complex and the White Sea series (SIDORENKO et al., 1971). The same work presents proofs of the primary sedimentary origin of both acid and basic (amphibolites) granulites. This, however, is a conclusion with which other investigators do not quite agree. The radiometric age of the granulitic complex is 1,950 ± 50 m.y. (according to Pb^{207}/Pb^{206}) and 1934 m.y. (by the K/Ar method); as in the case of the White Sea complex, it corresponds to the Karelian rejuvenation (individual datings of over 3,000 m.y. were obtained by the K/Ar method for pyroxenes).

However, these ideas on the White Sea complex cannot be regarded as sufficiently reliable. The structure of this complex is extremely complicated; the stratification given for it may in fact turn out to be tectonic rather than sedimentary i. e., it may be a result of interbedding of repeatedly deformed tectonic nappes composed of continental (gneisses) and oceanic (amphibolites and serpentinized ultrabasites) type rocks. Any correspondence between the White Sea and the Kola series, based on their symmetrical situation with respect to the Pechenga–Varzuga Trough system, and their combination into a single "Kola–White Sea complex" raises serious doubts, since they differ sharply both in composition and in structure. In particular, in contrast to the Kola series and the Karelian Lopian Complex (see below), jaspilites are totally absent in the White Sea complex, although they occur on either side of it; it is difficult to account for this by facies

---

[2] Thus it in fact coincided with the Saamian fold system as defined by POLKANOV & GERLING (1961).

differences alone. It is more likely that these formations are either older or younger than Late Archaean (analogues of the Keyvy series?), or that the latter are unconformably superposed on the former, with a missing iron-ore series. The first assumption seems to be most likely. It implies that the White Sea complex is an analogue of the basement of the Karelian and Kola megablocks, most extensively exposed on the surface and repeatedly remobilized till the Middle Proterozoic. This assumption is in maximum agreement with the similarity of the White Sea belt and especially of the granulite complex to granulite belts of other ancient shields.

The structure of the White Sea Megablock was in the Riphean complicated by the formation of the Kandalaksha Trough (Rift) superposed on its axial part. The deposits of this trough are now exposed on the surface along the Tersk shore of the White Sea in the form of gray shallow-water marine Turya, red alluvial Tersk, and again gray, probably marine, Upper Tersk clastic suites. The first, as suggested by recent data, may belong to the Middle Proterozoic, the second to the Middle Riphean, and the third to the Upper Riphean (A.L. RAGOZINA, see "Problems of Precambrian Geology ...", 1976). Subsidence of the Kandalaksha Graben resumed in recent times and led to the formation of the Kandalaksha Gulf and the present central area of the White Sea. The Kandalaksha Fault could, with the help of space photographs, be followed far to the WNW into the shield in Lapland; this fault marks there the southern boundary of the granulite complex.

The **Karelian Megablock** extends from the Caledonides front, across Swedish and Finnish Lapland into Karelia and Eastern Finland, emerging at the Ladoga and Onega lakes. The fault nature of the boundary between the Karelian and the White Sea megablocks has been confirmed by deep seismic sounding (DSS). It is most obvious geologically in the southeastern part of this boundary, in the region of the Vetreny Belt, where basic and ultrabasic Karelian magmatites are thrust over the White Sea complex (NOVIKOVA, 1971). In NOVIKOVA's opinion, the imbricate-overthrust structure is generally typical of the Karelian Megablock. The Karelian Megablock contains three unconformably overlapping structural complexes: pre-Karelian, Early Karelian, and Late Karelian (Yatulian). Initially, only the Upper Karelian was identified by ESKOLA (1925) under the name of Karelides. This concept of Karelian as equivalent to Yatulian (RAMSAY, 1907) has been retained in Finnish and Swedish literature, whereas in the Soviet Union, following K. O. KRATS (1958) investigations, there was (although not general) a wider meaning of Karelian, with a subdivision into lower and upper. GILYAROVA (1974) and others define KRATS' Lower Karelian as pre-Karelian; she believes it to be of the same age as the Kola (White Sea) complex, regarded as the geanticlinal and miogeosynclinal, terrigenous or terrigenous-volcanic equivalent of the eugeosynclinal volcanic pre-Karelian. The "pre-Karelian" is also identified by KRATS and his followers, though to a much lesser extent.

From this point of view, the pre-Karelian formations which constitute the basement of the Karelides, are of Archaean (Early Archaean) age (their radiometric age, probably rejuvenated, is 2,800–2,500 m.y.), and an analogue of the ancient basement of the Kola Megablock and of the "gray gneisses" of other shields. They are represented mainly by granite-gneisses which occur either in the form of blocks bounded by faults (GORLOV, 1967) or in cores of mantled domes (in most cases structures transitional between the second and first type are present). These outcrops of the ancient basement usually experienced vigorous remobilization and transformation into variously composed granitoids (from granodiorites to plagiomicroclinic granits) during the Early Karelian phases of diastrophism.

The Early Karelian complex can be divided into two distinct parts identified by KRATS (1958, 1963) under the names Lopian and Sumian. The Gimola iron-ore series in Western Karelia is a typical representative of the Lopian; the Bergaul, Parandovo, Tikshezero, Pebozero, and other series of Central and Eastern Karelia are its analogues. The Gimola series and some of its equivalents start with basal conglomerates, arkoses and quartzites unconformably overlying older granite-gneisses. Basic and some acid (mainly in the upper part of the sequence) including leptites and hälleflints, volcanics, metamorphosed in the greenschist or amphibolite facies, constitute the main part of these series. Many authors believe that these volcanics constitute a quite typical spilite-keratophyre formation; pyrite deposits, sometimes in minable quantities, are associated with it. Alongside volcanics, the Lopian contains various greenschists: micaceous, graphitic, chloritic, among which there occur magnetite greenschists and magnetite quartzites (jaspilites) in the Gimola series.

The accumulation of Lopian deposits ended with folding, metamorphism to the middle stages of the amphibolite facies, granitization, and migmatization. This phase of diastrophism was called Rebolian (KRATS, 1963); the formation of highly compressed, sometimes to isoclinal fold systems of meridional or northeasterly trend, is connected with it in Karelia. Deep metamorphism and granitization caused the rocks of the Gimola series and its analogues to be preserved in more or less indentifiable form only in individual synclines or synclinoria scattered through a vast field of granite-gneisses. The Lopian synclinoria obviously are equivalents of Archaean greenstone belts of other continents.

The early synkinematic granitoids of the Rebolian epoch belong to a diorite-plagiogranitic formation. More recent granitoids of the same epoch were already potassium-sodium or essentially potassium; they form plutons with clearly cross-cutting contacts.

The Rebolian tectono-magmatic epoch (phase) was previously dated as 2,300–2,200 m.y. old; the age of the granites which concluded it, as 2,180 m.y. ("Geochronological Boundaries ...", 1972). Higher ages, of about 2,800–2,700 m.y. and up to 3,020 m.y., were obtained recently. They indicate that the Lopian is Upper Archaean, and point out to its correspondence to the Kola series which is also iron-ore-bearing.

The upper part of the Early Karelian complex (and the Lower Proterozoic of the Karelian Megablock), which is quite independent in the structural-historical sense, consist of the Sumian–Sariolian series. Only the Sumian of KRATS (1958) was initially associated with this interval of the sequence; the Sariolian clastic suite was regarded as the basal part of the Upper Karelian, i.e., it was combined with the Yatulian[3]. Karelian geologists (V. A. SOKOLOV, L. P. GALDOBINA, and others; see SOKOLOV et al., 1973) have later proved convincingly the existence of a major break between the Sariolian and the Yatulian, and the alternation of Sumian volcanics and Sariolian coarse clastic formations.

---

[3] Some investigators, for example GILYAROVA (1974), believe that the Sariolian is coarser Yatulian facies associated with uplifts. On the Kola Peninsula, in the sequence of the Imandra–Varzuga Depression, the analogues of the Sariolian are identified in the lower part of the series, which is thought to be comparable to the Yatulian ("Problems of Precambrian Geology ...", 1971). It is possible that this contradiction is due to the fact the coarse clastolites of the Sariolian type in some cases conclude the Sumian sequence (PR_1), and in other cases begin the Yatulian sequence (PR_2).

The Sumian proper is a sequence of basic, less frequently acid volcanics: lavas (mainly diabases and diabase porphyrites), agglomerates, pyroclastolites which underwent greenschist, occasionally in near-fault zones, also amphibolite metamorphism. The Sariolian is characterized by granitic conglomerates (up to boulder or block size), gritstones, arkoses, and siltstones. Besides sequences with alternating volcanics and clastolites, there are sequences completely devoid of volcanic material and, on the other hand, sequences composed almost exclusively of such material. This is accounted for in the following way (SOKOLOV et al., 1973): sequences of purely Sariolian type are associated with slopes of ancient horst uplifts or depressions inside the latter, whereas Sumian sequences originated in the vicinity of eruption centers, obviously gravitating toward faults separating uplifts and depressions. Alternation of these two types of deposits, with a predominance of relatively fine- and thin-detrital rocks, is characteristic of central parts of depressions. On the whole, the Sumian-Sariolian formations fill simple or complex linear synclinal or graben-synclinal structures, usually bounded by faults on one or both sides.

K. I. HEISKANEN (see SOKOLOV et al., 1973) suggested a very interesting reconstruction of the tectonic conditions under which the Sumian-Sariolian deposits formed. He believes that during this epoch there existed in Central and Northern Karelia large arched uplifts, complicated by grabens and separated by troughs in which the Sumian–Sariolian series accumulated: coarser detrital or volcanic in uplifts, of mixed composition and finer detrital in interarch troughs. This implies that a quite high degree of consolidation had been attained by the central part of the Karelian Megablock by the Early Proterozoic. This region generally corresponds very closely to VAYRYNEN's "Yatulian continent" (1954) and to KHARITONOV's "Karelian Massif" (1966). These authors considered the basement of the massif to be older, i.e., pre-Svecofenno-Karelian. Finnish (SIMONEN, 1971) and Swedish (VELIN, 1972) geologists agreed with this view, on the basis of radiometric determinations of the age of the granite-gneisses in Eastern Finland.

Different tectonic conditions were typical, during this epoch, of marginal parts of the megablock, situated at the boundary with the adjacent White Sea (Belomorian) and Svecofennian megablocks. Here, the Northern Karelian and Northern Ladoga (Savo-Ladoga, according to SALOP (1971) who extends it through the apex of the Gulf of Bothnia into Sweden, Norbotten Province) troughs are distinguished.

The Upper Archaean Lopian formations in these two troughs are superseded more or less conformably by deposits of the Lower Proterozoic analogues of the Sumian which, in contrast to Central Karelia, are represented by typically geosynclinal formations.

Both in the narrower (up to 15–16 km) Northern and Eastern Karelian Trough (the Vetreny Belt Synclinorium is its further continuation to the east) and in the wider Northern Ladoga Trough, the lower part of the sequence is composed of volcanic-sedimentary formations; the upper part consists of flyschoid-terrigenous deposits. In the Northern Ladoga Trough, the former is called Sortavala series and the latter Ladoga series. The age of the Sortavala series is 2,600–2,500 m.y. on metabasalts, of the Ladoga series 1,880–1,810 m.y. (age of metamorphism). A break is observed between them, and a weak folding. However, the main deformations were most pronounced and led to isoclinal folding of a general northwestern trend with overthrusts and northeastern vergency, and to repeated deformations with the superposition of folds of different orders and strikes. They took place at the end of the Early Proterozoic, during the epoch of tectono-magmatic activity, called Seletsk epoch (1,880–1,810 m.y. B.P.) by KRATS (1958). Detailed structural studies, carried out by Karelian geologists (LAZAREV 1973, and others), showed that this epoch consisted of several (up to four) individual deformation phases. The first three phases were accompanied by synkinematic metamorphism which reached high grades of the amphibolite facies, as well as by ultrametamorphism, migmatization and granitization. The diorite-plagiogranite series of the Northern Ladoga zone is dated as 1,950–1,900 m.y. old (SUDOVIKOV et al., 1970).

A typical feature of the Northern Ladoga zone is the abundance of small (10–20 km across) mantled granite-gneiss domes of the Sortavala type in Karelia and Kuopio in Finland; this type of structure was first identified in the Ladoga region by ESKOLA in 1948.

The Seletsk deformations occasionally manifested themselves in a more consolidated massif of Central Karelia–Eastern Finland; however, as shown by LAZAREV (1973), they are of quite different character in the pre-Sumian – Sariolian basement and in the Sumian–Sariolian formations proper. This may partly be due to remobilization of the granitized substratum of the latter, although it may also be a consequence of the somewhat lower age of the Sumian–Sariolian in this massif. It is of interest that the Sumian–Sariolian, in contrast to the Ladoga and Tikshezero (Northern Karelia) series, is hardly intruded by granites, with the exception of individual veins. At the same time there exist in the Northern Ladoga zone, alongside the plagiogranite formation, also cross-cutting intrusions of Late Karelian potassium granites aged 1,800–1,700 m.y. This is yet another confirmation of the earlier consolidation of the Karelian Massif.

The uppermost structural complex of the Karelian Megablock, the Upper Karelian complex (Karelian s. str.), belongs stratigraphically to the Middle Proterozoic and consists of three series: sedimentary or sedimentary-volcanic Yatulian, volcanic Suisarian, and sedimentary Vepsian.

The Yatulian sedimentary formations of Karelia have been studied in great detail by V. A. SOKOLOV and coworkers (SOKOLOV et al., 1973). It was found out, first, that the Yatulian overlies, with a general structural unconformity, all older rocks, including Sariolian and Sumian (Tungudsk–Nadvoitsk series). The crust of weathering at the base of the Yatulian demonstrates the duration of the pre-Yatulian lacune. Secondly, the Segozero and Onega series, previously identified in the Yatulian sequence, are simply different facies (more specifically, formational) types with a definite structural affinity.

The Segozero type of the Yatulian is characterized "mainly by gritstone-sandy sediments with interlayers of quartzose conglomerates and clayey greenschists; carbonate rocks are less common and basic volcanics are present in various quantities" (SOKOLOV et al., 1973). Deposits of this type fill the large (up to 150 km wide) Karelian Depression. The latter extends from south of Segozero over a distance of 400 km northward on Soviet territory; it then passes into a trough which latitudinally traverses the territory of Finland from Lake Panajarvi to the Kemi River at the apex of the Gulf of Bothnia. The Yatulian here has a thickness of 2,000 m, although it is generally much less (400–600 m).

The Onega type of Yatulian is developed in the Southern Karelian Depression which encompasses the northern and western shores and the northwestern water area of the Onega Lake and is thus situated in the southeastern part of the Baltic Shield. The Onega type of Yatulian occurs in a thinner suite (up to 1,000 m); it not infrequently has an incomplete sequence which in some areas begins with the Middle Yatulian and in places also with Upper Yatulian deposits. The Yatulian of the Onega type is underlain by continental clastic rocks; carbonate rocks predominate higher up (dolomites and other rocks, with stromatolites); the top of the sequence is characterized by the famous shungites bitumen-bearing rocks. The Yatulian in the region of the Onega Lake is overlain conformably by the Suisarian series of basic volcanics (picritic basalts, etc.) associated with fills of gabbro-peridotites. The latter are particularly abundant in the extreme east of the Baltic Shield, in the region of the Vetreny Belt, in the zone bordering the Belomorides (White Sea Megablock); this may have certain tectonic implications (NOVIKOVA, 1971). The Suisarian series in the region of Petrozavodsk also plunges generally conformably under the Vepsian sedimentary deposits. These are the Kamennoborsk (Petrozavodsk) und Shoksha suites forming the southwestern shore of Lake Onega and the deepest part of the entire Onega Depression. This depression is bounded in the west by a fault with which centers of Suisarian lava effusions are associated.

Areas of the Karelian Megablock, outside these two depressions, are either completely devoid of Yatulian and generally Upper Karelian (Karelian proper) deposits, or are characterized by a sporadic occurrence of very thin strata of the continental sediments or basic lavas forming them. The Eastern Finland and the Northern and Southern Onega blocks are the largest salients of such pre-Yatulian formations.

Absolutely predominating among the fold structures of the Yatulian complex are synclines of two types (SOKOLOV et al., 1973). The first type are gentle brachymorphous synclines not complicated by faults, with flat bottoms and limbs inclined at angles of 15–30°.

These synclines are up to 20–30 km long. They are not infrequently complicated and divided into smaller brachysynclines by small, narrow, and steep anticlines, obviously connected with faults and often bringing pre-Yatulian rocks to the surface. The second type of Yatulian synclines are narrow (1 to 3 km) but long (up to 30–50 km) deep folds. Usually asymmetric (flat-dipping limbs inclined at angles of 15–20°, steep –60 to 80°–, sometimes overturned), complicated by faults. The latter in places cut off the steep limbs completely, thus converting synclines into monoclines or one-sided grabens.

As pointed out correctly by V. A. SOKOLOV and his colleagues (SOKOLOV et al., 1973), both types of Yatulian folds are undoubtedly connected with block deformations of the basement and therefore belong to the category of "reflected" folds. The first type belongs to the suprablock, the second to the suprafault or near-fault variant of these structural forms. The fold systems developed synsedimentarily but nonuniformly in time. Their structural plan changed gradually from NNW (submeridional) in the Early Yatulian to predominantly sublatitudinal in the south, at the end of the Yatulian and in the Suisarian time.

The Yatulian and Suisarian deposits are by most of their characteristics (the oligomictic, quartzose composition of conglomerates and sandstones, small thickness, the composition of the initial magma of volcanics and intrusive bodies, the nature of tectonic deformations, and the weak metamorphism) probably formations of the initial stage of eratonic consolidation in the central part of the Karelian Megablock. This is the interpretation of V. A. SOKOLOV and his colleagues (see SVETOV, 1979 for volcanics) who regard these deposits as protoplatform rather than orogenic formations as has been stated frequently. The Sariolian may be considered as Karelian molasse; the Sumian (Tungudsk–Nadvoitsk) volcanics exhibit traits of final rather than subsequent volcanism. The platform nature of the Vepsian has been generally recognized.

The Middle Proterozoic stage of evolution of the Karelian Megablock terminated in discordant intrusions of gabbro-diabases and dolerites aged up to 1,650 m.y., which traverse the Vepsian rocks in the southwestern Onega region[4]. In the southeastern part of the Northern Ladoga zone, near the boundary between the Karelian and the Svecofennian megablocks, there formed major intrusions of rapakivi granites (Vyborg and Salma) at that time. The latter, on the northeastern shore of Lake Ladoga is overlain by the clastic (conglomerates, gritstone, sandstones, etc.) Salma suite of the lower part of the Upper Proterozoic, which contains basic volcanics and tuffs. This is the only suite in Karelia, which may be regarded as true Jotnian, an analogue of the Swedish Jotnian (previously, deposits now called Vepsian were also referred to as Jotnian). The Salma suite underlies also the opposite shore of Lake Ladoga; it is hidden under the sedimentary cover of the Russian Platform in the southern half of the lake, but is known from the drilling data to fill the large and gentle Pish Depression (GARBAR, 1970).

All the above information on the Karelian Megablock concerned mainly its Soviet part which is covered by more detailed studies. Comparisons with the Finnish and Swedish parts of the megablock are still only tentative. That part of the megablock, belonging to Eastern Finland and reaching the Gulf of Bothnia near Kemi, is, according to Finnish investigators, composed mainly of pre-Karelian orthogneisses. These are overlain with a sharp unconformity, by platform (as determined by Finnish geologists such as SIMONEN, 1971, and others) Sariolian–Yatulian formations which are particularly widespread further north, west, and southwest (continuation of the Northern Ladoga zone) of this ancient massif. Analogues of the Western Karelian Gimola series wedge in at points between the Sariolian–Yatulian and the gneiss basement (SALOP, 1971). Finnish geologists also indentify the Kalevian above the Yatulian. It is composed of phyllites and micaceous greenschists. Still higher up, there is the Kumpu "formation" consisting of metaarkoses and conglomerates. The stratigraphic position of the Kumpu "formation" at the top of the sequence remains doubtful (SALOP, 1971).

The **Svecofennian Megablock** occupies a central position in the Baltic Shield and is the largest and the most peculiar one. It is situated practically completely outside the Soviet Union, in Finland, and Sweden (hence its name); however, its southeastern continuation plunges under the Russian Platform cover. In view of this, and the role of the Svecofennian Megablock in the structure of the Baltic Shield, it seems opportune briefly to describe its structure. Finnish geologists recently discovered that the boundary between the Karelian and the Svecofennian megablocks manifests itself in a broad fault zone (suture zone) which extends from the apex of the Gulf of Bothnia to Lake Ladoga. Numerous bodies of basic and ultrabasic rocks associated with sulfide-nickel mineralization are confined to this Bothnian-Ladoga Lineament. Towards the northwest, the lineament continues into Scandinavia and emerges in the Norwegian Sea north of the Lofoten Islands, where a major transform fault of the Northern Atlantic appears on its further trend. The Ladoga Depression lies on the continuation of the Bothnian-Ladoga suture zone in the southeast. The Pish Riphean Graben lies on the west and the Salma rapakivi granite pluton on the east. A possible continuation of the lineament in the region

---

[4] Older datings, up to 2,550 or even 2,455 ± 45 m.y. (KRATS et al., 1977) or, more recently (KRATS et al., 1984), 2,200–2,150 m.y. (U/Pb method on zircons) were obtained for the Yatulian diabases, confirming the Early Proterozoic age of this series.

of the Russian Platform will be discussed below. This suture zone is intruded by granites of the end of the Karelian–Gothian epoch.

The Svecofennian Megablock differs from the Karelian in several important features. The main ones are as follows:

The ancient Archaean granite-gneiss basement disappears here, at least in a tangible way, from the surface (with a possible exception of southern Finland). It may have subsided to great depths, as suggested by SALOP (1971) and undergone total anatexis, as MURATOV (1970) thinks. It seems, however, more likely, that the Svecofennian Trough originated directly on an oceanic crust (WATSON, 1977).

The second most important characteristic of the Svecofennian zone consists in the fact that the entire sequence of the Lower and Middle Proterozoic, including analogue of the Upper Karelian (proper), i.e., Yatulian-Suisarian and possibly also Vepsian, is represented here by typically geosynclinal formations. These underwent vigorous folding and deep metamorphism, as a rule, of the amphibolite facies. Even though analogues of the Yatulian-Kalevian, as well as of the Sumian and Lopian are undoubtedly represented in this sequence, it seems to be practically uninterrupted, i.e., development of the Svecofennian zone was generally of inherited nature. It is, however, quite possible that hiatuses and unconformities will later be revealed in this sequence at least in certain parts. Their presence is indicated by conglomerates and breccias with fragments of intrusive rocks, by abrupt changes in the degree of metamorphism, etc., as correctly pointed out by SALOP (1971). It is quite possible that the former division into slightly metamorphosed Bothnian and highly metamorphosed (gneisses) Svionian, now regarded by Scandinavian geologists as reflecting various intensities of metamorphism undergone by deposits of the same age, has a certain stratigraphic meaning.

A third characteristic of the Svecofennides is the appreciable role of various metavolcanics (including tuffs), both basic, close to spilites, and acid (the so-called leptites and hälleflintas) in the sequence. That these volcanics belong to an island-arc formation is quite obvious[5]. Quite conspicuous among the sedimentary rocks is the abundance and large thickness of graywackes, especially at the top of the sequence. The total (apparent) thickness of the deposits exceeds in some places (for example, in the Tampere region) 8–10 km. All this imparts to the sequence of the Svecofennian zone the typical appearance of eugeosynclines; most contemporary investigators, both Soviet and Scandinavian, regard this zone as such. One cannot, however, share their view of the adjacent Karelides zone as a miogeosyncline; this term could be appropriate only for the Savo–Ladoga marginal belt of the Karelian Megablock.

The fourth characteristic of the Svecofennian Megablock is the abundance of granites, i.e., a high degree of granitization. Some granitoid plutons are huge, for example, the Central Finland pluton is 250 km across. The Svecofennian granites have ages ranging from 1,900 to 1,500 m.y., and belong to at least two generations (VELIN, 1972): the earlier is synchronous with the Late Karelian folding and regional metamorphism (1,880–1,785 m.y.); the more recent generation (1,775–1,535 m.y.) corresponds to the time of transition from the orogenic to the cratonic stage of evolution.

Finally, the fold structure of the Svecofennian Megablock is very peculiar. It has a

---

[5] SALOP (1971) identifies a separate jaspilite-leptite subformation in the scope of this formation.

loop-shaped pattern with arcs convex alternately eastwards and westwards; sublatitudinal trends predominate in the south and submeridional ones in the north. As noted by
SALOP (1971), this complexity of the structural plan is due to the extensive occurrence of
major granitic plutons (especially the Central Finland one); he, like MURATOV (1970),
regards them as products of rheomorphism of the ancient granite-gneiss basement.
Highly stressed, usually isoclinal folds exist in the sedimentary-volcanic rock belts
preserved between the granitic plutons; the fold groups form synclinoria, less frequently
anticlinoria (KRATS et al., 1964).

The Svecofennian Megablock on the whole seems to be the region of a very vigorous
and prolonged manifestation of the geosynclinal process in the Early and Middle Proterozoic. It can by no means be considered as a median massif for this period of time, as
was suggested by MURATOV (1970).

After termination of the geosynclinal stage of evolution, at the beginning of the Later
Proterozoic, the central part of the megablock experienced uplifting; faults emerged,
relative subsidence and acid palingenetic magma eruptions took place on its western,
partly also on its southern and southeastern periphery. This led to the formation of a
series of the so-called sub-Jotnian in the west of Central Sweden. These consist mainly of
porphyrites, quartzporphyries and ignimbrites (the so-called Dala-porphyries) underlain by and alternating with conglomerates consisting of pebbles of Svecofennian rocks.
The overall thickness of the sub-Jotnian in a typical area is 2,000 m. It fills there the major
flat Malun-Idre Trough and is overlain unconformably by the Jotnian or Dala-sandstones
(quartzite-sandstones). Their overall thickness is also about 2,000 m; they are associated
with diabase and dolerite sills (BOGDANOV, 1967; GEYER, 1967).

Major granitic plutons of Smoland, Sorsele and Linn (younger part) are more or less
coeval with the sub-Jotnian volcanics; their radiometric age ranges from 1,745 m.y.
(Smoland) to 1,535 m.y. Rounded plutons of typical rapakivi granites formed during this
epoch (1,650 m.y. B.P.) They were found not only near Vyborg, but also on the Åland
Islands and in the north of Central Sweden.

Most Swedish geologists (GEYER, 1963; MAGNUSSON, 1965; LUNDEGÅRDH, 1971)
relate these formations to the Gothian cycle, setting the beginning of the latter at 1,750
m.y. B.P. VELIN (1972), however, is of the opinion that they complete the Svecofenno-
Karelian cycle. This is to a certain extent justified, since formation of these granitoids and
volcanics is practically a direct extension of the earlier synkinematic magmatism.

Conclusions. These are the principal data on the structure of the eastern Baltic Shield.
Each of the five main megablocks identified in it (the Southern Scandinavian is the fifth)
has a sequence of its own and is of Precambrian age; they also differ in crustal thickness,
which is minimal (30 km) in the White Sea and Murman Blocks of the Kola Megablock,
and maximal in the Karelian and Svecofennian megablocks (up to 40 km or more).
Transitional suture zones of especially complex structure were identified along all boundaries of the megablocks: the Imandra–Varzuga zone at the boundary between the Kola
and White Sea megablocks, the Northern and Eastern Karelian zone with its continuation
into the Vetreny Belt, at the boundary between the White Sea and Karelian megablocks,
the Savo–Ladoga zone between the Karelian and Svecofennian megablocks, and, finally,
the Gothian zone between the Svecofennian and Southern Scandinavian megablocks.
These sutures mostly represent overthrusts in the eastern and northeastern direction, as
was pointed out for Southern Scandinavia by BERTHELSEN (1976).

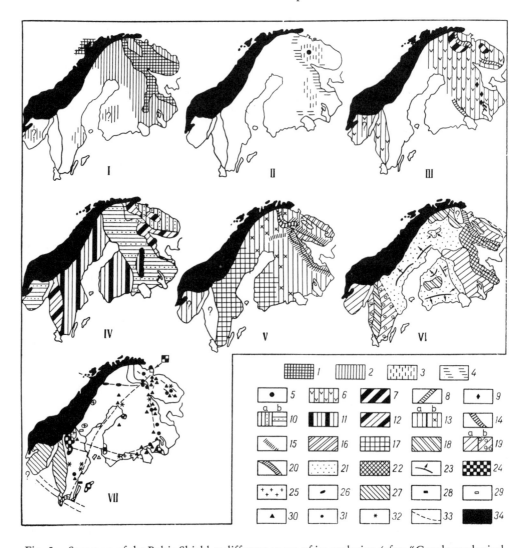

Fig. 5.   Structure of the Baltic Shield at different stages of its evolution (after "Geochronological boundaries . . . ", 1972).
Block with different age of the continental crust (m.y.): I = > 3,000, II = 3,000–2,800, III = 2,700 ± 100, VI = 2,500–2,150, V = 2,100–1,900, VI = (1,800–1,750) ± 50, VII = 1,700–1,000.
Areas of predominant development of: 1 = supracrustal rocks, 2 = granitoids; stabilized areas: 3 = proved; 4 = presumed; 5 = Moncha Pluton; areas of secondary reactivation: 6 = involved in reactivation; 7 = presumably not involved; 8 = zones of gneiss-granites, 2,500 m.y. old; 9 = intrusive granites, 2,500 m.y. old; 10 = areas of geosynclinal development (spilite-diabase and spilite-keratophyre associations): a) predominantly basic rocks, b) alternating linear zones formed of these associations and elevated basement blocks; 11 = areas of geanticlinal development (leptite-schist-porphyric associations and intense granite formation); 12 = elevated (stable) region of the median mass type; 13 = volcanic terrains (Sumian of Karelia and Kola Peninsula): a) andesite-dacite, diabase, spilite-diabase; b) ibid., with extensive acidic magmatism; 14 = zone of superimposed moderate-pressure granulite facies metamorphism 1950 ± 50 m.y.; 15 = ibid., high-pressure amphibolite facies 1950 ± 50 m.y.; 16 = blocks not involved in reworking of this age; 17 = areas of

Certain complexes, similar not so much by age as in the lithological respect, may be identified in several megablocks, if not the entire shield. The oldest of these complexes is basal for all the megablocks perhaps except the Svecofennian. This is a gray gneiss complex of tonalitic composition. The age of this ancient sialic basement must be at least 3.3–3.5 billion years (Fig. 5). It underwent metamorphism of the granulite facies as early as the Early Archaean; at the end of the Late Archaean it underwent diaphthoresis of the amphibolite facies and wide granitization.

There follows a typically eugeosynclinal sedimentary-volcanic, highly deformed, deeply metamorphosed and granitized complex corresponding to the Kola series of the Kola Megablock, Lopian of the Karelian Megablock, Bothnian, and Karelian (Upper) of the Savo-Ladoga zone and Svecofennian Megablock, and possibly represented by the Stura-Lö-Marstrand series in the Southern Scandinavian Megablock. Accumulation of this complex in the Kola and Karelian megablocks took place in the Late Archaean whose greenstone belts it strongly resembles; the deformations started in the Rebolian phase, about 2,700 m.y. ago. The upper boundary of this complex runs at the highest level in the Svecofennian Megablock where active geosynclinal (eugeosynclinal) conditions continued longest, up to 1,750 m.y. B.P., while noticeable stabilization began at the same time in adjacent megablocks. The Savo-Ladoga zone exhibits features of a miogeosyncline. Vergency in this zone, as in the Karelian Megablock, up to the boundary between the latter and the Belomorides, has a northeasterly direction, i.e., it is centrifugal with respect to the Svecofennian region.

The third complex, transitional from orogenic to early cratonic is of Early and Middle Proterozoic age and includes the Pechenga and Imandra–Varzuga series of the Kola Megablock, the Sumian–Sariolian–Yatulian–Suisarian, i.e., the "Middle" and "Upper" Karelian of the Karelian Megablock and the so-called Gothian supracrustal formations of the Gothian zone and the Southern Scandinavian Megablock. Accumulation of this complex had ended by the Late Karelian (Seletsk) epoch of tectonic deformations, metamorphism, and magmatism; it had all the features of a geosynclinal fold system with all accompanying phenomena only in the Svecofennian region and the Savo-Ladoga zone, and manifested itself in a general reactivation and radiometric rejuvenation of rocks in the remaining previously stabilized megablocks, especially the Kola and White Sea megablocks.

In the next stage, which corresponds to the interval 1,750–1,350 m.y. B.P. (the Early Riphean), the Kola, White Sea, and Karelian megablocks retained their elevated position and quiescent tectonic conditions. At the beginning of this stage, i.e., in the Hoglandian

---

sandstone-shale assemblages and intense acidic magmatism; 18 = very stable regions; 19 = orogenic zones, least mobile: a) quartzite-diabase assemblages; b) porphyric assemblages, granitoids; 20 = more mobile orogenic zones with predominant basic volcanics; 21 = very mobile orogenic zones (in deep sections – zonal high pressure metamorphism, granitoids); 22 = zones of secondary reworking; 23 = trend of metamorphic isograds, arrows showing direction of metamorphism increase; 24 = supracrustal series of platform type; 25 = rapakivi granites; 26 = intrusions of 1,500–1,450 m.y. age; 27 = regions of secondary magmatic reactivation dated 900–1,000 m.y.; 28 = pegmatites of 1,500 m.y. age; 29 = pegmatites of 1,380 m.y. age; K/Ar data in ancient rocks: 30 = 1,700–1,600 m.y. on amphibole and 1,600–1,500 m.y. on biotite; 31 = 1,600–1,450 m.y. on amphibole and 1,500–1,400 m.y. on biotite; 32 = 1,450–1,200 m.y. on amphibole and 1,400–1,200 m.y. on biotite; 33 = reactivation zones of 1,700–1,300 m.y. age; 34 = Caledonides.

of Polkanov (1947)[6], tectono-magmatic activity was high in the Svecofennian Mega-block, particularly the Gothian zone, and the Southern Scandinavian Megablock. It manifested itself in the formation of a late subsequent volcano-plutonic association of acid volcanics, ignimbrites, rapakivi granites, and similar rocks, as well as molassoids of the sub-Jotnian and its analogues. During the second half of this stage, i. e., after formation of the rapakivi, the Svecofennian Megablock entered, following the shield regions farther northeast, the cratonic stage of evolution proper, characterized by the accumulation of Jotnian sandstones. At the same time, partial regeneration of subsidence took place in the Southern Scandinavian Megablock. This produced the similar though not identical geosynclinal Dal and Seljur series. Movements of the Gothian tectono-magmatic epoch followed, causing quite pronounced folding of these formations. The Middle Riphean (1,350–1,000 m.y. B.P.) was a period of persistent tectonic activity and a high position of the Southern Scandinavian Block. It ended in uplifting, deformations, and intrusion of granites of the Bohus–Arendal complex in the Dalslandian tectonic-magmatic epoch. The Southern Scandinavian Block thereafter merged with the remaining part of the shield which had entered the phase of predominant upheaval.

Final separation of the Baltic Shield as independent and generally positive structural unit took place in the Late Riphean and Vendian. During this time a geosynclinal system formed in the northwest of Scandinavia, that produced the Scandinavian Caledonides in the Devonian. The Caledonian folded complex is underlain by a series of so-called sparagmite, i. e., sandstone, not infrequently coarse, with interlayers of carbonate rocks and clayey greenschists; an unconformity divides it into two parts: "gray sparagmite" with dolomite members and "red sparagmite" with limestone interlayers. A considerable part of the sparagmite formation is involved in the Caledonides nappe structure, forming their outermost (and lowest) nappes. Only in the extreme north, in Eastern Finnmarken, the Srednii Peninsula, and Kildin Island (USSR) does sparagmite occur relatively undis-turbed; however, even here the area of its occurrence is separated by a fault from the outcrops of the shield basement, and the contact retains its primary transgressive charac-ter only in some places.

The composition of the sparagmite formation, as was stressed by A. A. Bogdanov (1967), indicates strong uplifting of the source region which produced it. Had this region been the Baltic Shield, it would have undergone noticeable tectonic reactivation at the end of the Riphean. This, however, is not indicated by radiometric data: practically no datings give ages between 900 and 500 m.y. here (uplifts without a rise in the geoisotherms?)

Inland glaciation at the transition between the Riphean and Vendian, with an accumu-lation of tillites, was the culmination of these ascending movements. The situation stabilized gradually during the Vendian; this is indicated by the disappearance of tillites and the replacement of arkosic by quartzose sandstones at the top of the Varegian series, replacing the sparagmite series proper.

In the Cambrian–Ordovician–Silurian, the southern part of the shield underwent smooth subsidence associated with the development of the Grampian and Central European geosynclines. In the Devonian, during the final epoch of the Caledonian

---

[6] The Hogland Island in the Finnish Gulf is composed of porphyries similar to the Dala-porphyries of Sweden.

tectonic cycle, the Baltic Shield underwent general uplift. Its southeastern part separated from it, becoming involved in the subsidence of the Russian Platform (its Moscow Syneclise). The Baltic Shield has risen steadily since that time. This uplift was not large, judging by the preservation in its southern part, in Sweden and Finland, of outliers of a thin Cambrian–Silurian cover and in the north of the continental Devonian and pre-glacial weathering crusts. The Kola Peninsula was an exception. These uplifts intensified somewhat, especially in the north, only in the Devonian and Late Paleozoic, i. e., at the end of the Caledonian and Hercynian cycles. This is indicated indirectly by alkaline intrusion complexes and by the formation of the Oslo Graben. The main denudation of the shield, which exposed rocks formed at depths of up to 10–20 km, must have taken place before the Cambrian, during the Late Proterozoic; the quartzose sands of the Russian Platform Phanerozoic are the product of redeposition of Upper Precambrian sediments of the Baltic Shield and its nearest periphery. However, active erosion of the Kola Megablock continued also in the Phanerozoic.

2.3 The Ukrainian Shield and the Voronezh Massif

The Ukrainian Shield is, after the Baltic Shield, the second largest salient of the crystalline basement of the Eastern European Craton. It extends in the WNW–ESE direction in the southern part of the platform over a distance greater than 1,000 km, narrowing gradually toward the east; the band of Precambrian outcrops reaches 300 km in width in the west (Fig. 6). In the north and northeast, the Ukrainian Shield is truncated by boundary faults of the Pripyat–Dnieper–Donets trough system. In the west, the crystalline basement of the shield plunges smoothly under the deposits of the Dniester pericratonal downwarp, in the south under the Black Sea pericratonal downwarp. The Precambrian complex is exposed less in the Ukrainian Shield than in the Baltic Shield, mainly along river valleys. Young Cretaceous-Cenozoic deposits of the cover are exposed on water divides between these river valleys. The Konka–Yalyn Depression, east of the middle reaches of the Dnieper, is filled with these deposits. It separates the Azov Massif north of the Sea of Azov from the main part of the shield on the right bank of the Dnieper. The buried continuation of the Azov Massif, southeast of the Taganrog Gulf and the lower reaches of the Don River, is known as the Rostov Swell. The latter (DUBINSKY et al., 1978, consider it to be an independent block) is truncated in the east by the Sal'sk fault zone of NNE trend; however, further east, individual blocks of ancient granites and crystalline schists were revealed by drilling in a field of highly metamorphosed Carboniferous deposits. Consequently, the Ukrainian Shield, in its Azov–Rostov part, became a median massif separating the Donets Ridge and cis-Caucasian Hercynides, and then an anticlinorium between the Don–Caspian and cis-Caucasian Hercynide bands.

The block structure described above for the Baltic Shield is no less typical of the Ukrainian Shield, where it is particularly distinct in the magnetic field, and is confirmed by deep seismic sounding (DSS) data.

Generally speaking, the Ukrainian Shield consists of four or five principal blocks – megablocks of strongly granitized Archaean rocks. They are separated by narrow merid-ional suture zones along which lie synclinoria filled with Lower Proterozoic iron-ore bearing strata. In the magnetic field, the ancient granitized blocks are characterized by

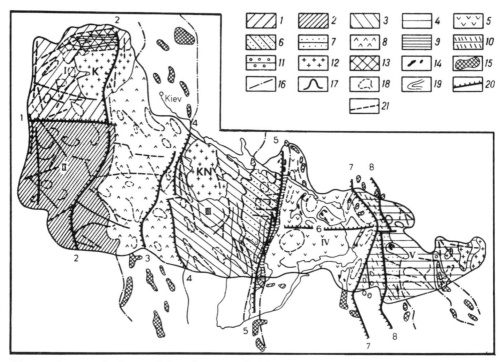

Fig. 6.   Structural scheme of the Ukrainian Shield (after "Tectonics of the Ukrainian Shield", 1972). 1–5 = blocks consolidated at the end of the Archaean: Volynian (I), Bug-Podolian (II), Kirovograd (III), Dnieprovian (IV), Azovian (V); 6–7 = marginal parts of Archaean blocks involved in geosynclinal subsidence in Early Proterozoic: 6 = West Ingulets, 7 = West Azov; 8–11 = Early Proterozoic protogeosynclines: 8 = Odessa–Kanev, 9 = Krivoy Rog–Kremenchug, 10 = Orekhovo–Pavlograd, 11 = Mangush; 12 = plutons of rapakivi granites and basic rocks; 13 = Ovruch Late Proterozoic intracratonic trough; 14 = outcrops of ferruginous quartzites; 15 = intense positive magnetic anomalies corresponding to buried deposits of ferruginous quartzites; 16 = boundaries of structural zones; 17 = contours of the shield; 18 = conventional contours of anticlinal structures in the basement; 19 = ibid., of synclinal structures; 20 = main deep faults; 21 = regional faults. Main faults: 1 = Andrushov, 2 = West Zhitomir, 3 = West Uman, 4 = Buz'ko–Mironov, 5 = Krivoy Rog, 6 = Devladovo, 7 = Orekhovo–Pavlograd, 8 = Azov–Pavlograd. Anorthosite–rapakivi plutons: K = Korosten', KN = Korsun–Novomirgorod.

mosaic anomalies of low intensity and alternate sign; the suture zones are characterized by strong linear positive anomalies. The ancient blocks, as shown by DSS data, are characterized by a crust of small or medium thickness (35–45 km), in some areas with a highly uplifted surface of Conrad discontinuity. The crustal thickness in suture zones was recently found to increase sharply to 50–55 km, mainly due to the increased thickness of the granitic layer (SOLLOGUB & CHEKUNOV, 1971).

The **Middle Dnieper** (Zaporozhye) **Block** occupies a central position in the structure of the Ukrainian Shield. This block is composed of the oldest rocks, with a radiometric age greater than 3,000 m.y. The inner structure of the Middle Dnieper Block resembles that of numerous other ancient blocks of the continental crust, in particular the Kapvaal and Zimbabwe massifs of South Africa. As was first shown by KALYAEV (1965), it is characterized by an abundance of gneiss domes between which variously oriented synclines are sandwiched. These are filled with the most ancient metabasite-jaspilite formation of the shield: the Konka–Verkhovtsevo series. Along the periphery of the block, as one approaches the faults limiting it, there is a tendency to the replacement of domes by arches (the Pyatikhatka Arch, etc.) extending parallel to these faults.

The Konka–Verkhovtsevo series, as believed by all its investigators, is a typical spilite-keratophyre or, more exactly, greenstone formation metamorphosed under the conditions of the amphibolite facies; its total thickness reaches 7–8 km. Two subformations are identified in the sequence of the formation from bottom upwards (KALYAEV, 1965): spilite-diabase, keratophyre-schistose; higher up there lies a jaspilite formation. Interlayers of ferruginous quartzites are present also in the spilite-diabase subformation; a considerable part of the keratophyre-schistose subformation is composed of metabasites. There is much volcanic material in the jaspilite formation; the sequence of the entire series is completed by volcanic terrigenous (schistose) formations metamorphosed only under conditions of the lowest grades of the amphibolite facies. This relatively weak metamorphism is typical also of supracrustal greenstone formations of the oldest massifs in other cratons.

The Konka–Verkhovtsevo series encloses numerous sills of serpentinized ultrabasic rocks.

Dome structures occupy the greater part of the territory of the Dnieper Block and are up to 60 km across (the Saksagan Dome). Their shapes vary from rounded to oval, with a transition to arches. The domes and arches are composed of migmatites among which relics of rocks of the Konka–Verkhovtsevo series were preserved; a major role belongs also to granitoids: older plagiogranites lying conformably in the structure of the domes (the Dnieper complex) and younger microcline, not infrequently aplite-like and pegmatoid granites, obviously cross-cutting with respect to the inner structure of the domes.

The domes and arches are separated by narrower synclines whose shape is determined by that of the domes. The Saksagan Dome and the Demurin Dome further north are separated only by a latitudinal fault zone along which there are numerous bodies of serpentinized gabbro-peridotites. This basic to ultrabasic formation is clearly younger than ultrabasic rocks occurring conformably with the rocks of the Konka–Verkhovtsevo series.

Radiometric determinations of the age of minerals from the Konka–Verkhovtsevo rocks furnished figures of about 3,100–3,000 m.y. This proved the Lower Archaean age of this series. It had previously been compared with the Lower Proterozoic Krivoy Rog series which also contains jaspilites, however, in association with sedimentary rather than volcanic rocks. The Dnieper granitoid complex has been dated radiometrically as 2,800 to 2,600 m.y. old, i. e., it is also Archaean.

It has been proved recently that the Middle Dnieper block comprises older rocks than the greenstone Konka–Verkhovtsevo series, such as tonalites, metaporphyries and metadacites, i. e. gray-gneiss-type rocks (The oldest granitoids …", 1981). Thus, the Middle Dnieper Block, like the Murman and Kola blocks of the Baltic Shield, is one of the few relics of the oldest complexes of the basement of the Eastern European Craton. Geological data make it possible to trace the continuation of the Middle Dnieper Block north of the Dnieper–Donets Depression to the Sumy–Belgorod –Kharkov region (GAFAROV, 1971).

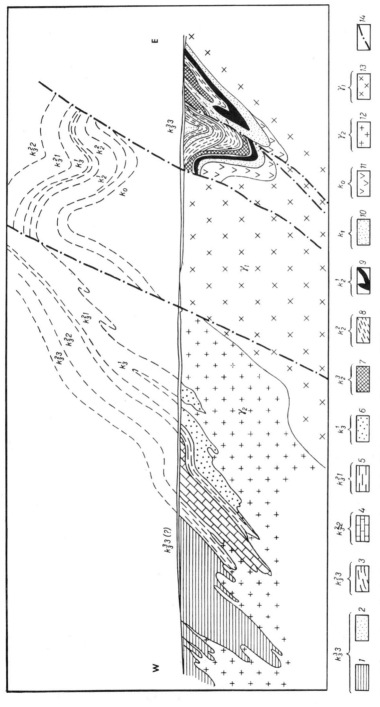

Fig. 7. Schematic geological cross-section of the Krivoy Rog Basin and West Ingulets zone (after DOBROKHOTOV, 1969). 1 = gneisses and "microgneisses", plagioclase-biotite; 2 = metasandstones and metaconglomerates; 3 = quartz-sericite-graphite and quartz-feldspar-biotite schists; 4 = dolomite marbles and calciphyres; 5 = quartz-sericite-graphite schist, locally plagioclase-bearing; 6 = metasandstones and white quartzites with schist intercalations; 7 = ferruginous quartzites, phyllitoid schists; 8 = phyllitoid schists; 9 = ferruginous quartzites; 10 = metaconglomerates, metasandstones, phyllites and talc schists; 11 = amphibolites; 12 = Ingulets granites and migmatites; 13 = Saksagan plagiogranites; 14 = faults.

The Dnieper Block is bounded in the west by the meridional Krivoy Rog (Kremen-chug-Krivoy Rog) near-fault synclinorium. Being just up to 10 km wide, it can be traced directly by exposures over 50 km; drilling data make it possible to follow it over 220 km. Magnetic anomalies enable one to extend this zone northward, to the Belgorod–Mikhailov zone of KMA (Kursk Magnetic Anomaly), and in the south to the Kherson Magnetic Anomaly which extends it further to 1,000 km. Together with the Western Ingulets marginal zone of the Kirovograd Block, parallel to it in the west, the Kremenchug–Krivoy Rog zone reaches 70 km in width (against 170 km for the Dnieper Block).

The Krivoy Rog Synclinorium proper is a narrow syncline of Lower Proterozoic formations of the series of the same name, dipping east, complicated by higher-order fold systems. The western limb is reduced (initially, as believed by KALYAEV, 1965), and truncated by a deep fault, it has a more quiescent eastern limb (Fig. 7). The Krivoy Rog Fault has the character of an upthrust-overthurst, with the surface dipping steeply westward, under the meridional Ingulets gneiss arch. The faults around it complicate the inner structure of the synclinorium. The eastern limb of the Krivoy Rog Synclinorium is at the same time the western limb of the Saksagan gneiss-migmatite zone.

The Krivoy Rog series is underlain by a metabasite series which is shown by radiometric evidence to be of Archaean age. It is probably an analogue of the Konka–Verkhovtsevo series of the Middle Dnieper Block. The Krivoy Rog series itself is subdivided into three suites. The lower suite is of terrigenous composition, starts with conglomerates, and consists mainly of quartzite-sandstones, quartz-sericite, chlorite, and graphite schists (phyllites); schists predominate at the top of the suite; conglomerate (gritstone) interlayers occur throughout its lower part. The middle suite contains major deposits of ferruginous quartzites of the Krivoy Rog Basin. Ferruginous quartzite and jaspilite members are separated by schist members, forming a rhythmic alternation with them. KALYAEV (1973) considers the jaspilite formation of the Krivoy Rog region to be an analogue of the flysch formation of younger geosynclines. The upper suite of the Krivoy Rog series, especially its upper part, has a molasse character; it starts with quartzite-sandstones overlain by carbonaceous-graphite and sericite schists; there follows an alternation of these schists with quartz-carbonate schists and dolomitic marble; the sequence is completed by conglomerates, coarse-grained quartzite-sandstones with minor members of the same schists and marble. This coarse upper molasse is up to 1.7–1.9 km thick. The overall thickness of the Krivoy Rog series reaches up to 8 km. Its metamorphism proceeded under the conditions of higher grades of the greenschist – lower grades of the amphibolite facies.

The Krivoy Rog series is intruded by aplite-like and pegmatoid granites aged 2,000–1,800 m.y., which belong to the so-called Kirovograd–Zhitomir complex. This shows that formation of the Krivoy Rog Synclinorium ended in the Early Karelian epoch of diastrophism. A comparison between the synclinorium structure with that of adjacent migmatite-gneiss domes and arches shows that growth of the latter structures, which had started in the Archaean, continued or rather recommenced during this epoch.

The Western Ingulets zone was produced by subsidence of the eastern margin of the Kirovograd Block linked with the closure of the Kremenchug–Krivoy Rog Trough. Deposits of the upper suite of the Krivoy Rog series are common here. They occur in many cases with a gentle slope, filling synclines which are parallel to the Krivoy Rog Fault in the east and assume a northwesterly trend further west. The synclines are separated by minor granite-migmatite domes and arches. The boundary with the main part of the Kirovograd Block is indistinct, since the outer contours of the Krivoy Rog series are sinuous.

The Orekhov–Pavlograd suture zone, which borders the Dnieper Block in the east and separates it from the Azov Block, is generally less distinct than the Krivoy Rog–Kremenchug zone. As suggested by KALYAEV (1973), it rather resembles the Western Ingulets zone. Only individual synclinal remains of analogues of the Krivoy Rog series, including its middle iron-ore suite, are preserved in this zone. As in the Western Ingulets zone, the series thickness is much less than in the Dnieper Block. A major fault lies west of this zone; it also has the character of a steep upthrust-overthrust with the surface dipping west.

The Azov Block is characterized by various paragneisses of the amphibolite facies, migmatites, and granitoid rocks. Individual radiometric datings point to the Archaean age of the gneisses; the western part of the block contains relics of jaspilite formations of the Konka–Verkhovtsevo series. However, the majority of radiometric determinations indicate an age of 2,300–2,100 m.y., proving pronounced alteration of the Azov Block rocks in the middle of the Early Proterozoic. This tectonomagmatic epoch about 2,300 m.y. ago, called Azov–Volyn epoch (KOZLOVSKAYA et al., 1971), was very important in the evolution of the Ukrainian Shield structure. It was also revealed in the Voronezh Massif and may correspond to the Seletsk phase of Baltic Shield diastrophism (see ch. 2.2). The gneiss complex has a general NNW trend; however, against this background there are dome-like and arch-like structures much less distinct than in the Dnieper and Bug regions. A group of alkaline plutons of Middle Proterozoic age (about 1,700 m.y.) was found in the eastern part of the block.

Magnetic anomalies and drilling data suggest the existence, within the western boundary of the Rostov salient (block), of a Lower Proterozoic zone similar to the Krivoy Rog–Kremenchug and Orekhovo–Pavlograd zones. Data of aeromagnetic surveys indicate that this zone extends far to the south, in the direction of Krasnodar or even Maikop. The Rostov Block is composed of Archaean gneisses and Proterozoic (Middle Proterozoic?) granites.

The **Kirovograd Block** occupies a central position in the Ukrainian Shield. It is situated between the Krivoy Rog–Kremenchug suture zone in the east and the similar Odessa–Kanev zone in the west. Like the Azov Block, it is composed mainly of paragneisses of the amphibolite facies, partly Archaean and partly Lower Proterozoic (the Teterev series) having a radiometric age of 2,300–2,000 m.y. and a general NNW trend; however, linear structures here predominate over dome-like ones, except in the Western Ingulets zone which was discussed earlier. Rocks in individual blocks have Archaean isotope datings. The block contains major granitic massifs (the Kirovograd Massif and others); in its northwestern part there is the huge Middle Proterozoic Korsun'–Novomirgorod stratified pluton of gabbro and rapakivi granites.

The western part of the Ukrainian Shield, west of the Odessa–Kanev Deep Fault, is subdivided into two blocks by a sublatitudinal fault; the Volyn Block in the north and the Bug–Podolian Block in the south.

The **Bug–Podolian Block** is composed entirely of Archaean formations having a radiometric age of 2,600 m.y. (individual determinations give ages of up to 3.5 b.y.!), metamorphosed in the granulite facies. These are pyroxene-plagioclase gneisses and charnockites with sheet-like intrusions of biotitic granites containing garnet and cordierite (the so-called Chudnov-Berdichev granite). The structural pattern of the block is typically nonlinear, with numerous but relatively small domes whose cores usually contain aplitoid and pegmatoid granites.

The **Volyn Block** is similar to the Kirovograd Block in terms of age and composition of the constituent rocks, and its structure; it is separated from the latter block by the Odessa–Kanev Fault zone. There also is a similarity with the Azov Block. Ancient Archaean rocks altered during the Azov–Volyn epoch (2,300 m.y. B.P.), predominate in its southwestern part which is transitional to the Bug–Podolian Block. Farther east and north lie rocks of the lower part of the Lower Proterozoic; these are mainly gneisses and migmatites. The predominant structural orientation is NW–SE, i.e., typical of the Azovo-Volynides. There also occur quite large granite plutons of the same epoch (Zhitomir, Uman, etc.). In the northern part of the block, there is the major Korosten, much differentiated, pluton of rapakivi granites and gabbroids; they formed, like the Korsun'–Novomirgorod Pluton, at the end of the Middle Proterozoic, during the cratonization stage. It is probably no coincidence, as noted by SEMENENKO (1972), that these two plutons, as well as coeval alkaline intrusions of the Azov region, are confined to a band of the northeastern boundary of the Ukrainian Shield. DSS investigations showed that the Korosten Pluton is of a bed-like shape, with the lower interface lying at a depth of 17–18 km (SOLLOGUB et al., 1963). The formation sequence of the Korosten Pluton rocks was studied by POLKANOV (1938). According to him, basic rocks intruded first, and then granites which occupied a position between the enclosing rocks and the gabbro massif, as well as above the latter. Some recent works, however, substantiate the metasomatic origin of the Korosten and Korsun'–Novomirgorod rapakivi granites.

The extreme northwestern part of the shield is occupied by the peculiar Osnitsk zone of sublatitudinal-northeasterly trend. It is separated from the Volyn Block by the Sushchany–Perzhansk Fault of the same trend. The Osnitsk zone differs considerably from the remaining complexes of the shield in the composition of its constituent rocks, the extent of their metamorphism, and the relatively simple structure. It is the youngest formation of the shield basement. The Osnitsk complex is a late subsequent volcano-plutonic association consisting of quartz porphyries, metamorphosed tuffs of acid composition (similar to leptites), migmatites, and granites, with some amounts of intermediate and basic rocks. The relationships with the Korosten Pluton further east show that the Osnitsk complex is older than this pluton; on the other hand, it must be younger than the Krivoy Rog series and therefore it most likely belongs to the Middle Proterozoic, possibly being an analogue of the Scandinavian sub-Jotnian. It is not ruled out that the formations of the Osnitsk complex are the fill of an intracratonal geosyncline.

The Ovruch series is younger and of more obvious cratonic nature. It fills the latitudinal syncline of the same name, which is about 90 km long and up to 20 km wide in the extreme north of the Volyn Block. The southern limb of this syncline overlies the eroded granite surface of the Korosten Pluton; its northern limb borders the same granites as well as the Osnitsk complex (in NW) along a major fault. Consequently, the Ovruch series is younger than the Korosten Pluton and must have originated at the beginning of the Late Proterozoic, at about the same time as the Swedish Jotnian. The Ovruch series is composed of quartzites, quartzose sandstones, shales, porphyrites, orthophyres, and diabases; these rocks are in the initial stage of regional metamorphism and are deformed into small folds which, alongside faults, complicate the structure of the Ovruch Syncline.

Younger Riphean and Vendian deposits are not involved in the shield structure, but lie transgressively on its slopes; they fill the Dniester, Black Sea, and partly the Pripyat troughs.

The **Voronezh Massif.** The top of the Precambrian basement lies above sea-level over a considerable area in the highest part of the Voronezh Anteclise, and at elevations of up to 500 m in an even larger zone. This and especially the Earth's biggest deposits of iron-ores in the Precambrian of the Voronezh Massif (the region of the Kursk Magnetic Anomaly), cause the latter to be traversed by large quarries and numerous boreholes. It has been covered by extensive geophysical studies; this has made it possible to prepare a geological map of its surface and to identify the principal features of its stratigraphy, structure, and magmatism (GORBUNOV et al., 1969; GORBUNOV et al., 1973; KRASOVITSKAYA & PAV-LOVSKY, 1970). The Voronezh Massif was thus an important intermediate reference region permitting comparison between the Precambrian of the Baltic and Ukrainian Shields and preparation of a map of the Russian Platform basement.

The Precambrian basement of the Voronezh Massif, as in other areas of the Eastern European Craton, has a pronounced block structure; the faults separating the blocks and their constituent rocks have a trend ranging from NW–SE to meridional. The formations of the Oboyan series are assumed to be oldest (Lower Archaean). The lower part of this series has a metabasite composition and contains charnockite-like rocks; the main part of the series is composed of gneisses with amphibolite interlayers; it experienced strong migmatization associated with plagiogranites. A complex of small lenticular bodies of gabbro-amphibolites, pyroxenites, peridotites, and serpentinites, called Besedin, is also connected with the Oboyan series. The Mikhailov series, a spilite-keratophyre formation containing magnetite quartzites, also belongs to the Archaean. Typical komatiites were found in its lowermost part (KRESTIN, 1980). Both series, Oboyan and Mikhailov, are considered now as corresponding to the Konka–Verkhovtzevo greenstone series of the Ukrainian Shield. The Mikhailov series also encloses small bodies of ultrabasic rocks and much larger bodies of plagioclase granites, as well as plutons of microcline-plagioclase and younger Early Proterozoic aplite-pegmatoid granites. The Early Proterozoic structure contains consolidated, partly reworked blocks of Archaean rocks bounded by faults.

The Lower Proterozoic in the massif is represented by the Kursk series which is an analogue of the Krivoy Rog series of the Ukrainian Shield; the main deposits of the KMA (Kursk Magnetic Anomaly) ferruginous quartzites are associated with it. Coarse detrital rocks, as well as micaceous and carbonaceous shales, phyllites, etc. lie at the base of the series; ferruginous quartzites are concentrated in its middle; the upper part has a metater-rigenous psammite-pelite composition. Metamorphism is pronounced in the southeastern part of the massif; its rocks pass into micaceous shales and gneisses. The deposits of the Kursk series form synclinorial bands which, by tracing magnetic anomalies and by geological comparisons, can be correlated with the Krivoy Rog–Kremenchug and Orekhovo–Pavlograd zones of the Ukrainian Shield. The synclinoria consist of isoclinal fold systems of western vergency.

Folding of these formations, which occurred about 2,000–1,950 m.y. B.P., may have been coeval with formation of the Stoilov–Usman gabbro-plagiogranite intrusive complex, followed by the Pavlovsk normal granitoid complex.

The eastern part of the massif, separated from its remainder by a major deep fault, is composed of the Vorontsovo series of Lower and Middle Proterozoic age (2,300–1,750 m.y. B.P.). This series contains clasts of the rocks of the Kursk series and is of terrigenous composition, not infrequently with appreciable amounts of volcanics. Its analogues were also found in the KMA region, where there are also carbonate rocks in the middle part. Metamorphism there is generally very slight (phyllite stage). Simultaneously with deposition of the series in the principal area occupied by it, a marginal volcanic belt superposed on an Archaean block formed on the other side of the fault bounding it. The rocks of this belt vary in composition from basalts to dacites. The Vorontsovo series is deformed to broad folds complicated in places by smaller ones, without persistent vergency, but with highly sinuous axes. It encloses numerous relatively small stratiform intrusions of basic to ultrabasic composition, with copper-nickel sulfide mineralization (the Trosnyansk-Mamonovo complex). The younger granitoids range from quartz diorites to binary granites.

The youngest formations of the massif include early cratonic metavolcanic rocks of basic-acid composition, as well as a complex of basic rocks and granitoids (granite- and syenite-porphyries, microcline granites and pegmatites) having an age of 1,450–1,200 m.y.; this corresponds to the Gothian tectonomagmatic epoch. This prolonged magmatic activity of the Voronezh Massif corresponds to a similar activity in the northwestern part of the Ukrainian Shield.

2.4 The Russian Platform

2.4.1 Inner structure of the basement

There is no doubt whatsoever that the basement of the Russian Platform is composed mainly of the same Archaean, Lower and Middle Proterozoic metamorphic and intrusive rocks which are exposed in the Baltic and Ukrainian shields and at the summit of the Voronezh Anteclise. These rocks have by now been traversed by several thousand deep boreholes; by 1972 there had been 2,500 such boreholes in the Volga–Uralian region alone. The material on the core samples was processed by VARDANYANTS & TIKHOMIROV (1968), who published the first petrographic map of the basement and also by VESELOVSKAYA (1963) and LAPINSKAYA (1966), and later with participation of S. V. BOGDANOVA (see NEVOLIN et al., 1968). Much of this material was studied radiogeochronometrically; this made it possible to determine more accurately the age of the rocks traversed by drilling. Reports on these determinations were published by VINOGRADOV, TUGARINOV et al. (1960) and SEMENENKO (1970). Geophysical observations, in particular magnetometry in combination with gravimetry, contributed much to the coordination of all these data and reconstruction of the general structural pattern of the basement. It was shown as far back as 1937 by ARKHANGEL'SKY, ROZE et al., and then in 1946 by SHATSKY that the character of magnetic anomalies indicates two types of structural elements in the platform basement: some of them are characterized by relatively weak mosaic anomalies, others by much stronger band-like anomalies. In the shields, magnetic fields of the first type correspond to Archaean blocks, those of the second type to Lower and Middle Proterozoic fold zones. Correlation with drilling data and radiometric dating supports these assumptions. This method was further developed successfully by E. E. FOTIADI (1958) and then by GAFAROV (1963, 1971), NEVOLIN (1968), DEDEEV (1972), and others. GAFAROV, NEVOLIN, and DEDEEV published recent maps of the basement of the Eastern European Craton (NEVOLIN and DEDEEV within the borders of the USSR). All these maps, which differ appreciably even in most essential details, show that the platform basement consists of numerous (28 on GAFAROV's map, 19 on the map of DEDEEV!) relatively small isometric blocks (massifs) of epi-Archaean consolidation, as well as of Proterozoic (PR_{1-2}) Early and Late Karelian geosynclinal fold-zones separating them. The Archaean rocks are in the granulite, less frequently amphibolite grade of metamorphism; the Proterozoic rocks experienced metamorphism mainly in the amphibolite facies. The largest of the Archaean blocks is mapped by GAFAROV and DEDEEV on the southeastern continuation of the White Sea Megablock of the Baltic Shield (Fig. 8); in the south, GAFAROV (1971) identifies another major block beneath the North Caspian Depression. Alongside Archaean blocks proper, the investigators identify Archaean blocks which were reworked in the Early Proterozoic. This was as a rule accompanied by regressive metamorphism of the amphibolite or epidote-amphibolite facies.

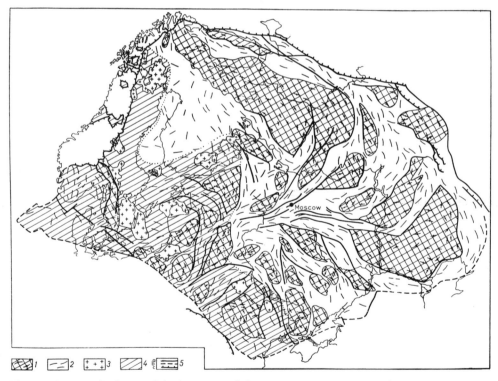

Fig. 8.   Structural scheme of the basement of the East European Craton (after R. A. GAFAROV, T. A. LAPINSKAYA, S. V. BODGANOVA, simplified). 1 = crustal blocks consolidated at the end of the Archaean; 2 = Karelian fold systems; 3 = plutons of rapakivi type granites; 4 = area of Gothian – Dalslandian reworking; 5 = main faults; a) vertical and subvertical, b) ibid. presumed, c) over-thrusts and nappes.

The Karelides and the basement rocks in general, as viewed by GAFAROV (1971) and DEDEEV (1972), form an arc in the northern half of the platform which is convex to the east. The trend changes from northwesterly, typical of the Baltic Shield, to northeasterly and latitudinal at the Moscow latitude. Southeast of Moscow, southeasterly trends predominate, meridional trends, characteristic of the Ukrainian Shield, predominate south of Moscow. The Moscow region, as shown on the diagrams of GAFAROV (the most structural of all the three) and DEDEEV, is the center of virgation of the Karelides which fans eastward. NEVOLIN's map interprets these relationships in a somewhat different manner: he shows, east of Moscow, an intersection of structures of northwesterly and northeasterly trends, regarding these two diagonal trends as most typical of the crystalline basement; a sublatitudinal trend was revealed only in the eastern part of the platform.

Neither of these three maps, based principally on geophysical evidence, gives sufficient attention to the processes of basement tectonomagmatic reworking at the beginning and in the middle of the Late Proterozoic. These are distinctly indicated by extensive intrusions of the same age (their radiometric dating indicates ages between 1,750 and 1,100 m.y.). Plutons of rapakivi granites and associated gabbroid rocks, including anorthosites, formed at the beginning of the Late Proterozoic (about 1,650 m.y. B.P.); they were found to be typical not only of the southern periphery of the Baltic Shield and the western part of the Ukrainian Shield, but were also revealed by drilling in the area between these two regions: in the Baltic region (the huge Riga Pluton) and in Northern Poland (ZNOSKO et al., 1972). This indicates that extensive reactivation processes occurred at the time throughout the western part of the platform. "Young" radiometric datings show a tendency to "spread" from here eastwards and southeastwards to the Volga–Uralian region along major fault zones in the basement. The "material bearers" of these Gothian–Grenville datings are zones of crushing, metasomatic alteration of ancient rocks, and intrusions of basic rocks, mainly gabbro-diabases. Reworking of the basement must have proceeded simultaneously with the formation and development of the system of Riphean aulacogens "imprinted" in the base of the sedimentary cover of the Russian Platform.

A recent scheme of the interior structure of the East European Craton has been compiled by a group of geologists from Leningrad, headed by KRATS & ZAPOLNOV (see "Main problems ...", 1979). Main structural units distinguished in this scheme, like in the previous ones and of approximately similar shape, are angular-isometric "fields" and linear zones called "field boundaries", but their age and genesis are interpreted differently. It is believed that they originated before Archaean times (> 3.5 b.y.) or in Early Archaean, the "boundaries" being, in fact, granulite belts, especially zones of basic granulites separating granite-gneiss fields. It is assumed that these structures successively developed in Early Archaean. In Late Archaean-Early Proterozoic, when in the central parts of the "fields" "vortex massifs" originated (apparently corresponding to gneiss ovals of SALOP, 1971), the "boundaries" were transformed either into geosyncline-type fold systems, or into zones of structural-metamorphic reworking. The authors emphasize the connection between the aulacogens of the Russian Platform and the above mentioned mobile zones of the basement, which, to their opinion, is indicative of close inheritance in the development of the craton's structure at early stages of its development.

2.4.2 The structure of the sedimentary cover

If the structure of the Russian Platform sedimentary cover is represented most schematically, i.e., according to the hypsometry of the basement surface (see Fig. 2), the following pattern will result: The Baltic Shield is encircled in the south, southeast, and east by a broad subsidence zone which includes the Baltic Syneclise in the west, the Moscow Syneclise in the center, and the Mezen' Syneclise in the east (northeast). This Baltic–Mezen' zone, which originated in the Riphean and developed further in the Early Paleozoic, where the basement lies at a depth of 4–5 km, separates the Baltic Shield from the zone where the basement lies higher. This zone has a general northeasterly trend and includes the Ukrainian Shield, and the Mazurian–Belorussian, Voronezh, and Volga–Uralian anteclises. The basement surface in this zone lies at increasingly greater depths in the easterly direction, ranging from positive elevation marks to minus 1–1.5 or even 3 km in the region of Perm. The Ukrainian–cis-Uralian uplift zone (the buried ancient Sarmatian Shield identified by SHATSKY, 1946) gives way in the southeast to the very deep North Caspian Syneclise. The latter appears to be the antipode of the Baltic Shield occupying the opposite corner of the Eastern European Craton.

Besides this SW–NE zonality, the structure of the platform basement surface exhibits a somewhat different zonality. The Baltic Shield is linked via the Latvian Saddle, where the basement lies at depths of less than 0.5 km, with the Mazurian–Belorussian Anteclise; the latter is linked, via the similar Polessie Saddle, with the western part of the Ukrainian Shield, thus forming the meridional Baltic–Ukrainian uplift zone. East of this zone, there is a general subsidence in the direction of the Paleozoic Uralian geosyncline; it runs to the west, over a shorter distance, toward the Central European Paleozoic geosyncline. To put it differently, this meridional axis of the Eastern European Craton plays a role similar in many respects to the Nemaha–Boothia axis of the North American Platform. The main evolutional stage was Hercynian for both geosynclines; activity on this axis became most pronounced from this particular stage on to the Recent epoch, when the Baltic–Black Sea water divide was confined to it over a considerable distance.

Below, we present a more detailed description of the main structural elements of the Russian Platform, proceeding from west to east and from north to south.

The **Baltic Syneclise** occupies the northwestern part of the platform. The slope of the Baltic Shield forms its northern boundary. In the west, it is separated from the Moscow Syneclise by the Latvian Saddle. In the south, the syneclise is adjacent to the Mazurian–Belorussian Anteclise, in the southwest to the Denmark–Poland suture fold zone which formed along the edge of the ancient craton. The Baltic Syneclise developed practically throughout the Phanerozoic; however, its structural pattern changed repeatedly (SUVEIZDIS, 1972), being characterized by rather moderate rates of subsidence. The syneclise acquired its present shape mainly at the end of the Paleozoic and beginning of the Mesozoic, i.e., during the Late Hercynian–Early Cimmerian stage of evolution. Its greatest depths, of about 5 km, occur in the Gdansk Gulf of the Baltic Sea; the axis extends in the northeastern direction, in which its basement generally becomes flat. The basement surface and partly also the bottom of the sedimentary cover are in some places dislocated by faults of mainly latitudinal trend (Riga and Neman); there may also exist a submeridional (NNE) fault along the eastern shore of the Baltic Sea. Local uplifts complicate mainly the southern part of the syneclise; here, oil deposits were found at the base of the sedimentary cover (Kaliningrad region).

The **Moscow (Central Russian) Syneclise** is, after the North Caspian Syneclise, the largest negative structure of the Russian Platform; it is situated in the center of the Eastern European Craton, between the Baltic Shield in the northwest, the Belorussian Anteclise in the west, the Voronezh Anteclise in the south, and the Volga–Uralian Anteclise in the southeast. The boundary with the Mezen' Syneclise is drawn quite arbitrarily, approximately along the Northern Dvina. Evolution of the Moscow Syneclise was preceded by formation, in the Riphean, of a system of echelon-like grabens and graben-like troughs, ranging from Valdai to Soligalich, which together form the so-called Central Russian Aulacogen of ENE trend. This aulacogen borders and cuts off the salients and troughs, of meridional to northwesterly trend, which complicate the Baltic Shield slopes. Another deep Riphean Graben – the Moscow Graben – was revealed in the south of the syneclise; a borehole drilled in it near the town of Pavlov Posad, southeast of Moscow, had not penetrated the base of the Riphean deposits at a depth of 4,783 m. On the whole, however, the deepest part of the syneclise is in the region of Yaroslavl'–Kotlas.

In the Vendian and Early Paleozoic, the Moscow Syneclise constituted a single structure with the Baltic Syneclise; it became separated from the latter only at the end of the Devonian. Downwarping of the Moscow Syneclise continued, with interruptions at tectonic stage boundaries (end of Silurian–beginning of Middle Devonian, Middle Triassic), nearly to the end of the Cretaceous, after which deposition here ceased. However, the zone retained a tendency toward subsidence relative to adjacent structures of the platform.

Since the Vendian, the Moscow Syneclise has developed as a typical platform depression; it therefore possesses rather flat slopes (less than 1°); the strata within the syneclise lie practically horizontally. Besides, it is relatively little complicated by dislocations of higher orders (flexures, arches, and local uplifts); the existing structures of this type also dip very gently. The Sukhon Arch in the central part of the syneclise is the most remarkable structural complication of the sedimentary cover. It formed over the Soligalich Graben-Trough of the base of the sedimentary cover; this arch is composed of Permian deposits at the surface.

The **Mezen' Syneclise,** situated in the extreme northeastern part of the Russian Platform between the eastern slope of the Baltic Shield, the northern plunge of the Volga-Uralian Anteclise, and the Timan, is closely connected with the Moscow Syneclise; it is often regarded as part of the latter under the name of Mezen' Basin. The deep structure of the Mezen' Syneclise is still little known; the recently begun seismic profiling has shown that the basement in the southern part of the syneclise lies much deeper (more than 4 km) than was believed hitherto. In the northeast, the syneclise goes over into the peri-Timan Trough which is up to 4–5 km deep; in the northwest, it merges with the Kandalaksha Graben which splits the eastern slope of the Baltic Shield. The wider southeastern continuation of this graben is divided into two independent troughs by the Arkhangel'sk Swell. The Mezen' Syneclise, like the Moscow one, originated in the Vendian; Riphean aulacogens may occur at its base. This development continued with interruptions to the end of the Jurassic and beginning of the Cretaceous, after which the Mezen' Syneclise became involved in the general uplift of the northern part of the Russian Platform.

The inner structure of the cover of the Mezen' Syneclise is still little known; it may be quite complex.

The **Mazurian–Belorussian Anteclise** forms the westernmost link in the central uplift zone of the Russian Platform. It has a sublatitudinal trend, but a promontory issues from its northern limb in the northeastern direction. It runs across the Latvian Saddle to link with the Loknov Promontory of the southern slope of the Baltic Shield. In the south, the anteclise borders the Pripyat Aulacogen along a major fault with an amplitude of up to 3 km; in the west it borders the Brest Trough; in between lies the Pinsk Saddle which links the Belorussian Anteclise with the Ukrainian Shield. In the west, this anteclise is truncated by the meridional Olsztyn Fault, behind which the basement subsides abruptly, from 2 to 10 km toward the Baltic–Podolian Lineament (pericratonal downwarp). In the northwest, a gentle slope links the anteclise with the Baltic Syneclise; in the east it is separated from the Voronezh Anteclise by the shallow but ancient (Riphean!) Orsha Trough which is closed in the south by the Zhlobin Saddle forming a bridge between the two anteclises.

At the apex of the Mazurian–Belorussian Anteclise, in the upper reaches of the Neman River, the Precambrian basement lies above sea-level, in some places directly beneath a thin Quaternary cover. In the Riphean, the anteclise formed the southern salient of the Baltic Shield; in the Caledonian stage it had a northeasterly trend, and its entire western part lays within a region of marine sedimentation. The anteclise acquired its present sublatitudinal, more specifically WNW–ESE trend in the Hercynian stage.

The **Voronezh Anteclise** is a major positive structure of the Russian Platform. It extends in the NW–SE direction. Its northeastern and northwestern limbs slope gently, the southwestern and southeastern limbs are steeper. In the north, the Voronezh Anteclise merges with the Moscow Syneclise, in the northeast, it borders the Ryazan–Saratov (Pachelma) Trough (Aulacogen) along faults (in the basement); in the southwest, it borders the Ukrainian Syneclise with the Dnieper–Donets Aulacogen at the base; further east, it borders the Donets fold system. In the southeast, the Voronezh Anteclise subsides stepwise toward the North Caspian Syneclise from which it is ultimately separated by the Volgograd Fault.

In the highest part of the Voronezh Anteclise, southeast of Kursk and south of Voronezh, the Precambrian basement lies above sea-level in a considerable area; its rocks were traversed by iron-ore mines in the region of the Kursk Magnetic Anomaly and form outcrops near Pavlovsk and Boguchar in the Don Valley. In the northeastern limb of the anteclise, the basement is overlain by Devonian deposits, in the southwestern limb by Carboniferous ones. The Voronezh Anteclise was initially part of the Sarmatian Shield, but as early as the Early Riphean it was separated from the Volga–Uralian part of the shield by the Pachelma Aulacogen. At that time, the Orsha Trough separated the future Voronezh Anteclise, which was part of the Ukrainian–Voronezh Shield, from the Belorussian Anteclise. The Riphean was also the formation time of the southeastern slope of the Voronezh Anteclise and, probably, of the initial separation of the Voronezh Anteclise from the Ukrainian Shield. Ultimate separation of the anteclise took place in the Middle Devonian, when the Dnieper–Donets Aulacogen formed; since that time it exists as an independent uplift which is obviously of residual nature.

The southeastern slope of the Voronezh Anteclise is complicated by the extended (more than 300 km long) system of Don–Medveditsa dislocations which form a complex arch of the same name. Carboniferous, Jurassic, and Cretaceous strata outcrop in it among Paleogene deposits at the top of several local box-shaped uplifts.

The complex **Volga–Uralian Anteclise,** which is separated from the Voronezh Anteclise by the Ryazan–Saratov Trough, is the easternmost submerged link of the Ukrainian – cis-Uralian uplift zone. It is almost directly adjacent to the Urals whose bend near the Ufa Plateau is, as assumed already by KARPINSKY (1919), linked with a salient of this anteclise. Its separation took place as far back as the Riphean; at that time there formed narrow grabens (aulacogens) which divided the anteclise into individual uplifts (arches). Further fragmentation of the anteclise and the beginning of its general subsidence took place at the end of the Middle and beginning of the Late Devonian. Subsidence continued in the Late Paleozoic; as a result, the whole anteclise became a buried structure whose existence was revealed only by means of deep drilling (SHATSKY, 1945). The western part of the syneclise still experienced slight subsidence in the Mesozoic, the southeastern part in the Paleogene.

The following structural elements are identified in the present structural pattern of the anteclise along the basement surface. In the west, on the right bank of the Volga, this is the Tokmov Swell whose basement top lies less than 1 km deep; in the eastern limb of the swell at the boundary of the Moscow Syneclise, there extends the Oka–Tsna Arch, one of the most typical structures of this type in the Russian Platform. East of the Volga and on both sides of the Lower Kama lies the Tatarian Swell with a minimal basement depth of about 1.5 km. It is separated from the Tokmov Swell by the Kazan Saddle; in the south from the Voronezh Anteclise by the Ul'yanovsk–Saratov Trough which is filled with Mesozoic and Paleogene rocks. The Kotel'nich (Kotel'nich–Sysol) Uplift (more precisely, Ridge), is the northern, or rather north-eastern salient of the Tokmov Swell; the Komi–Permian Uplift (Ridge) has the same relation to the Tatarian Swell. Both these uplifts abut against the peri-Timan Trough in the north and are separated by the Kirov–Kazhim or Kazan–Sergiev Aulacogen. The latter developed actively first in the Riphean, then in the Middle Devonian, and later slightly subsided until the Jurassic. Subsequently, the complex Vyatka Arch emerged above them, extending from Kirov to Kazan over a distance of about 400 km.

The most easterly part of the Volga–Uralian Anteclise is known as the Perm Swell, and the southernmost part as the Zhiguly-Orenburg Swell. The former extends in the NNW – SSE direction; the latter runs in the latitudinal direction parallel to the rim of the North Caspian Syneclise which it borders with a system of flexures and faults. In the north, the Zhiguly–Orenburg Swell is also bounded (along the basement) by a fault which, according to recent drilling data, is an overthrust to the north. It manifests itself on the surface as the Samarskaya Luka flexure known from the work of PAVLOV (1887) which can be traced for 350 km, and also as the parallel to it Zhiguly Arch, which in the east gives way to the Bol'shekinel' Arch. This faults forms the southern boundary of the Sernovodsk–Abdulino Aulacogen, which originated in the Riphean and separates the Zhiguly–Orenburg Swell from the Tatarian Swell. The northern boundary of the aulacogen (in upper horizons of the trough cover) is also accompanied by local uplift chains forming arches (Sok–Sheshma and Tuimaza).

Arches (Kerensk–Chembar and others) complicate also the limbs of the Ryazan – Saratov Trough which formed over the Pachelma Aulacogen. The latter has also a very complex structure and consists of several partial grabens and horsts separating them.

Especially complex is the Saratov area, where the aulacogen links with the rim of the North Caspian Syneclise. Its manifestation in the cover are dome-like uplifts with limbs dipping unusually steeply for a platform structure (angles of up to 20° or more).

The complex **Pripyat–Dnieper–Donets Aulacogen** is a most remarkable structure of the Eastern European Craton. It originated in the Middle Devonian as a rift on the summit of the Ukrainian–Voronezh Shield. It is also part of a system of fault dislocations which can be traced to the southeast across the Caspian and Mangyshlak to Soviet Central Asia, and also to the northwest across Warsaw and Poznań to Berlin, as well as along the northern front of the Central European Hercynides. Existence of this aulacogen was first noted by KARPINSKY (1883) (the famous "Karpinsky line"); its significance was subsequently shown by Soviet geologists (KHAIN, 1958; USPENSKAYA, 1961; AIZBERG et al., 1971). The Pripyat–Dnieper–Donets Aulacogen extends for about 1,500 km; its width is up to 250 km.

The western link of this complex aulacogen is the sublatitudinal Pripyat Aulacogen which separates the Belorussian Anteclise and the northwestern part of the Ukrainian Shield. The Pripyat Aulacogen is up to 4–4.5 km deep (along the basement) and has a very complex inner structure. It is complicated, in particular, by a longitudinal horst in whose western part (the Mikashevichi salient) the basement rises to sea-level. Two salt-bearing series were found in the Devonian of the Pripyat Trough: one of them of Frasnian and the other of Famennian age; embryonic manifestations of salt tectonics (salt pillows) are associated with them.

The Dnieper–Donets Aulacogen is separated from the Pripyat Aulacogen by the Bragin transverse salient (horst). In contrast to the Pripyat Aulacogen, the Dnieper – Donets Aulacogen's trend is NW–SE and it is much deeper (8 to 10–12 km in the extreme southeast, at the joint with Donets Basin). This aulacogen is also known to contain two salt-bearing series; the lower is of the same age as the upper Famennian salt-bearing series of the Pripyat Trough; the upper is of Lower Permian age. Vigorous manifestations of salt tectonics are associated with the lower series with rather typical salt domes. When viewed in detail, the inner structure of the Dnieper–Donets Aulacogen is very complex (CHIRVINSKAYA et al., 1968). Besides marginal faults, there are faults bounding the central graben and continuing, according to DSS data, into the upper mantle. These faults manifest themselves in the sedimentary cover by relatively short, echelon-like discontinuities. Chains of salt domes are associated with the longitudinal faults. In addition to longitudinal faults, the structure of the depression is complicated by transverse faults which usually lie in the continuation of the deep faults separating the megablocks of the Ukrainian Shield (Krivoy Roy, Kremenchug, Kirovograd, etc.). Segments of the aulacogen, separated by these faults, differ in structure. The zone of the Orekhovo–Pavlograd Fault coincides with the transition from the aulacogen to the Donets fold system and to the coal basin of the same name. This transition is along the trend and is quite gradual: the basement depths increase regularly along the aulacogen; individual stratigraphic units, as well as anticlinal and synclinal zones could be traced continuously from the Dnieper–Donets Depression (DDD) to the Donets Basin. On the other hand, appreciable changes take place over a relatively short distance. They consist of the following:

1) the basement depth and the thickness of the sedimentary fill double, from an average of 10 km in the Dnieper–Donets Depression to 20 km or more in the Donets Basin; this increase is due mainly to the Carboniferous in which there appears a thick paralic coal-bearing formation;

2) there disappear manifestations of salt tectonics probably in connection with a facies replacement of the Devonian evaporitic series; the fold structure acquires a linear character, and the so-called Main Anticlinal occupies a central position in it, suggesting inversion of the overall structure; fault dislocations are represented by overthrusts;

3) the degree of sedimentary rock catagenetic alterations increases sharply and there appear minor basic intrusions and a mercury-polymetallic mineralization. All these features, as well as the fact that the Donets (Donets–Caspian) fold system in its southeastern continuation links with the cis-Caucasian Hercynides, do not permit us to regard the Donets Basin ("Donbas") as a simple intracratonic structure of aulacogen type or talk of a single aulacogen or trough of the Greater Donbas.

At the same time, it is quite obvious that the Pripyat–Dnieper–Donets Aulacogen and the Donets–Caspian Geosyncline originated simultaneously in the Middle Devonian over an ancient epi-Karelian cratonic basement, along a single fault system. It has been repeatedly suggested that the entire Pripyat–Dnieper–Donets Aulacogen formed as early as the Riphean. This was apparently confirmed by the presence of Riphean deposits in the Pripyat Depression. But the Riphean here occurs in the zone of the Volyn–Orsha Trough wich cuts obliquely across the Pripyat Graben; consequently it has nothing to do with the formation of this graben. However, recent DSS studies did reveal a relatively narrow graben (narrower than the Devonian graben by a factor of 1.5–2) at the base of the eastern part of the Dnieper–Donets Depression and the Donbas. This graben is filled with thick (up to 10–12 km!), clearly pre-Devonian, most likely Riphean or possibly Lower Paleozoic strata (SOLLOGUB & CHEKUNOV, 1971). Thus, the Riphean age of initial formation is quite likely for the central and eastern parts of the aulacogen. Possibly the western prolongation of this Riphean aulacogen is represented by the Ovruch Syncline (SOLLOGUB, 1978).

Formation or regeneration, in the Middle Devonian, of the Dnieper–Donets zone of subsidence (less in the northwest and quite intense in the southeast) was accompanied by considerable alkaline-basaltic (volcanic rocks) and alkaline-ultrabasic (intrusive bodies) magmatism. Its maximum occurred in the west of the Dnieper–Donets Depression, in the region of Chernigov. More acid derivatives appeared at the southern edge of the Donets Basin. Generally speaking, this volcanism is quite characteristic of continental rift zones. The marginal faults of the Pripyat–Dnieper–Donets Aulacogen remained active up to the middle of the Viséan age (even later in the Donbas). Since then, the region of subsidence has expanded noticeably; the aulacogen gradually became transformed into the flat Ukrainian Syneclise (Fig. 9). In the Late Cretaceous, the subsidence involved also slopes of adjacent uplifts (Ukrainian Shield, Belorussian and Voronezh anteclises); however, later the area occupied by them decreased successively.

Many oil and gas fields are associated with the Pripyat–Dnieper–Donets Aulacogen. They occur in deposits ranging from Devonian to Permian (the latter in the east of the Dnieper-Donets Depression). Major deposits of potassium salts were found in the Pripyat Depression.

The **North Caspian (Pericaspian) Syneclise,** as noted above, belongs to a quite specific type of cratonic depressions. It was proposed to call them nodal syneclises (KHAIN, 1964), exogonal syneclises (ZHURAVLEV, 1970), bathysyneclises (STAVTSEV, 1965), phialogenes (NALIVKIN, 1962), or amphiclises (MURATOV, 1972). However, none of these terms has so far been generally accepted. A specific feature of this structure is not

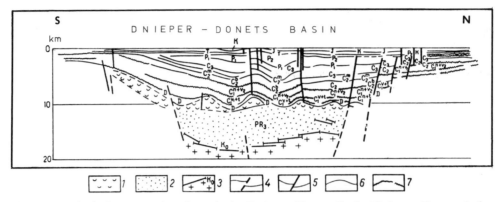

Fig. 9. Geological cross-section through the Dnieper–Donets Basin (Dnieper–Donets Aulacogen and Ukrainian Syneclise) along the line Sinelnikovo–Chuguev (after SOLLOGUB &
CHEKUNOV, 1971). 1 = Devonian evaporitic series; 2 = Riphean; 3 = surface of the crystalline
basement; 4 = faults according to geophysical data; 5 = ibid., according to geological data; 6 =
stratigraphic boundaries according to geological data; 7 = ibid., according to geophysical data.

only its size (0.5 million km^2), but mainly the enormous thickness of the sedimentary
cover (up to 20–23 km!) in the center of the depression where the crust is only 30 km
thick. This is accompanied by thinning-out of the basaltic layer and pinching out (in the
same area) of the granite-gneiss layer (Fig. 10). The latter is to a certain extent substituted
in the center of the depression by a layer having longitudinal-wave velocities from
6.7–6.8 to 7.0 km/sec., whereas the basaltic layer proper exhibits velocities from 7.0–7.2
to 7.5 km/sec. This "suprabasaltic" layer may be interpreted in different ways, but is, in
all probability, a relict of the degranitized lower part of the granite-gneiss layer. The
regions of absence of a typically granitic layer exhibit strong gravity highs (Aralsor and
Khobda).

The North Caspian Basin is separated from the main part of the Russian Platform by
peripheral scarps of flexure-fault nature. The surface of the subsalt Paleozoic subsides
along the northern scarp from 1.2–2.5 to 3–4 km, the western scarp has an amplitude of
up to 2 km. In the south, the basin is separated from the Donets–Caspian Hercynian fold
zone by the Astrakhan Swell whose basement surface has a maximum depth of 4 km. This
uplift must be separated from the depression by a fault along the lower reaches of the
Volga. In the southeast, a fault along the axis of the Southern Emba Uplift separates the
North Caspian Basin from the southeastern branch of the Zilair–In zone of the Southern
Urals. Finally, in the area south of Aktyubinsk, the degenerating termination of the
Uralian Foredeep enters the depression; in the west, it is accompanied by a subsalt
Paleozoic uplift, probably also by an uplift of the basement.

The North Caspian Basin is the region of most spectacular manifestation of salt
tectonics associated with accumulation of a thick stratum of rock salt at the end of the
Early (Kungurian age) and beginning of the Late (Kazanian age) Permian. The initial
thickness of this series was about 3 km. At present, the salt forms numerous (more than
thousand) stocks over which salt domes formed; some of these are of giant size (Chelkar
and Sankebai), up to 100 km across and 10,000 km^2 in area. Investigations by SOKOLOV
(1970) and KRICHEVSKY (1963) have shown that the salt stocks merge into extended ridges

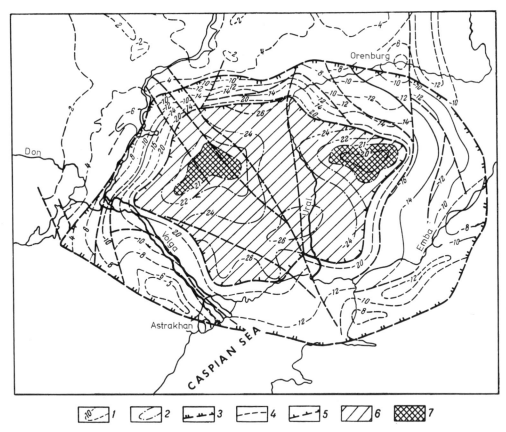

Fig. 10.   Scheme of the surface relief of the crystalline basement of the North Caspian Syneclise (after FOMENKO, 1972, simplified). 1 = isobaths of the surface of the Precambrian "granitic" crystalline basement; 2 = isobaths of the surface of the "basaltic" layer; 3 = zones of deep faults of suture type; 4 = other faults; 5 = presumed limit of the zone devoid of the granitic layer; 6 = area of "basaltic" basement of the North Caspian Basin; 7 = most elevated segments of the "basaltic" basement.

via narrow salt bridges at a depth of 1,000–1,500 km; at a depth of 2,000–2,500 m these ridges acquire a honeycomb or cellular structure in which, however, one can observe elements of some distinct orientation (Fig. 11). For example, in the east, near the Urals (Mugodzhary), the salt ridges are mainly of meridional trend but are diagonal in the central part (NW and NE); ridges of latitudinal and northeasterly trend respectively predominate along the northern and western edges. A similar pattern is typical of the Gulf Coast in North America.

Salt is generally completely squeezed out of the space between the ridges and domes. The surface of the subsalt Paleozoic lies at the depth of 9–10 km in the central part of the basin (Fig. 12). It is generally subhorizontal, but in some places dislocated by faults. Subsidence is maximal in the Volga–Uralian interfluvial region, on the continuation of the Ul'yanovsk–Saratov Trough (the Novouzensk Graben) and also in the east, in the region of the Khobda gravity high, south of the Buzuluk Trough. The Permo–Triassic

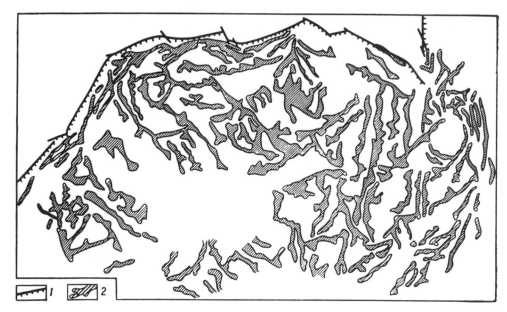

Fig. 11. Salt walls of the North Caspian Syneclise (after Sokolov & Krichevsky, 1966). 1 = flexure-fault boundary of the basin; 2 = salt walls.

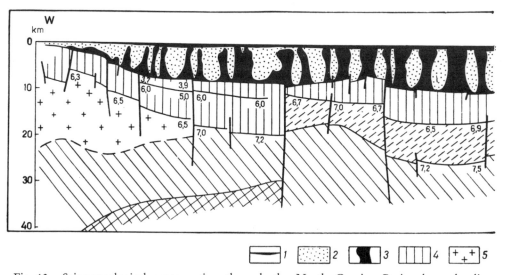

Fig. 12. Seismogeological cross-section through the North Caspian Basin along the line Kamyshin–Aktiubinsk (after Sokolov, 1970, modified). 1 = Upper Pliocene; 2 = Paleogene, Mesozoic and Upper Permian; 3 = salt plugs and walls; 4 = subsalt strata; 5 = "granitic" layer of the consolidated crust; 6 = "basaltic" layer of the crust; 7 = upper mantle; 8 = layer with

molassoid formation accounts for the greater part of the thickness of the suprasalt stage. It is overlain by much thinner Jurassic, Cretaceous, and Paleogene deposits of typically platform nature. There is a gap from the end of the Miocene to the Middle of the Pliocene, during which the roofs of the salt domes experienced considerable erosion. As a result, the salt stocks are often directly overlain by a cover of Upper Pliocene–Quaternary deposits. However, wherever Mesozoic–Paleogene rocks are still preserved in their roofs, they are usually confined to grabens. Besides, the domes are often accompanied by compensation rim synclines.

The origin and the early stages of evolution of the North Caspian Basin are still insufficiently known, although the picture has become clearer in recent years. Analysis of the magnetic, and partly of the gravitational fields has shown that the basin is a super-posed structure; the continuation of the Russian Platform structures in the relief of the subsalt bed and even of the suprasalt stage is quite clear. GAFAROV (1971) stresses the similarity of the smooth and mosaic magnetic field of the depression with the fields of the White Sea and Northern Dvina massifs, and suggests the existence of a buried Archaean massif here. It is very likely that the Pachelma Aulacogen extended into the depression during the Riphean and that another, meridional aulacogen, existed in its eastern part. More definite information is available on the Middle and Late Devonian, and from the Carboniferous onward. The lithologic-facies variations of the respective formations in the direction of the North Caspian Basin point to the probable accumulation in the latter, on the one hand, of relatively deep-water sediments enriched with organic matter and, on the other hand, of sandy-clayey strata corresponding to epochs of gaps in the platform surroundings. The structural differentiation of the future depression must have been considerable. Zones of maximum subsidence must have coincided, according to

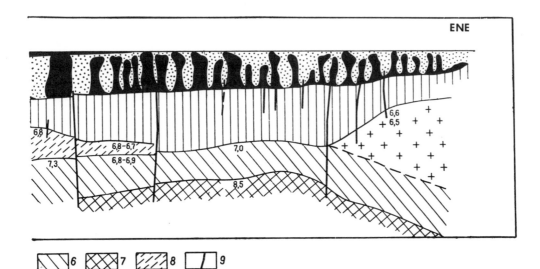

intermediate seismic velocities between "granitic" and "basaltic" layers (probably the lowermost part of the sedimentary cover); 9 = faults. Numbers in the figure = boundary velocities of longitudinal (P) waves.

SOKOLOV (1970), with present gravity highs, i. e., with zones where the crust is thin. This should mean that the crust of the North Caspian Basin had a "suboceanic" character even at that time.

Very important information on the next stage of depression evolution, which covers the Viséan–Artinskan time, was furnished by drilling at the depression rim in the area of Krasny Kut, southeast of Saratov. Here, the thickness of the carbonate formation of the Middle Carboniferous-Lower Permian age decreases sharply toward the depression (from 1,200 to 50 m). This formation is replaced by an essentially terrigenous dark-colored series having an overall thickness of more than 1,500 m. Reef structures of barrier type, accompanied by a clastic apron occur in the transition zone. A steep slope, 1,500 m in height, formed in the relief of the subsalt bed, due to sharp decrease in the carbonate-formation thicknesses; a small-amplitude fault corresponds to it in the basement and in the older Paleozoic. These and other data caused SOKOLOV (1970) to conclude that at the latest in the Middle Carboniferous the North Caspian Basin acquired its present shape and became a deep-water basin with uncompensated subsidence. However, elements of an older structural pattern were preserved, although in attenuated form.

In the Kungurian–Kazanian time, rising of the Hercynian fold system in the southern and eastern surroundings of the depression, as well as of the Voronezh Anteclise west of it, caused intense salt accumulation in the North Caspian Basin. At the end of the Permian and in the Triassic it gave way to the deposition of a thick variegated molassoid formation, replaced by true molasses in the east ond south. At the end of the Triassic and beginning of the Jurassic, the depression was filled with sediments, dried out and became an alluvial plain; however, from the Middle Jurassic onward, subsidence resumed and reached the greatest amplitude in the Late Cretaceous epoch. In the Paleogene, subsidence became localized in the Volga–Uralian interfluvial, in the Neogene subsidence stopped for a long time, resuming at the end of the Pliocene–Pleistocene. Sediments of this uppermost structural stage of the depression have small thickness, excepting the zones where they fill sub-Akchagyl' erosion cuts, in particular, the valley of the paleo-Volga along the Novouzensk Trough inherited from the Riphean(?).

Salt ridges started developing, according to SOKOLOV (1970), already in the Late Permian; salt domes continued to grow, especially during periods of regional breaks of sedimentation (EISENSTADT, 1966), throughout the Mesozoic and Cenozoic. Geodetic observations on the Baskunchak test site and geomorphological studies in the valley of the Lower Volga show that the domes continue to grow at a rate of several cm / year.

The North Caspian Basin contains numerous relatively small oil and gas fields. It is likely that its main, not yet explored reserves (mainly gas) are associated with large buried uplifts of the subsalt Paleozoic in peripheral parts of the depression, as well as with the Permo–Triassic rocks of interdome depressions in its central part (SOKOLOV, 1970).

2.5 Main stages of evolution

2.5.1 The Archaean stage

Recent data on the stratigraphy of the Baltic Shield Early Precambrian show that the oldest formations exposed here at the surface are granite-gneisses of tonalitic composition. These constitute the basement of the Kola series on the Kola Peninsula and of the Gimola, Parandovo and equivalent Karelian series. A possible analogue of this ancient continental crust within the Voronezh Massif is the Oboyan series; in the Ukrainian Shield this is the Bug complex. In the Dnieper Megablock, the Konka–Verkhovtsevo series with its radiometric age of about 3.0–3.5 billion years may be a relict of an even older oceanic crust, which escaped granitization. But it is more probable that the granite-gneiss domes and arches separating the areas occupied by this series represent its remobilized substratum. In any case, the spatial relationships of this greenstone series and the granite-gneisses strongly resemble those observed in other ancient shields. The Gimola series of Karelia and its analogues (Lopian of KRATS, 1958), as well as the Mikhailov series of the Voronezh Massif with its recently discovered komatiites fully correspond to Archaean (Late Archaean) rocks of the greenstone belts of other continents (Canada, South Africa, Western Australia). Such a similarity was recently proved by French investigators (BLAIS ET AL., 1977) for analogues of the Gimola series in adjacent Finland.

On the other hand, the Murman and White Sea megablocks of the Baltic Shield, the Bug–Podolian Megablock of the Ukrainian Shield, and formations revealed by drilling in the Baltic region, Belorussia, and the Volga region, similar to them in the grade of metamorphism (attaining the granulite facies) and in composition (occurence of charnockites), may be regarded as representatives of the granulite belts common in practically all shields. Judging by radiometric datings, the age of granulite metamorphism must be at least Archaean. Anyhow, extensive granitization, which spared only individual relics of greenstone belts (especially those containing jaspilites), caused the continental (or nearly continental) crust to form by the end of the Archaean in the whole area of the future craton. This epoch saw the formation (beginning of the formation) of the numerous granite- (magmatite-)gneiss domes responsible for the patchy-mosaic patterns of the gravitational and magnetic fields. True potassium (microcline) granites formed later, in the Early Proterozoic.

At the boundary between the Archaean and Proterozoic, the protocontinental crust experienced intensive destruction. This produced the block mosaic reflected in the present structure of the platform basement (see Fig. 8).

2.5.2 The Early Proterozoic (Karelian) stage

By the beginning of this stage, the territory of the future Eastern European Craton had been separated into numerous relatively consolidated massifs (see Fig. 8) with relatively narrow geosynclinal systems (protogeosynclines) of various trends between them. All these systems in the craton considered are still of fully eugeosynclinal type, i. e., they are characterized by strong basic volcanism, at least in the early evolutional stages. The presence, alongside metaspilites and diabases, of metaporphyries and keratophyres indicates the occurrence of a typical spilite-diabase-keratophyre association. Somewhat later, abundant jaspilites (Krivoy Rog and KMA) formed. Quite common are metamorphosed gabbro and ultrabasic rocks. Metamorphism is mainly of amphibolite, less frequently of epidote-amphibolite grade. The granitoids may already be divided (SHURKIN, 1968) into two formations: an earlier autochthonous and synkinematic plagiogranite formation and a more recent allochthonous formation of plagio-microcline granites. Regional metamorphism and granitization of the Karelian (Early Karelian) tectonomagmatic epoch completed the formation and consolidation of the continental crust over the greater part of the Eastern European Craton. Thus, quite large protocratons had emerged by the beginning of the Middle Proterozoic. These include the entire northeastern part of the Baltic Shield, almost the entire Ukrainian Shield, the western part of the Voronezh Massif, etc.

2.5.3 The Middle Proterozoic (Svecofennian) stage

The Svecofennian zone of the Baltic Shield, which reaches Lake Ladoga (Ladoga series) in the southeast, is a typical geosyncline of this stage. Another geosynclinal zone (continuation of the first?) could be detected by the occurrence of the Vorontsovo series in the eastern part of the Voronezh Anteclise. The flyschoid appearance of both the Ladoga and the Vorontsovo series and the general predominance of terrigenous over volcanic rocks are of interest. A geosyncline of this age may also exist in the northwestern part of the Ukrainian Shield (the Osnitsk complex). Geosynclines of the same type could exist in central areas of the Russian Platform. However, geophysical data do not make it possible to differentiate between the Karelides proper (Early Karelides) and the Svecofennides (Late Karelides), and thus to determine the relative extent of the latter. The Middle Proterozoic geosynclines completed their evolution by relatively strong folding, metamorphism from the amphibolite to the greenschist facies, intrusion of basic rocks (gabbro, gabbro-diorites, and gabbro-diabase) and, much less frequently, ultrabasic rocks, and, finally, granitization; the granites are mainly allochthonous, plagiomicroclinic.

Sediments of the cover (the Yatulian and its analogues) accumulated within relatively low areas of the protocraton, for example, in Central Karelia; basic magmatism also manifested itself there (Suisarian volcanics, Onega gabbro-diabases, etc.).

As this stage came to an end, practically the entire territory of the Eastern European Craton experienced complete consolidation. Moreover, there is every reason to believe that this consolidation initially encompassed the entire area of the future European continent and even adjacent continents. The present shape of the platform formed much later (see below).

2.5.4 The beginning of the Late Proterozoic (Gothian) stage

The western periphery of the future platform, from Swedish Lapland to the Azov region, exhibited increased mobility and magmatic activity at the beginning of the Late Proterozoic (see Fig. 8). The most typical manifestation of the latter was the formation of a volcano-plutonic association having an average age of 1,650 m.y. It includes large differentiated plutons of rapakivi granites, gabbro-labradorites, anorthosites, as well as acid lavas (Dala-porphyries, Ovruch quartz porphyries, etc.) and ignimbrites (sub-Jotnian). This peripheral crustal magmatism, which points to an increased heat flow and which promoted homogenization of the crystalline platform basement, was connected spatially (marginal volcano-plutonic belts) with areas where future geosynclinal belts formed. In the east, is manifestations were found in the Bashkirian Urals (the Berdyaush Pluton).

After formation of the rapakivi granites and the accompanying magmatites, there continued general uplifting of the craton, with its being split into individual blocks. The faults reached the upper mantle; the magmatites acquired an exclusively basic composition. They alternate in some places with quartzose sandstones of the Jotnian type, deposited in flat and more or less isometric depressions.

2.5.5 The middle and the end of the Late Proterozoic (Dalslandian und Early Baikalian stages)

At the beginning of this stage, the future craton assumed its present configuration in connection with the formation of the geosynclines around it (Timan–Uralian and Galician–Caucasian). A possible exception was the northwestern boundary, since sediment accumulation in the geosyncline of Scandinavian Caledonides began only 800 m.y. B.P. The northwestern salient of the craton (the Southern Scandinavian Massif and its continuation beneath the Caledonides) at that time underwent partial regeneration; however, it ended soon with the Dalslandian folding and the intrusion of granites of the Bohus–Arendal type.

At the same time a system of early aulacogens began to form on the future Russian Platform: the stage of cratonization was superseded by the aulacogen stage. The aulacogens were filled with sediments of the Lower(?) to Middle Riphean (Fig. 13). The longest aulacogen band dissected the platform approximately along the Lvov–Kotlas diagonal. A second band, more discontinuous, runs in the direction from Lake Ladoga to Saratov, with a possible continuation along the Novouzensk Trough. Parallel to this NW–SE diagonal there is a shorter diagonal from the Onega Peninsula (White Sea) to Northern Bashkiria. Here, this aulacogen band merges with the Kama pericratonal downwarp; the Volyn–Kotlas zone, in its turn, links with the Dniester downwarp of similar type. Both these pericratonal downwarps pass, in a direction away from the platform, into the miogeosynclinal zones of the Uralian and Galician (cis-Carpathian–Dobrogean) geosynclinal systems respectively. A pericratonal downwarp could well exist also along the Timan Geosyncline, and could link with the Volyn–Kotlas subsidence zone.

The Volyn–Kotlas aulacogen zone divided the basement of the craton into the Baltic and the Sarmatian shields. The Pachelma Aulacogen additionally divided the latter into the Ukrainian–Voronezh and the Volga–Kama shields. Finally, the Volga–Kama Shield was in its turn dissected by the latitudinal Sernovodsk–Abdulino Aulacogen in the south and by several longitudinal grabens in the north.

According to POSTNIKOVA (1976), a red clastic formation evolved extensively at the beginning of the Riphean. It was mainly continental and rudaceous in the lower part. By the end of the Early Riphean, gritstones and sandstones had been replaced by siltstones and clays; the red color had partly been replaced by gray or even black. The appearance of glauconite suggests that the continental environment gave way to littoral-marine conditions. Tectonic activity resumed at the boundary between the Early and Middle Riphean. It was accompanied by gabbro-diabase sill intrusions both at the western (Volyn–

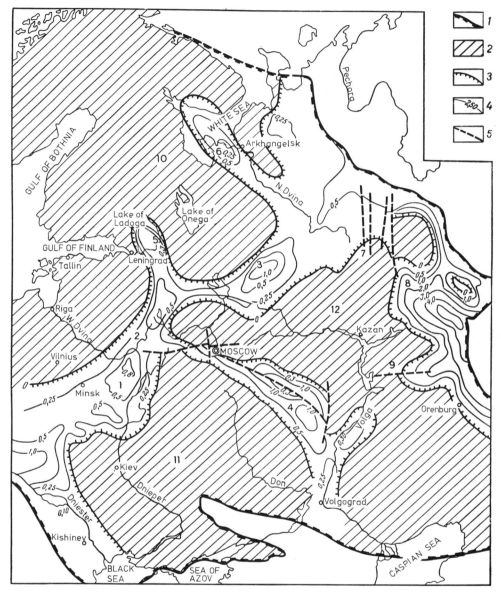

Fig. 13. Scheme of extension and distribution of thicknesses of Riphean and Lower Vendian in the Russian Platform of the East European Craton (after IGOLKINA, KIRIKOV & KRIVSKAYA, 1970, modified and supplemented after POSTNIKOVA, 1976; OSTROVSKY et al., 1975). 1 = actual boundary of the Platform; 2 = areas devoid of sediments; 3 = limit of sediment distribution; 4 = isopachs; 5 = main faults. Main structures: aulacogens: 1 = Orsha, 2 = Krestets, 3 = Middle Russian, 4 = Pachelma, 5 = Ladoga, 6 = Belomorian, 7 = Kazhim, 8 = Kaltasa, 9 = Radaevo; shields: 10 = Baltic, 11 = Sarmatian, 12 = Volga–Kama.

Podolia) and the eastern (Kama Depression) margins of the craton, in some places even in its central part (Ladoga region and Krestsy Depression).

The Middle Riphean sedimentation cycle was a near repetition of the Early Riphean one; it, too, ended in a phase of tectonomagmatic reactivation with gabbro-diabase intrusions.

The thickness of the Lower-Middle Riphean formations in the Kama pericratonal depression reaches 4.5 km; in the aulacogens of the central part of the Russian Platform (Central Russian, Moscow, Pachelma) it is 2–3 km. The overall thickness of the Riphean in the Dnieper–Donets Aulacogen must be considerable.

Upper Riphean deposits, according to Postnikova (1976), are less common in the Russian Platform than Lower and Middle Riphean deposits. Their thickness is less (up to 1 km only in the Kama pericratonal downwarp, and not greater than 0.5–0.6 km in central parts of the Russian Platform). This must be largely the consequence of pre-Vendian erosion. At the same time, the decay of some graben-like structures of an earlier period cannot be ruled out either. In its structural composition the Upper Riphean is, in contrast to older formations, almost exclusively represented by marine formations (sometimes terrigenous, mainly siltstone, also with glauconite, sometimes carbonate). Barrier reefs developed in the Kama pericratonal downwarp. The end of the Late Riphean was marked by new uplifts and gabbro-diabase intrusions.

2.5.6 The Late Baikalian–Caledonian stage (Vendian–Early Devonian)

At the boundary between the Riphean and Vendian, more specifically during the Early Vendian, the aulacogen stage of Russian Platform development was superseded by the platform stage. This manifested itself primarily in the expansion of sedimentation regions beyond the aulacogens. By the end of the Early Vendian, there formed the huge Baltic-Moscow Syneclise which stretched from Southern Belorussia to Timan. It was bounded in the west by the Baltic Shield slopes which at that time reached Riga and Grodno. Within this syneclise, there formed by the beginning of the Cambrian the Loknov Uplift; it was later transformed into the Latvian Saddle separating the Baltic Syneclise from the Moscow one.

The Lower Vendian sediments in the lower part are represented by a marine terrigenous or terrigenous-carbonate formation, red or variegated below, gray-colored, with glauconite, higher up. Some investigators relate this part of the sequence to the Riphean; it was limited by aulacogens in its occurrence. In the second half of the Early Vendian, the area of the Russian Platform underwent intense glaciation. It manifested itself in the accumulation of tillites and tilloids (marine-glacial deposits). Basaltic volcanism was quite strong, especially at the western margin of the craton; volcanic material was found also among sediments of the central areas of the Russian Platform. It is, in all probability, not a coincidence that glaciation and the outburst of volcanism were synchronous with a considerable rearrangement of the overall structure of the platform.

In the Late Vendian, sedimentation spread to the entire Baltic–Moscow Syneclise. The latter was widely linked with the Timan, cis-Uralian and Visla-Dniester pericratonal downwarps and formed a gulf in the direction of Ryazan and Saratov (Fig. 14). A marine gray terrigenous formation with glauconite is predominant; it is replaced by a variegated

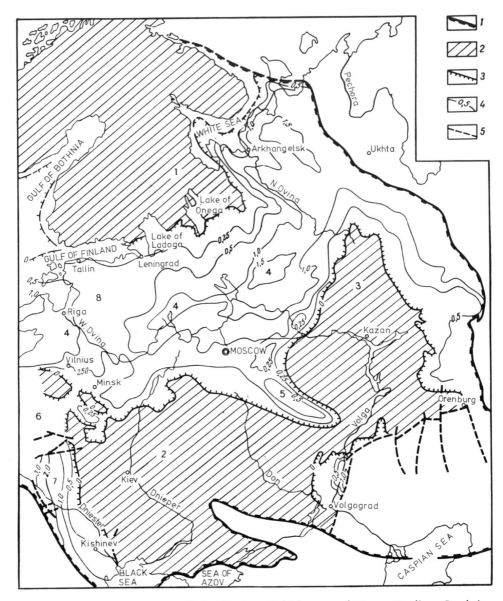

Fig. 14. Scheme of extension and distribution of thicknesses of Upper Vendian, Cambrian, Ordovician and Silurian in the Russian Platform (after IGOLKINA, KIRIKOV & KRIVSKAYA, 1970, modified and supplemented after POSTNIKOVA, 1976; OSTROVKSY et al., 1975). 1 = actual boundary of the Platform; 2 = areas devoid of sediments; 3 = limit of sediment distribution; 4 = isopachs; 5 = main faults. Main structures: 1–3 = shields: 1 = Baltic, 2 = Sarmatian, 3 = Volga–Kama; 4 = Baltic–Moscovian Syneclise; 5 = Pachelma Trough; 6 = Brest Basin; 7 = Lvov Trough; 8 = Loknov Uplift.

formation at the periphery of the uplifts. Volcanism attenuated gradually, together with a general reduction of tectonic activity. Transition to the Cambrian was not accompanied by any appreciable changes in sedimentation; however, a noticeable rearrangement took place in the structural pattern: the Baltic–Moscow Syneclise expanded greatly to the west, absorbing the southern part of the Baltic Shield and merging with the Grampian Geosyncline. Conversely, its area was reduced in the northeast, probably due to uplift of the Timan. Thus, the plunge of the axis of the regional subsidence of the syneclise became reversed. The cis-Uralian pericratonal downwarp ceased to exist for a long time after the beginning of the Cambrian.

In the Late Cambrian, the territory of the platform experienced general, though weak upheaval which is indicated by a hiatus. In the Ordovician, subsidence resumed and continued during the Silurian, but encompassed only the Baltic region, the western part of the Moscow Syneclise, and the Visla–Dniester Downwarp. The main Ordovician–Silurian formation is shallow-water carbonate, with the appearance of regressive evaporites at the top of the Silurian. Toward the western edge of the ancient craton, i. e., in the pericratonal zone of Polish Pomerania, the Ordovician–Silurian carbonate sediments are replaced by graptolitic shales of a quite considerable (more than 1,000 m) thickness (continental slope deposits).

At the end of the Silurian and beginning of the Devonian, a new general emergence of the craton took place synchronously with intensive uplifts in the Caledonian Geosyncline of Scandinavia, and with slighter uplifts in the Urals. Red continental deposits of the Lower and lower part of the Middle Devonian were encountered only in the Visla-Dniester pericratonal trough where in some places they overlie older deposits, including Silurian with considerable unconformity. This suggests Caledonian deformations of the cover in the marginal part of the ancient craton.

The total magnitude of subsidence at this stage of craton evolution was relatively small. The thicknesses exceed 1,000 m only in the central part of the Moscow and Mezen' syneclises and in pericratonal downwarps (Visla–Dniester, more than 3,000 m!, and Timan). Vendian deposits account for the greater part of this thickness.

2.5.7 The Hercynian stage (Middle Devonian–Middle Triassic)

The onset of the Hercynian stage was marked by a considerable rearrangement of the craton structure. By that time it had acquired fold system surroundings (Baikalian in the northeast and southwest, and Caledonian in the northwest); in the east and south, subsidence along the craton periphery intensified sharply; considerable areas previously forming its parts, in particular the Donets–Caspian zone, became involved in the geosynclinal subsidence. The Ukrainian–Voronezh Shield was split by the Pripyat–Donets Aulacogen which emerged in the continuation of this band. Further west, the shallower Podlyas-Brest Trough, opening into the Central European Hercynian Geosyncline, formed in the same subsidence zone. Had the Ukrainian Shield been less stable and had the Polessie bridge between it and the Mazurian–Belorussian Anteclise been destroyed, the entire southwestern part of the craton would have become part of the Mediterranean geosynclinal belt. This, however, did not happen. Those investigators were wrong who, like Sobolev (1939) and Tetyaev (1935), tended to classify the

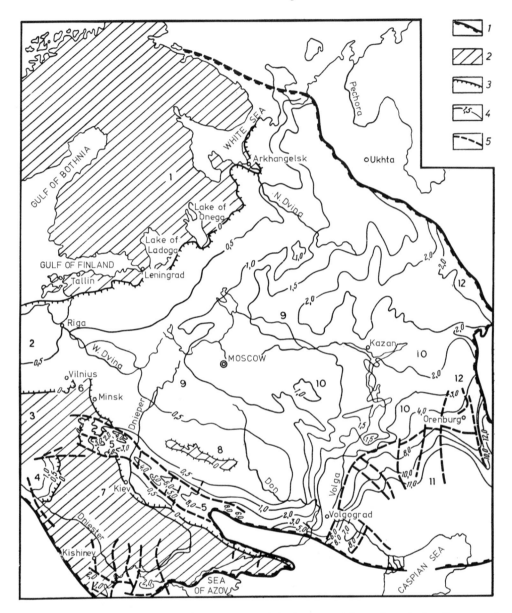

Fig. 15. Scheme of extension and distribution of thicknesses of Devonian, Carboniferous and Permian in the Russian Platform (after IGOLKINA, KIRIKOV & KRIVSKAYA, 1970, simplified). 1 = actual boundary of the platform; 2 = areas devoid of sediments; 3 = limit of sediment distribution; 4 = isopachs; 5 = main faults. Main structures: 1 = Baltic Shield, 2 = Baltic Syneclise, 3 = Brest Basin, 4 = Lvov Trough, 5 = Pripet–Dnieper Aulacogen, 6 = Belorussian Anteclise, 7 = Ukrainian Anteclise, 8 = Voronezh Anteclise, 9 = Moscow Syneclise, 10 = Volga–Kama Anteclise, 11 = North Caspian Syneclise, 12 = Uralian Foredeep.

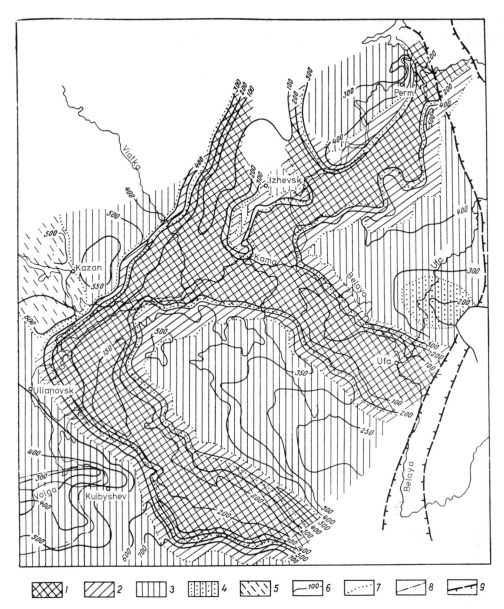

Fig. 16.   System of Kama–Kinel troughs and distribution of thicknesses and facies of the Devonian carbonate complex in the Volga–Urals region (after MIRCHINK, KHACHATRIAN et al., 1965,
simplified). Relatively deep marine facies: 1 = carbonaceous siliceous-argillaceous limestones,
marls and shales of "domanik" facies; shallow marine facies: 2 = predominantly massive carbonate
rocks of reef facies; 3 = layered limestones and dolostones with subordinate reef facies; 4 = layered
limestones and dolostones with a member of terrigenous Orlovka series; 5 = layered limestones
and dolostones with coal-bearing terrigenous Uslon series; 6 = isopachs; 7 = facies boundaries; 8 =
limits of areas devoid of Middle Frasnian strata in the Samarskaya Luka–Chapaev and Karatau
regions; 9 = boundaries of the Uralian Foredeep.

Pripyat–Donets Trough together with the Donbas as a geosyncline (the "Amodecian Geosyncline" of SOBOLEV, 1939) and the Ukrainian Shield as a structure of geoanticline, median massif, or anticlinorium type.

The Mazurian–Belorussian and Voronezh anteclises had acquired their present configuration at the beginning of the Hercynian stage (Fig. 15). The Baltic Syneclise again became, in the Devonian, as in the Vendian, the western centrocline of the Moscow Syneclise; in the Carboniferous its independent existence ceased temporarily. The Moscow Syneclise (together with the Mezen' Syneclise) after the Carboniferous became a large and deep (sediment thickness of up to 2 km) depression with an axis dipping toward the Uralian Geosyncline. The entire band adjacent to this geosyncline became involved in subsidence; this ultimately (in the Late Devonian) caused the Volga–Kama Shield to sink below sea-level and become the Volga–Uralian Anteclise. This subsidence was accompanied by splitting of the shield, causing the formation, particularly in the north, of the submeridional Kazan–Sergiev Aulacogen. In the Late Devonian–Early Carboniferous, the peculiar system of Kama–Kinel troughs (Fig. 16) developed in the Volga–Uralian region. The primarily tectonic features of these troughs were much intensified by sedimentation-erosion processes. Relatively deep-water deposits settled in the axial parts of these troughs. The so-called domanik bituminous facies of the Upper Devonian can be identified there. Erosion predominated in the Middle Viséan, during a period of continental hiatus. Later, the deficit of sediments was compensated by rapidly accumulating alluvial facies. Chains of bioherms stretch along the rims of the troughs; beyond them there lie shallow-water marine carbonate deposits of small thickness (Fig. 17).

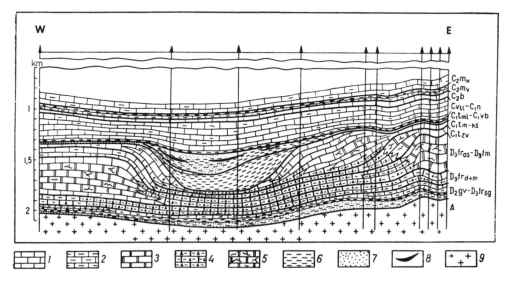

Fig. 17.   Schematic cross-section through the Ust-Cheremshan part of the Kama–Kinel system of troughs along the line Mordovino–Nurlat (after MIRCHINK, KHACHATRIAN et al., 1965, simplified). 1 = limestones; 2 = shaly limestones; 3 = dolostones; 4 = carbonaceous shaly-siliceous limestones and argillites; 5 = massive, predominantly reef limestones and dolostones; 6 = shales and argillites; 7 = sandstones; 8 = coal and coaly shale; 9 = granite-gneisses.

Existence of the North Caspian subsidence zone can be established quite definitely for the first time in the Devonian. This region was a zone of accumulation of deep-water sediments. However, as pointed out in the previous paragraph, this subsidence was differentiated; in the west, there is a continuation of the Lower Volga Trough which separated the Voronezh and Volga–Uralian anteclises; in the east the continuation of the Buzuluk Trough situated between the Zhiguly–Pugachev and the Orenburg swells of this anteclise. Between them there may exist a zone of relative uplift.

Replacement of the Caledonian structural pattern by the Hercynian was accompanied by renewal of movements along the deep faults in the craton basement and by a considerable outburst of magmatic activity. The latter attained a maximum at the end of the Middle and especially at the beginning of the Late Devonian. The products of this magmatism, represented in both volcanic and intrusive (subvolcanic) forms belong mainly to a trap (tholeiite-basaltic) formation, but vary much in composition (from ultrabasic rocks to alkaline basaltoids). Devonian magmatism was most pronounced in aulacogens, especially in the Pripyat–Donets one, and where they link with shields (Ukrainian) and anteclises (Voronezh and Volga–Uralian).

The sequence of sedimentary formations in this stage starts with a variegated lagoonal-continental formation with a rather extensive presence of quartzose sands – products of erosion of the crystalline basement, mainly of the Baltic Shield. At the end of the Middle and beginning of the Late Devonian, this formation was superseded by a carbonate formation, represented in the east and center of the platform mainly by a subformation of biogene limestones, in the south and west by micritic limestones and marls. A thick evaporite Upper Devonian formation is associated with the Pripyat–Dnieper–Donets Aulacogen; formations of bituminous clays, limestones, and oil shales (domanik) are associated with the North Caspian Syneclise, the system of Kama–Kinel troughs, and the cis-Uralian pericratonal downwarp. The "domanik" formation accumulated under conditions of uncompensated subsidence. The predominance of carbonate formation lasted until the beginning of the Permian inclusively, but was interrupted over a considerable area in the Middle Viséan, by short-term accumulation of a limnic coal-bearing formation in the central and eastern areas of the Russian Platform. A paralic coal-bearing formation was widespread in the south in the Late Viséan and Namurian (Bashkirian). From the Sakmarian–Artinskian time onward there appear evaporites which were most common in the Kungurian stage in the North Caspian Syneclise. Evaporite accumulation there continued also in the Kazan age; deposition of evaporites proceeded at the same time also in the south of the Baltic Syneclise which since the Permian constituted the eastern centrocline of the huge North Sea–Baltic (Central-European) Basin. The evaporite formation was superseded by a variegated lagoonal-continental Upper Permian-Lower Triassic formation which concludes the Hercynian formational sequence. In the Middle Triassic, nearly the entire area of the Eastern European Craton was dry land. Only in central parts of the deepest depressions were also variegated, continental sediments of the Middle Triassic found; along the southwestern and southern edges of the ancient craton in Poland, in cis-Caucasia, and in the North Caspian Syneclise, ingressions of a brackish marine basin with Central European Muschelkalk fauna were observed.

Certain changes in the structural pattern took place during the Hercynian cycle of craton evolution. For example, from the second half of the Carboniferous onward, upheaval intensified along the meridional axis which connected the Baltic and Ukrainian shields; this led to total separation of the Baltic and Moscow syneclises whose depths increased in the westerly and easterly directions respectively. At the same time, the eastern part of the Russian Platform adjacent to the Uralian Geosyncline subsided more noticeably. The geosyncline had turned into a highland by the end of the Carboniferous and beginning of the Permian; the cis-Uralian Foredeep formed in front of it. The Pripyat–Donets Aulacogen completed its most active development in the Viséan age, when movements along the faults bounding it ceased and more than half the sedimentary fill had accumulated; at that time it began to turn into a syneclise. The Baltic Shield experienced the effect of Hercynian movements through formation of the Oslo Graben in its southern part. This graben contains Lower Permian red beds and a somewhat younger volcano-plutonic association. Major alkaline and ultrabasic plutons of the Khibiny and Lovozero tundras were formed in the northern part of the Baltic Shield.

The most important structural transformations of the Eastern European Craton occurred during the Hercynian cycle. The majority of arches and local uplifts of the Russian Platform, in particular of the Volga–Uralian region, formed during this cycle.

2.5.8 The Cimmerian and Alpine stages (Late Triassic–Quaternary period) (Fig. 18)

In this stage, the principal structural transformations and subsidence affected the southern part of the craton, under the obvious effect of the high tectonic activity of the Mediterranean geosynclines (Carpathian and Caucasian). The main structural elements continued their inherited development; this applies to the Baltic (Polish–Lithuanian), Moscow, and Mezen' syneclises, the Mazurian–Belorussian and Voronezh anteclises, the Ukrainian Syneclise which formed in place of the Pripyat–Donets Aulacogen, and the North Caspian Basin. However, while the northern syneclises gradually "left the scene", slowing down abruptly or even discontinuing their downwarping by the end of the Cretaceous, the Ukrainian Syneclise expanded markedly at the expense of the adjacent uplifts; the Black Sea pericratonal downwarp and the North Caspian Basin subsided considerably. In the east, within the buried Volga–Uralian Anteclise, there formed the Ul'yanovsk–Saratov Trough which continued to the North Caspian Basin. The Don-Medveditsa dislocation system formed at the junction between the North Caspian Basin and the Voronezh Anteclise. Growth of salt domes, which had begun at the end of the Hercynian stage, continued in the Ukrainian and North Caspian syneclises.

Deposits of this stage are less than 0.5 km thick throughout the northern half of the platform; thickness is maximum in the North Caspian Basin (more than 3.0 km) and reaches 1.0–1.5 km in the Ukrainian Syneclise, and the Visla–Dniester and Black Sea downwarps.

The Cimmerian–Alpine structural sequence of the Russian Platform starts with a basal gray continental, partly coal-bearing formation, which predominated in the depressions during the Liassic and the first half of the Dogger (to the middle of the Bajocian). This formation was gradually superseded by a gray, terrigenous, but already marine Upper Jurassic formation consisting mainly of dark, in some places bituminous clays and glauconite sandstones. In the axial parts of the deepest depressions (the Ul'yanovsk–Saratov Trough and the North Caspian Basin), these sediments are partly replaced by

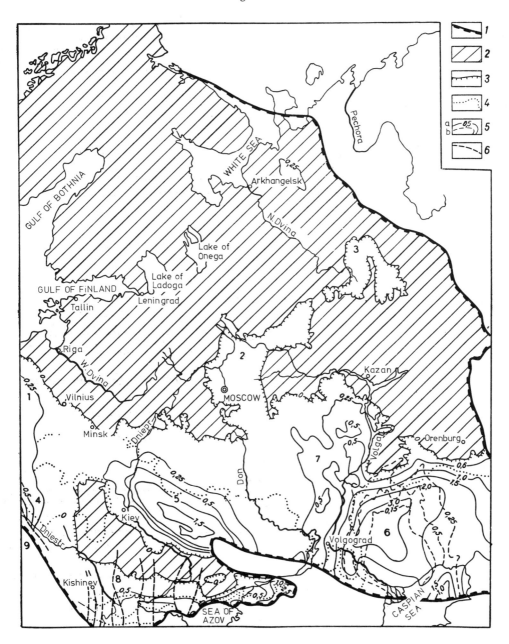

Fig. 18.   Scheme of extension and distribution of thicknesses of Mesozoic and Cenozoic in the Russian Platform (after IGOLKINA, KIRIKOV & KRIVSKAYA, 1970, simplified). 1 = actual boundary of the Platform; 2 = areas devoid of sediments; 3 = limit of development of Mesozoic sediments; 4 = ibid., of Cenozoic sediments; 5 = isopachs: a) Mesozoic, b) Cenozoic; 6 = major faults. Main structures: 1 = Baltic Syneclise, 2 = Moscow Syneclise, 3 = Viatka–Kama Depression, 4 = Lvov Basin, 5 = Ukrainian Syneclise, 6 = North Caspian Syneclise, 7 = Ul'yanovsk–Saratov Trough, 8 = Peri–Black Sea Depression, 9 = Carpathian Foredeep.

marls and oil shales. In the Early Cretaceous, accumulation of the marine terrigenous formation continued. The quartzose-glauconite sands are typical of shallower water areas. Clays with siderites and ammonite fauna are typical of deeper water areas associated with axial parts of depressions in the southern half of the Russian Platform. Marine sediments alternate with continental ones in its northern half. The Late Cretaceous was a time of maximum transgression and predominance of a carbonate, more specifically marl-chalk, formation throughout the southern and western parts of the platform. On the uplift slopes, it is replaced by glauconite-quartzose sands. Approximately the same sedimentation conditions prevailed in the Paleocene–Eocene, but purely carbonate rocks were replaced by marls and calcareous clays; added to the glauconite-quartzose sands at the depression periphery were siliceous rocks – gaizes. In the Oligocene and Miocene, the glauconite-quartzose sand formation was replaced, due to the general regression, by a continental formation of quartzose sands and kaolin clays, most typical of the Ukrainian Syneclise (Poltava series). In the Black Sea Downwarp, a marine, relatively deep-water, dark, bituminous clayey formation became widespread. It was also found in the Caucasus where it is known under the name of Maikop series. Finally, in the Neogene (from the Middle Miocene), the same thin strata of continental quartzose sands and clays accumulated in residual depressions. In the Dniester and Black Sea downwarps and in individual areas of the North Caspian Syneclise the deposits consist of shallow-water-coastal sands, clays, and shell limestones. At the southwestern periphery of the platform, alongside the Carpathian Uplift in the Miocene, there developed the cis-Carpathian Foredeep. Of great interest is the recently discovered ingression of Neogene seas deep into the Don Basin. This seems to be connected with embryonic rifting in the continuation of the African–Arabian rift system via Western Caucasus (MILANOVSKY, 1976). During the Quaternary period, the entire northern and central parts of the craton were involved in glaciation and became a region of deposition of a tillite, sand, and varved-clay ice sheet formation. The south remained as area of erosion, with the exception of the North Caspian Syneclise where the same shallow-water marine regressive formation continued to be deposited.

The history of the Eastern European Craton is, on the whole, one of gradual fragmentation and subsidence. It reached a maximum in the Hercynian cycle (hence the most intensive magmatism) and was superseded by renewed uplifting at the end of the Alpine evolutional stage.

2.6 Conclusions on the history and structure of the Eastern European Craton

The Eastern European Craton, alongside the Northern American one, is among the best explored ancient cratons of the world. At the same time, both cratons are the most typical representatives of stable cratons; their example can give a general characterization of this type of structure.

Formation of the crystalline basement of the Eastern European Craton was completed at the chronological boundary of 1,650 m.y. B.P. It was marked by the intrusion of major differentiated plutons of gabbro-labradorites and rapakivi granites. A more or less equivalent event in North America is the formation of anorthosite massifs of the Nain province and Grenville Belt, etc., dated at about 1,350 m.y. B.P. Considerable areas of the basement of both platforms had consolidated much earlier, 2,000 m.y. B.P., some of them even earlier. The beginning of formation of a continuous sedimentary cover of the Eastern European Craton is estimated at about 650 m.y. B.P. (Early Vendian); therefore, the interval between the end of basement formation and the beginning of accumulation of the cover amounts to the enormous figure of a billion years! This is the duration of the aulacogen evolutional stage of the Eastern European Craton, which superseded the relatively short stage of cratonization (1,750–1,650 m.y. B.P.). The aulacogen stage in Europe, in contrast to North America, manifested itself quite typically. The main aulacogen system extends across the central part of the platform in a SW–NE direction (Volyn–Orsha and Central Russian aulacogens), i.e., parallel to the Scandinavian and Southern Embian boundaries of the platform. In the area of Moscow, a southeast-trending system branches off from it and runs via Ryazan and Saratov to the North Caspian Syneclise. It is parallel to the Timan and Carpathian edges of the craton. Geosynclinal systems originated along all these boundaries of the platform.

The development of aulacogens was accompanied by plateau-basalt magmatism whose manifestations are quite common also in North America (Keewenawan and Coppermine River traps, etc.). Outbursts of this magmatism are associated with the Gothian, Grenville, and Baikalian epochs of tectonomagmatic activity, i.e., to the chronological boundaries of 1,350, 1,000 and 650 m.y. B.P. The increased activity of endogenic processes was characteristic not only of aulacogens but also of the uplift areas between them. This is attested to by the "radiometric rejuvenation" of the rocks, which accompanied the processes of fissuring, diaphthoresis, and hydrothermal activity. Interestingly, the basement rocks have radiometrically a lower age in the interval of 1,650–600 m.y. B.P. mainly within the Russian Platform and pericratonal downwarps, whereas in the Baltic and Ukrainian shields, and even in the Belorussian and Voronezh massifs, only older rocks were preserved. This means that the part of the platform, which was more fragmented in the aulacogen evolutional stage, was prone to subsequent subsidence and became the site of syneclise formation in the platform stage proper. A similar pattern is observed also in the Northern American craton whose basement beneath the sedimentary cover has a generally lower radiometrical age of 1,350–1,000 m.y.

The platform stage of development of the Eastern European Craton started somewhat earlier than for the Northern American Craton (in the Vendian and not Late Cambrian), but rather later than in the Siberian Craton, where the basement, too, is generally much older. This stage is quite clearly divided into distinct individual tectonic stages which quite well correspond to stages of development of adjacent geosynclinal systems: Baikalian Timan–Uralian and Galician, Caledonian Scandinavian, Hercynian Uralian and Central Dobrogean–Caucasian, Cimmerian Northern Dobrogean–Mangyshlak, and Alpine Carpathian and Caucasian. The boundaries between these stages (Middle–Late Cambrian, Early Devonian, Middle Triassic) were marked by nearly complete emergence of the craton, with the exception of the pericratonal downwarps and the North Caspian "exogonal" (ZHURAVLEV, 1972) or nodal syneclise. The structural pattern of the craton underwent noticeable rearrangement from stage to stage. This applies primarily to the distribution of the principal subsidence zones (syneclises). It generally followed the pattern established by KARPINSKY (1894), i.e., preferably occurrence near and parallel to the geosynclines particularly active at that stage. This pattern is associated with the manifestation of movement trends inherited from previous stages of evolution.

In the Baikalian stage, i. e., in the Vendian–Early and Middle Cambrian, the principal subsidence affected the Timan, Uralian and cis-Carpathian regions (Dniester pericratonal downwarp), but spread also by inheritance to the zone of the Central Russian and Pachelma aulacogens, having formed the Baltic–Moscow Syneclise in the center of the future Russian Platform.

The Caledonian stage saw active development of the Baltic–Moscow Syneclise whose trend is generally the same as that of the Scandinavian Geosyncline from which it is separated by the Baltic Shield. The Dniester Downwarp and the North Caspian Syneclise developed not as rapidly. This corresponded to the slower subsidence in this stage, of the eastern, southern, and southwestern geosynclinal surrounding of the platform.

The platform underwent considerable reactivation and rearrangement at the beginning of the Hercynian stage, in the Middle and Late Devonian, simultaneously with intensified subsidence in its eastern and southern surroundings. This coincided with formation of the Pripyat–Dnieper–Donets Aulacogen zone, of the Donets–Caspian superposed miogeosyncline, and of the Belorussian–Voronezh Uplift zone north of the miogeosyncline. This uplift zone had the shape of a narrow basement ridge. In the east, the ancient Volga–Uralian Shield was intersected by the channel of the Kama–Kinel troughs paralleling the Urals; this may well have a tectonic implication. Ultimately, at the end of the Hercynian stage, the entire eastern part of the platform became involved in strong subsidence after forming the huge Eastern Russian Depression whose southern part merged with the North Caspian Syneclise. Downwarping of the latter intensified from the Kungurian onward; it became the site of accumulation of a thick evaporite, and then (Permian–Triassic) of a red clastic formation. Thus, the entire southern and eastern parts of the Eastern European Craton underwent much alteration in the Hercynian stage under the obvious influence of the most active development of the Central European, Dobrogean–Caucasian, and Uralian geosynclines. However, also the remaining northeastern part of the craton became reactivated judging from the manifestation of Permian alkaline magmatism in the Oslo Graben and on the Kola Peninsula. This Hercynian reactivation was preceded by a Caledonian, less vigorous reactivation coeval with the Devonian orogenesis of the Scandinavian Caledonides.

In the Cimmerian stage, which is identified quite arbitrarily, the strongest subsidence took place along the southwestern and southern margins of the craton, including the North Caspian Syneclise; in the south, it involved the zone affected by the geosynclinal or nearly geosynclinal troughs of the Crimea Mountains, cis-Caucasia, Greater Caucasus, and Mangyshlak. In Western Europe, however, the geosynclinal region was separated from the Eastern European Craton by a large zone of inherited Hercynian uplifts. Farther north, there formed the huge North Sea–Baltic (Central European) region of intensive subsidence, whose northern rim was superposed on the edge of the ancient platform. In the east and northeast of the platform, there continued weak subsidence inherited from the Hercynian stage. This subsidence partly continued or recommenced also in the Alpine stage, until the Late Pliocene transgression in the Volga and Kama regions; the latter proceeded along a weakened zone of nearly longitudinal trend which can be traced back to the Devonian (KIRSANOV, 1971).

The greatest subsidence of the Alpine stage affected the southern periphery of the craton, i. e., the Ukrainian and North Caspian syneclises and the Black Sea pericratonal downwarp.

The influence of the most active adjacent geosynclines was decisive in forming the craton structural pattern during individual stages of its evolution. At the same time, zones of subsidence or uplifting which had formed in earlier, especially in the immediately preceding evolutional stages, were preserved (inherited). In some cases, subsidence zones are not directly adjacent to the geosynclines, but are separated from them by uplift zones. However, the agreement in the trends of all three types of structures points to their interrelated formation (Scandinavian Geosyncline–Baltic Shield–Baltic-Moscow Syneclise in the Caledonian stage; Dobrogean-Caucasian Geosyncline–Ukrainian Shield–Dnieper-Donets Aulacogen in the Hercynian stage, etc.). The Ukrainian Shield underwent strong reactivation in the orogenic period of the Hercynian cycle.

Subsidence on the craton lagged by half a phase behind its start and culmination in geosynclines, as noted by RONOV in 1949. However, the highest uplifts in mobile zones and on the platform generally coincide: by the end of the stage the craton seems to catch up the geosyncline, due to its faster emergence; it then lags behind, due to the longer period of uplifting. As a result, the boundaries between the cycles (Middle Cambrian, Early Devonian, Middle Triassic, Quaternary) are shared by the craton and adjacent geosynclinal-orogenic systems.

The finer cyclicity in the alternation of transgressions and regressions was discussed by KAUTSKY (1949) for the Early Paleozoic of the Scandinavian Geosyncline, the Baltic Shield, and the western slope of the Baltic Syneclise, by TIKHOMIROV (1951) for the Late Cretaceous of the Caucasus and Russian Platform, by FEDOROV & KUTUKOV (1950) for the Middle and Late Paleozoic of the Urals and Volga region. They proved the synchronism of short-term "interstage" regressions on the Russian Platform and in adjacent geosynclines.

The present author (KHAIN, 1951), after an analysis of the relationship between the movements of the Eastern European Craton and adjacent geosynclines, arrived at the conclusion that oscillatory movements proper are synchronous (eustasy!); wave-like movements (dictyogenesis according to S. BUBNOV, 1954) on platforms lag behind movements in adjacent geosynclines. Subsidence of geosynclines begins in an environment of continuing uplifting of adjacent continental platforms which supply abundant detrital material. Subsidence gradually involved the pericratonal downwarps of the platform continental margins; however, only the alternation of subsidence and uplifting, i. e., inversion in the geosynclines, induces a subsidence wave which then penetrates deep into the craton, until it is overtaken by a more powerful upheaval wave from the former geosyncline (now orogen).

Another problem repeatedly discussed with reference to the Russian Platform is the relationship between the sedimentary cover structure and the inner structure of the basement. At first glance, there is no connection between them, due to the sharply discordant structural patterns. The trends of the major structures of the Russian Platform (Baltic and Moscow syneclises, Belorussian–Voronezh Anteclise, Dnieper–Donets Aulacogen, Ukrainian Shield) which are oriented either WSW–ENE or WNW–ESE, i. e., in sublatitudinal directions, are sharply discordant with respect to the predominantly longitudinal trend of the linear, mainly Karelian structures of the basement. This discrepancy seems quite natural, in view of the enormous gap between the end of the basement formation and the beginning of deposition of the continuous sedimentary cover.

It is, however, noteworthy that most of the above cover structures were shaped (in their present outlines) under the influence of Hercynian movements, mainly of the Mediterranean Belt, reinforced by Alpine movements of similar orientation. Older structures of the cover, especially Riphean aulacogens, their ancestors, are not so sharply discordant with respect to the basement structures. On the contrary, as noted in several publications, especially by BOGDANOV (1964), GAFAROV (1971), and POSTNIKOVA (1976), aulacogens formed in most cases above Lower or Middle Proterozoic geosynclinal channels, even though in individual cases they have cross-cutting positions, with respect to these channels. Syneclises in turn form above aulacogens, if not at their intersection; consequently, this inherited character is acquired also by syneclises, at least in the initial stages of their development. Only in more recent stages was the structure of the sedimentary cover shaped almost completely independently of the inner structure of the basement. But even in these stages und until the recent tectonic epoch, movements and deformations continued, associated with renewed movements along the faults bounding the Karelian suture geosynclinal troughs. The important role of such movements was demonstrated by Ukrainian geologists for the Black Sea slope of the platform (ERMAKOV, 1972; and others).

3. The Timan–Pechora epi-Baikalian Platform

3.1 Boundaries and principal structural subdivisions

The Timan–Pechora Platform occupies the extreme northeast of the Russian Plain, which is separated from its main part by the low (up to 463 m high) Timan Ridge with a NW–SE trend (Fig. 19). In the east and southeast the platform passes into the northern segment of the Uralian Foredeep, bordering with the Northern and Polar Urals and Pai Khoi. Its boundary with the foredeep has the character of a quite gradual subsidence in the southern half; however, a system of quite abrupt dislocations appears along it more to the north, which morphologically appear as the Chernyshev, Gamburtsev, Sorokin, and Chernov ridges. According to the composition of their sediments these structures belong to the platform, but their deformations are genetically connected with the formation of the Ural–Novaya Zemlya geosynclinal fold system in the Late Hercynian–Early Cimmerian epoch. These zones, according to their location and origin are obviously equivalents of the Jura Mountains folds in Western Europe, which are located on the boundary between the Alpine Foredeep and the epi-Hercynian platform.

The convergence of Timan and Urals in the south causes pinching out of the Timan–Pechora Platform in this direction. On the other hand, it becomes wider in the north (northwest) and continues to the Pechora Sea, occupying its entire area and practically the whole space between the Kola Peninsula in the southwest and the Southern Island of Novaya Zemlya and Vaigach Island in the northeast. In the direction of the central part of the Barents Sea the structures of the Timan–Pechora Platform, which have a NW–SE trend, give way to structures of different trend, originating on another basement. This change probably occurs along a fault zone of WSW–ENE trend. However, the upper horizons of the cover of the Pechora Basin merge into the South-Barents Depression (Syneclise).

One of the main structural units of the Timan–Pechora platform is the **Kanin–Timan Ridge,** a large uplift of the Baikalian (Riphean) basement of the platform, extending from the Kanin Peninsula in the northwest to the Polyudov Ridge in the southeast where it links with the Urals. The Kanin–Timan Ridge borders upon the ancient Eastern European Craton and the Russian Platform along a fault (boundary suture). This is a fault which, on the basis of seismic and magnetometric data is a large and flat overthrust along which the Baikalian complex of the Timan and Kanin overlaps the basement of the Russian Platform for several tens of kilometers. Its magnetic field appears through this considerably weaker magnetic complex. The seismic profile indicates that the surface of the thrust slopes at an angle of 30 to 40°.

The fault along the boundary of the Timan and Kanin Baikalides with the ancient craton can be traced to the northeast in the sea at the Murman coast of the Kola Peninsula. It appears again on land between the Srednii and Rybachi peninsulas in the USSR, and on

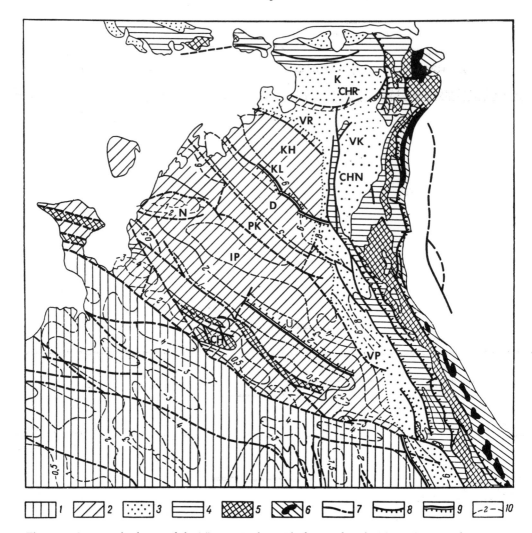

Fig. 19.   Structural scheme of the Timan—Pechora Platform (after the Tectonic Map of Europe, 1:10,000,000, 1977). 1 = sedimentary cover of the Russian Platform; 2 = ibid., of the Timan — Pechora Platform; 3 = Upper Paleozic and Triassic of the Uralian Foredeep; 4 = miogeosynclinal Lower and Middle Paleozoic of the Novaya Zemlya, Vaigach, Pai Khoi and Urals; 5 = salients of the Baikalian fold complex; 6 = eugeosynclinal Lower and Middle Paleozoic and outcrops of ophiolites (shown in black); 7 = faults (mainly vertical); 8 = overthrusts; 9 = flexure; 10 = isobaths of the basement surface. CH = Chetlassky Kamen Uplift; U = Ukhta Flexure; IP = Izhma — Pechora Basin; N = Naryan Mar Swell; PK = Pechora—Kozhva anticlinal zone; D = Denisov Trough; KL = Kolva anticlinal zone; CHN = Chernyshev Ridge; CHR = Chernov Ridge; K = Karataikha Basin; VK = Vorkuta Basin; VP = Upper Pechora Basin; KH = Khoreiver basin; VR = Varandei Ridge (Sorokin Ridge).

the Varanger Peninsula in Norway (DEDEEV et al., 1974; SIEDLECKA, 1975). Intensely dislocated geosynclinal sediments of the Middle Riphean–Lower Vendian are here thrust over their platform equivalent, lying monoclinally and transgressively on the ancient Precambrian basement of the Baltic Shield. These Late Precambrian platform sediments fill the Kil'din Trough which runs parallel to the boundary suture (overthrust) (they form Kil'din Island). The equivalent of this trough to the southeast, on the continent, is the **fore–Timan Trough** whose basement lies at a depth of 5 to 6 km. The Riphean sediments composing the basement of this cover are evidently formations of a zone of pericratonal subsidence of the Eastern European Craton; only the Vendian sediments can be considered to be filling a foredeep. They, in their turn, are overlain by platform deposits.

The second very large structure of the Timan–Pechora Platform is the **Pechora Syneclise** which on the continent attains a width of 400 km along the coast. The maximum depth of the basement in this region is 7 to 8 km on land and 8 to 9 km at sea. The basement of most of the depression, if not throughout its area (see below), is by age and constitution identical with the Baikalian fold complex of the Timan. The inner structure of the depression is rather complex, especially on the northeast. It consists of a series of large and complicated arches and troughs separating them.

The Timan–Pechora Platform, especially its northeastern part, the Bol'shezemelskaya Tundra, remained poorly explored until 10 to 15 years ago. An exception was the Ukhta oil- and gas-bearing region in the southeast. This changed quickly when it became clear that the entire Pechora Basin is an oil- and gas-bearing region and that bauxite deposits occur in the Timan. The territory of the platform has now been completely explored by means of geophysical methods; drillings have been carried out over most of it, including Kolguev Island. The most important summaries of work in this area were given by ZHURAVLEV (1972), MATVIYEVSKAYA (1974), and TSZYU (1964).

3.2 The Kanin–Timan Ridge and the nature of the basement of the Timan–Pechora Platform

Along the surface of the basement the Kanin–Timan Ridge forms a flat and wide (up to 150 km) asymmetric uplift with steeper and narrower southwestern, but wider and stepped northeastern limbs. They are joined along faults respectively with the fore-Timan Trough and the Izhma–Pechora Depression. In the longitudinal direction the ridge is split into inadequately uplifted blocks by transverse (northeasterly) and obliquely transverse (latitudinal) faults (BASHILOV & KAMINSKY, 1975). This causes the discontinuous nature of the outcrops of basement rocks on the surface. About ten such basement outcrops are known between the Kanin Peninsula and the Polyudov Ridge; they form the rocky uplands-"Kamni". The largest is the Chetlas kamen in the Central Timan, where the relatively most complete sequence of the Baikalian fold complex has been ascertained.

A fault, of a steep (?) overthrust type, separates the northeastern limb of the meganticline of the Chetlas kamen from the southwestern limb and forms the actual boundary between two basically different structural-facial zones of the basement (Fig. 20). The southwestern zone is built up very smoothly and consists of very gradually (between 2 to 4 and 15 to 20°) southwesterly dipping terrigenous (quartzite and schistose) formations of the Middle and Lower Upper Riphean and of a carbonaceous formation of the Upper

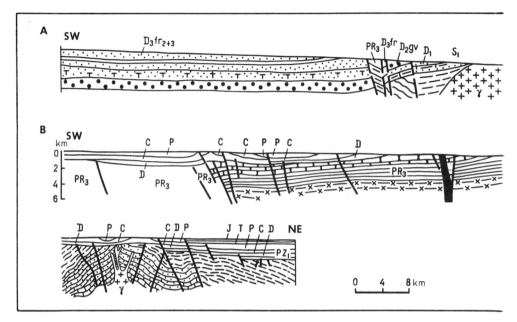

Riphean (the latter is, as regards stromatoliths, the equivalent of the Minýar series of the Bashkirian Anticlinorium of the Urals). Metamorphism of these rocks is not stronger than the lowest stages of the greenschist facies. The monocline of the southwestern limb is complicated by small upthrusts along which vein intrusions of derivatives of alkaline-basic magma are found (lamprophyries, diorites, plagioclasites, carbonatites, etc.). Near the main fault forming the zone boundary the steepness of the dip of the strata increases to between 30 and 60°, and its direction becomes inverse.

The terrigenous, basically schistose sediments of the Middle-Upper (?) Riphean, which constitute the northeastern zone, exhibit much stronger metamorphism which on the Kanin Peninsula reaches the amphibolite stage, as well as deformation into steep, locally isoclinal folds; they were also subject to intensive cleavage. A fairly large number of small intrusive rock bodies are found here; they belong to three different formations, namely gabbroic (earliest), granitoid (whose radiometrically determined age is between 625 and 525 m.y.), and monzonite-syenite which includes nepheline syenites (600 to 500 m.y.).

The Baikalian basement has been explored by drilling at several sites of the Izhma–Pechora Depression. Exploration has recently been carried out much further to the east, south of the Kol'va Arch and in the southern part of the Khoreiver Depression (DEDEEV, ZHURAVLEV & ZAPOL'NOV, 1974). This basement is represented by highly dislocated metamorphic schists of the same type as in the northeastern zone of the Kanin–Timan Ridge; it also contains granites which here, according to the probably correct suggestion of ZHURAVLEV (1972), form larger massifs, having the scale of batholiths, than in the Timan and Kanin peninsula.

These data in general corroborate the suggestions, made by STILLE (1955) and SHATSKY and supported by BOGDANOV, that the Timanides were formed within a narrow intracratonic trough (aulacogen according to SHATSKY & BOGDANOV, 1961) dividing two

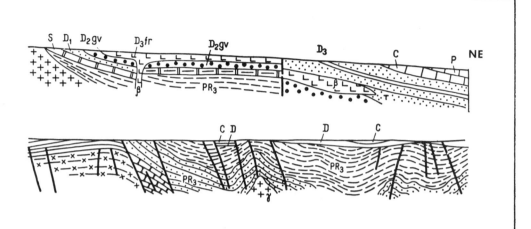

Fig. 20.   Cross-sections of the Timan Ridge: A: in the north (after GETSEN, 1975); B: in the middle part – Chetlassky Kamen (after SMIRNOV in ZHURAVLEV, 1972); γ = granites; β = basalts.

ancient blocks–plates–Russian and Barentsian (Barentsia according to STILLE). Many scientists, in agreement with these suggestions, thought that the basement of the north-eastern part of the Pechora Syneclise, the Bol'shezemelskaya Tundra, is not Baikalian (Riphean) but Early Precambrian. In fact, the magnetic field of the Bol'shezemelskaya Tundra differs from those of the western part of the Pechora Syneclise and of the Timan; it is, in contrast to the latter, characterized by high maxima with predominantly northeasterly trend, but with quite sharp virgations. Some geophysicists assume this magnetic field to coincide with that of the Russian Platform, while ZHURAVLEV & GAFAROV stated in 1959 that an eugeosynclinal zone of Baikalides continued here from the region of the Polar Urals. ZHURAVLEV & GAFAROV thus assume, in agreement with earlier suggestions made by SHATSKY, that the Kanin, the Timan, and the western part of the Pechora Syneclise were formed on the base of a miogeosynclinal zone of Baikalides, while the eastern part was formed on top of an eugeosynclinal zone.

The discovery of dislocated and metamorphosed Riphean on the Kolva Arch and in the Khoreiver Depression, and especially of acid volcanics in the latter region (similar to the volcanics of the Upper Riphean of the Polar Urals), to some extent confirmed the view of ZHURAVLEV & GAFAROV, at least as far as showing that the Timan–Pechora Platform as a whole obviously belongs to the Baikalian geosynclinal fold region. It may, however, be that this region includes some older blocks, i. e., median massifs with an Early Precambrian basement transformed during the Baikalian era. This assumption is most probably correct for the Bol'shezemeskaya Uplift buried beneath the Khoreiver Depression and for its possible continuation in the Pechora Sea. Riphean acid volcanics by themselves do not prove the eugeosynclinal nature of this region in the Baikalian cycle; the characteristic encircling of the Bol'shezemelskaya Uplift by the Polar Urals and the Pai Khoi indicates its considerable, probably earlier consolidation.

An important though indirect confirmation of the latter view is provided by new data on the age of the metamorphics of the Kharbei Anticlinorium core of the Polar Urals, which in the southeast borders the eastern part of the Timan–Pechora region. These metamorphics were considered to be Riphean and were assumed by PERFIL'EV (1968) to extend to the Bol'shezemelskaya Tundra. However, data obtained by PODSOSOVA (1981) indicate that they are pre-Riphean, obviously Early Precambrian. This changes the situation radically.

Let us now again consider the Kanin–Timan Ridge. We note that the sedimentary cover on top of its slopes and in transverse subsidences begins in the north with the Silurian, but in the south with the Middle Devonian and includes the Upper Devonian, the Carboniferous, and the Permian. The cover of the most deeply subsided parts, especially in the southeastern Timan, exhibits large, flat, brachyanticlinal and dome-shaped folds. These include, in particular, the Ukhta Brachyanticline with which the oil field of the same name, known since the 18th century, is associated; this was the forerunner of the entire Timan–Pechora oil- and gas-producing province.

3.3 The Pechora Syneclise

The Pechora Syneclise occupies the remaining large part of the Timan–Pechora Platform. It is filled with an enormous sedimentary cover varying in thickness between 2 to 3 and 6 to 8 km, which includes almost the entire sequence of the Phanerozoic. This sequence is quite clearly subdivided by regional unconformities into several structural layers (or rather complexes). The age of the lowest layer, the so-called Izhma–Omra terrigenous complex, is still disputed (some scientists relate it to the Vendian, others to the Cambrian, still others to the Ordovician which is the most likely, or consider it to form elements of all three or two of these systems). The complex lies on a rough basement surface and is with some unconformity overlain by the second layer (counted from below) which includes Silurian, and in the east also Lower Devonian carbonaceous formations. This Late Caledonian level is with a clear unconformity overlain by a third, Hercynian, terrigenous-carbonaceous layer comprising sediments of the Middle and Upper Devonian, Carboniferous, and Lower Permian. An unconformity, evident in particular in the west inside the Lower Carboniferous, between Tournaisian and Viséan, divides this layer into two sublayers (or rather, divides the complex into layers). The main oil fields in this region are associated with this layer. The remainder of the Permian, which is evaporitic-carbonaceous-terrigenous, and the terrigenous Triassic form a fourth, Late Hercynian–Early Cimmerian structural layer. They are unconformably overlain by Jurassic and Lower Cretaceous sediments, and near the coast also by Upper Cretaceous and Paleogene sediments, which form the fifth, Late Cimmerian–Alpine layer. Lastly, a thin nappe of Pliocene–Quaternary glacial-marine sediments represents the sixth, Late Alpine layer which is practically not yet dislocated.

There is a sharp nonconformity in the attitudes of the various structural layers. The most contrasting structural forms appear in the Devonian–Permian (Triassic) layer and are due to Late Hercynian–Early Cimmerian movements. The largest of these structures are, from SW to NE, the wide and flat **Izhma–Pechora Depression** whose basement lies at a depth of up to 4 km, with very gentle uplifts on its terraces; the pronounced,

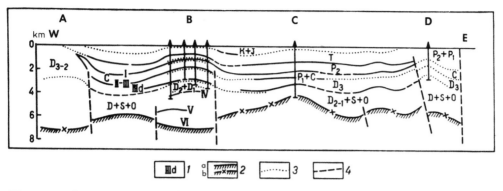

Fig. 21. Seismic cross-section through the northeastern part of the Timan–Pechora Platform (after Tarbaev, Mokrushin & Konovalov, 1973). 1 = seismic reflectors; 2 = basement surface: a) ascertained, b) presumed by gravimetric data; 3 = geological boundaries; 4 = faults. A = Pechora–Kozhva anticlinal zone; B = Denisov Trough; C = Kolva anticlinal zone; D = Khoreiver Trough and Chernyshev Ridge.

complicated **Pechora** (Pechora–Kozhva) **Ridge Arch,** bounded along the basement and the pre-Carboniferous part of the cover by faults; the narrow (40 to 60 km wide) **Denisov Trough,** the complicated **Kolva Arch,** the **Khoreiver Depression;** and finally the **Sorokin Arch** which already forms a transitional structure to the fold zone of the **Gamburtsev–Chernov Ridge** bordering the Pechora Syneclise in the northeast. A similar zone of the **Chernyshev Ridge** borders the northeastern part of the Pechora Depression in the east. Like the Izhma–Pechora Depression in the west, it gradually goes over, through a flexure, into the Upper Pechora Depression of the Uralian Foredeep.

The structure of the syneclise is quite different on the surface of the basement, and partially in the two lower structural layers. The surface of the basement subsides quite abruptly under the Pechora–Kozhva and Kolva arches, where it attains the greatest depth in the whole syneclise (7 to 8 km). The thickness of the pre-Carboniferous, especially of the pre-Middle Devonian sediments is here, correspondingly, maximal. On the other hand, the basement is raised slightly in the region of the Denisov Trough and especially the Khoreiver Depression where the Lapian Salient and the Bol'shezemelskaya Anticline stand out (Fig. 21). The Pechora–Kozhva and Kolva arches were thus formed on the site of inverted aulacogens.

The Jurassic–Cretaceous (Paleogene) layer occurs mainly in depressions and, highly attenuated, repeats the structural relief of the Triassic surface. The thin nappe of Pliocene–Quaternary has a subhorizontal attitude.

3.4 Main stages of evolution

It is very likely that in the pre-Riphean (pre-Middle Riphean?) time there existed in place of the present Timan–Pechora Platform a continuation of the ancient Eastern European Craton with its Early Precambrian granitic-gneissic basement. In any case, this must be so for the entire miogeosynclinal zone of the Timanides up to the Pechora Fault zone (the western boundary of the Pechora–Kozhva Arch), and is quite likely for the presumed Bol'shezemelsky median massif. Recently it was confirmed by seismic soundings.

3.4.1 Dalslandian–Baikalian stage

Formation of the Timan–Uralian Late Precambrian geosynclinal system took place in the Middle Riphean. It is howewer, not impossible that the Varanger–Timan segment was formed earlier, already in the Early Riphean. The eugeosynclinal zone, which in the future Timan–Pechora Platform most likely occupied a place between the Pechora–Kozhva and Kolva arches, must have appeared as a result of destruction of the ancient continental crust, by whose break-up the Bol'shezemelsky Massif may have been formed.

The Timan geosynclinal underwent intensive subsidence during the **Middle and Late Riphean;** in the northwest, along the shore of the Kola Peninsula, this also took place during the **Early Vendian.** A layer having a thickness of the order of 10 km, composed mainly of clastic sediments, accumulated. This material was mainly derived from the ancient Eastern European Craton. As regards the zone between the Pechora–Kozhva and Kolva arches, it may be assumed from the intensive magnetic anomalies that considerable ophiolitic magmatism occurred. The first upheaval in the Timan zone occurred already before the Late Riphean, in the Dalslandian (Grenvillian) epoch. A certain attenuation of tectonic activity must have occurred during the second half of the Late Riphean in view of the accumulation of a carbonaceous formation of the Bystra series and its analogues. According to GETSEN (1975), however, this formation is a barrier reef at the boundary of two longitudinal zones of the Timan.

The main diastrophism of the outer zone of the Timanides most likely occurred in the **Late Vendian.** Their sedimentary filling at that time underwent intensive folding and overthrust deformations, being in general displaced in the direction of the ancient craton. These deformations were accompanied by regional metamorphism reaching the deeper levels of the amphibolite stage, as well as by the formation of granitic batholiths in the more interior part of the miogeosynclinal zone and of smaller plutons of granitoids nearer to its periphery. Granitic plutonism was preceded by the intrusion of basic magma. Baikalian diastrophism may have begun earlier to the east of the present Kanin-Timan Ridge (already at the end of the Riphean and beginning of the Vendian), the more so, if the Vendian really forms part of the Izhma–Omra molassoid complex with which the platform cover begins. Granites 780 m.y. old are known here. Uplift of the Kanin–Timan Ridge, which began after the folding, was accompanied by compensatory subsidence of the fore-Timan Trough.

3.4.2 Caledonian stage (Ordovician–Late Devonian)

The Izhma–Omra complex (Ordovician?) in any case already belongs to the Caledonian stage of regional evolution. Its extension towards the east from the Kanin–Timan Ridge and its composition prove that this ridge continued to undergo uplift. Formation of the Pechora–Kozhva and Kolva aulacogens took place simultaneously along the ancient tectonic sutures of the platform basement, which in the past had separated the miogeosyncline, the eugeosyncline, and the median massif.

Tectonic activity decreased considerably in the second half of the Caledonian stage, the Silurian. This is evident from the accumulation of a marine shallow-water carbonaceous (mainly dolomitic) formation. Sediments of this formation overlaid the northern part of the Kanin-Timan Ridge, while the southern part remained an erosion zone. Accumulation of this formation occurred in the east also during the Early Devonian. The presence of sulfate admixtures proves the gradual closing of the basin.

Thus, on the border between Early and Middle Devonian, the southwestern part of the Timan–Pechora Platform underwent a general upward movement, this being most pronounced in the Kanin–Timan Ridge. This movement and the accompanying deformations were undoubtedly repercussions of Late Caledonian tectogenesis. They were accompanied by an outburst of basaltic volcanism, especially at the very beginning of the Late Devonian.

3.4.3 The Middle Devonian and early Late Devonian

gave rise to a terrigenous sandy-clayey formation with some basalts. Substitution of the continental and coastal-marine facies by purely marine facies occurs east of the Timan. Erosion centers remained during all this time in various parts of the Timan Ridge. The thickness of the sediments also increases towards the east, attaining a maximum of 1,200 m in the Pechora–Kozhva Aulacogen. No uplift or rearrangement of the structural pattern took place here, and in general in the northeastern part of the platform at the Early-Middle Devonian boundary (MELKUMOV, 1975). The sediments of the Silurian and Middle Devonian form one structural layer. Its thickness exceeds 4 to 5 km in the Kolva Aulacogen and the analogous structure of the future Sorokin Ridge. The Bol'-shezemelsky Swell is clearly distinguished by its thickness minimum.

3.4.4 Hercynian stage (Late Devonian–Early Permian)

The terrigenous D_2–D_3^1 formation gives way to a clayey-carbonaceous D_3^1–C_1^1 formation, also developed to a maximum thickness (> 1.5 km) in the Pechora–Kozhva Aulacogen, which underwent general subsidence by 7 km (!) at the beginning of the Carboniferous. A characteristic "Domanik" deep-water bituminous-carbonaceous subformation can be traced at the base of this clayey-carbonaceous formation, that is common to the Timan–Pechora and Russian platforms. Reef structures, also similar to the latter and associated with this formation, developed on the uplifts. The rate of vertical movement and the corresponding change of the type of sediments were in general practically the same for both platforms from the Middle Devonian onward. They combined into one plate already as a result of the Baikalian tectogenesis.

This common character of development is evident also from the brief episode of upheaval and regression at the beginning of the **Viséan age of the Early Carboniferous.** This upheaval was most pronounced also in the Kanin–Timan Ridge where it was linked with the formation of bauxite. Interruptions in the accumulation of sediments occurred repeatedly in the Northern and Central Timan during the Carboniferous. Early Viséan regression caused wedging of a sandy-clayey terrigenous member of Lower Viséan – lower Upper Viséan into the carbonaceous, calcareous-dolomitic formation of Upper Devonian–Carboniferous–Lower Permian, thus dividing it into two. The magnitude of the Early Viséan disconformity becomes less in the northeast, while the terrigenous sediments of the Lower Viséan are replaced by carbonaceous ones. Sulfates were deposited here at the end of the Viséan, while sediments of the Middle and Upper Carboniferous are often missing in the sequence.

The thickness distribution of the upper part of the carbonaceous formation of the Hercynian stage differs in principle from that characteristic of all underlying sediments of the cover; the thicknesses are least in the Pechora–Kozhva and Kolva zones, but increase in the depressions bordering and separating them. This proves the beginning inversion of the structural pattern of the Pechora Syneclise, which came to an end in the Middle Permian.

3.4.5 Late Hercynian–Early Cimmerian stage (Middle Permian–Triassic)

A new interruption in the accumulation of sediments can be noticed in the sequence of the platform in the Early Permian. It is accompanied by a complete or partial disappearance of the Artinskian and Kungurian levels and by the change in the type of formation. The carbonaceous D_3^1–P_1^2 formation gives way to a red, lagoonal-continental, molassoid formation of **Upper Permian and Triassic.** The Kungurian transitional, evaporitic (sulfatic-)-carbonaceous-terrigenous subformation, which is up to 100 m thick, is located at its base. Its thickness is the least on the slopes. The detrital material of the overlying red subformation was derived from the Urals which at that time underwent considerable uplift. Certain strata with marine fauna, which appear in the northern part of the Pechora Syneclise, indicate ingressions from the north, and indirectly the predominance of marine conditions in the present area of the Pechora Sea. The maximum thickness of the formation is about 2 km; it can be clearly associated with the Denisov Trough and the Khoreiver Depression. The Pechora–Kozhva and Kolva arches and the Sorokin Arch finally appear at the end of this stage in the form of uplift zones. The structural pattern of the syneclise, even of the entire platform becomes close to that existing now. Shifting of the Hercynian final movements to the end of the Triassic, is similar to a phenomenon observed in the Northern and Polar Urals and the Pai Khoi. These movements were quite pronounced and caused a lengthy (during the whole Early Jurassic) interruption in the accumulation of sediments, and unconformable deposition of sediments of the following structural stage on the strata dating from the Triassic to the Devonian. It is quite natural that the amplitude of the hiatus and unconformity is maximum on the northeastern slopes of the Timan.

3.4.6 Late Cimmerian–Alpine stage (Jurassic–Paleogene)

The end of evolution of the adjacent Uralian geosynclinal-orogenic system caused the tectonic regime of the Timan–Pechora Platform, now united with the West Siberian

Platform into one zone of consolidation through the Polar Urals, to become even more quiescent. Movements along the basement faults practically ceased already at the end of the preceding stage. The formational sequence of the Middle Jurassic to Lower Cretaceous consists of a gray, continental, coal-bearing formation of the **Middle Jurassic,** a marine sandy-clayey formation of the **Upper Jurassic–Neocomian,** and the gray continental, sandy-clayey **Aptian–Albian** formation. None of these formations is thick. The maximum thickness is 800 to 1,000 m in the northeast (northern part of the Bol'-shezemelskaya Tundra), the minimum occurs in the Timan slopes and the Chernyshev Ridge, i. e., on the periphery of the syneclise. The thickness reduction and pinching out of the stage in the Pechora–Kozhva and Kolva arches indicates uplifts of these structural elements of the syneclise, which were inherited from the preceding stage.

Almost the entire Timan–Pechora Platform (in the area of the present dry land) became a region of erosion during the **Late Cretaceous and Paleogene.** The only exception was the northern part of the Pechora Syneclise, where thin strata of continental-marine sediments of the Upper Cretaceous and Paleogene occur[1].

3.4.7 Late Alpine stage (Neogene–Quaternary)

The Timan–Pechora Platform continued to undergo general slight uplift during the **Miocene and Early–Middle Pliocene.** This uplift probably extended also to its water-area part, being slightly stronger in the zone of the Timan–Kanin Ridge. Uplift of the ridge intensified during the **Late Pliocene and Quaternary,** thus giving it its present low-hill relief. Accumulation of sediments resumed at the same time in the depressions of the northern part of the Pechora Syneclise. These, however, belong to glacial-marine sandy-clayey formations. Their thickness only slightly exceeds 100 m.

Geophysical information and the tendency to facies and thickness changes of the sediments in the dry-land part of the Timan–Pechora Platform enable us to predict that the overall thickness of the sedimentary cover increases in the area of the Pechora Sea, primarily due to post-Triassic sediments.

We may state in conclusion that four main stages can be distinguished in the evolution of the Timan–Pechora Platform, as compared to adjacent regions. The first stage belongs to the beginning of Late Precambrian, when the territory of the platform formed part of an ancient craton which probably united the Eastern European and North-Asian cratons. During the second, Riphean, stage it became part of the Ural–Timan geosynclinal belt. The Baikalian tectogenesis removed the Timan–Pechora Platform from this belt and it again joined the Eastern European Craton, causing in the third, Paleozoic–Triassic stage a uniformity of movements and accumulation of sediments with the Russian Platform. However, the reduced extent of basement consolidation and the proximity of the Uralian Geosyncline (later orogen) caused the greater mobility of the Timan–Pechora Platform and its larger deformations during the Paleozoic and Triassic. Finally, from the Jurassic onward, the Timan–Pechora Platform again became, as before the Riphean, part of a huge consolidated territory encompassing all of northern Eurasia.

---

[1] Members with marine fauna occur in it near the Pechora Sea.

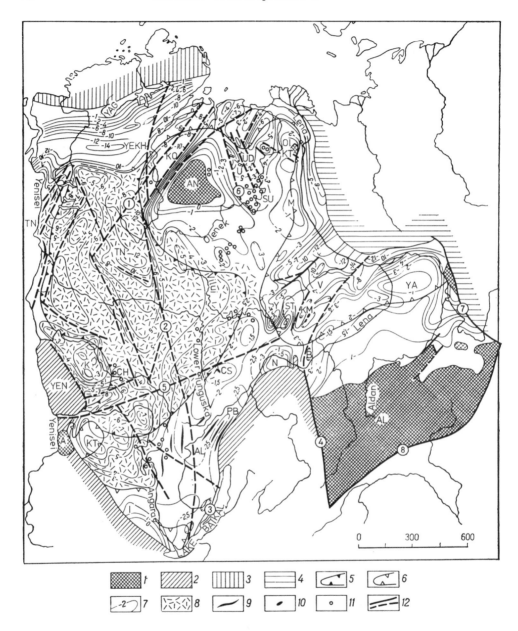

Fig. 22. Tectonic scheme of the old Siberian Craton (isobaths of the basement surface after SAVINSKY, 1971). 1 = salients of the Early Precambrian basement; 2 = Baikalian fold framing; 3 = Late Hercynian–Early Cimmerian framing; 4 = foredeeps of the Verkhoyansk fold system; 5–7 = platform cover: 5 = Upper Paleozoic–Triassic complex; 6 = Jurassic–Cretaceous complex; 7 = isobaths of the basement surface; 8 = Tungussian trap formation; 9 = folds of the platform cover; 10 = salt domes; 11 = ultrabasic-alkaline ring intrusions and kimberlite pipes; 12 = faults. YAG = Yangoda–Gorbit Uplift; YEKH = Yenisei–Khatanga Trough; KO = Kotuy Trough; TN = Turukhansk–Norilsk zone of dislocations; U = Udzha Trough; UD = Udzha Swell; SU =

4. The ancient Siberian Craton

4.1 Boundaries and principal subdivisions of the Siberian Craton

The Siberian (Central Siberian) Craton forms the oldest consolidated element of Northern Asia. It has a mainly Archaean basement and occupies a central position in the structure of the north of the Asian continent. It is 2,500 km across. Its area is 4.4 million km², not including the Baikalides on its western and southern periphery, in view of their complete independence from the structural aspect; historically the formations constituting them belong to the Ural–Okhotsk Belt.

The boundaries of the Siberian Craton (Fig. 22) are generally quite undisputed, except for the northwest and north, although there are differences in detail in tracing the other boundaries (SAVINSKY et al., 1971). The craton boundaries in the southeast coincide with the Northern Tukuringra deep fault running almost east to west, which can be traced from the shores of the Okhotsk Sea to the zone between the Olekma and Chara rivers, if we include the Stanovoi Range in the craton, or rather in the Aldan Shield. It will be shown below that there is enough reason for this. The boundary here turns north (NNE) and continues along the submeridional Zhuia Fault separating the Aldan Shield of the platform from the Baikal–Patom Baikalian fold system; its role is then taken over by the Primorye Fault (overthrust), whose general trend is SW–NE, and along which this Baikalian system is in contact with the southern edge of the craton. At the source of the Angara the craton boundary again changes direction abruptly, running from there in a northwesterly direction along the Biryusa Fault which borders the Baikalian structure of Eastern Sayan and the Kan Block of the Yenisei Ridge from the east. The craton boundary shifts slightly to the east at the lower reaches of the Angara; it then runs again in the same northwesterly direction along the periphery of the Baikalides of the Yenisei Ridge until it crosses the Yenisei River at the mouth of the Podkamennaya Tunguska at the northern end of the ridge. A series of boreholes west of the Yenisei, as far as north of the Igarka, have shown chlorite, quartzose-sericite, and micaceous schists and phyllites, most likely of Riphean age, i. e., similar to the rocks of the Yenisei Ridge, while platform Paleozoic rocks were found nearer to the Yenisei (BOCHKAREV & POGORELOV, 1973). It may therefore be assumed that the craton boundary north of the Podkamennaya Tunguska runs almost directly to the west of the Yenisei and obviously coincides with a fault. The

Sukhan Trough; OL = Olenek Anteclise; KT = Kutundy Graben; M = Muna Swell; V = Vilyui Syneclise; Y = Ygyatty Trough; ST = Suntar Uplift; KM = Kempendyay Trough; YA = Yakutsk Uplift; Al = Aldan Shield; B = Berezov Trough; N = Nuya Downwarp; CS = Central Siberian Anteclise (Nepa–Botuoba Uplift); AL = Angara–Lena Syneclise; PB = Peribaikalian Trough; I = Irkutsk Jurassic Trough; KT = Kansk–Tasseyevo Syneclise; A = Angara–Kan Uplift; YEN = Yenisei Range; K = Baykit Anteclise (Kama Uplift); CH = Chadobets Dome; AN = Anabar Anteclise; TN = Tunguska Syneclise. Deep faults: 1 = Kotuy; 2 = Taymyr–Baikal; 3 = Primorsk; 4 = Zhuya; 5 = Angara–Vilyui; 6 = Udzha; 7 = Nelkan; 8 = North Tukuringra.

zone of the Turukhansk–Igarka Dislocations must be included in the platform as its folded edge, since the Upper Riphean, the Vendian, and the Paleozoic are here represented by platform formations.

The Paleozoic cover of the craton north of the Dudinka dips beneath the deep Mesozoic–Cenozoic Ust'–Yenisei Trough. The northern boundary of the craton runs between the lower reaches of the Yenisei and the Laptev Sea north of the axial zone of the Yenisei–Khatanga (fore-Taimyrian) Trough, on whose other side lies the Late Hercynian–Early Cimmerian fold structure of the Southern Taymir Mountains (Byrranga Range).

The boundary of the craton turns south almost at a right angle in the Lena Delta and follows the edge of the Verkhoyansk Foredeep, first in a meridional and then, along the lower reaches of the Aldan, in an east-west direction. At the northern end of the Sette–Daban Range, where the Aldan Valley turns south, the craton boundary also turns sharply to the south and again assumes a submeridional direction coinciding with a fault (the Nel'kan Fault in the south) which separates the craton from the folded Southern Verkhoyansk Range, including the Sette–Daban Range and the Yudoma–Maia Trough. The junction, again at a right angle, between the Nel'kan boundary suture and the eastern end of the Northern Tukuringra Fault should run near the Okhotsk Sea coast, but is masked by a large granite pluton belonging to the southeastern end of the young Okhotsk–Chukotka volcanic-plutonic belt.

If follows from the above that the Siberian Craton has the polygonal contours characteristic of old cratons. It is bounded by faults (boundary sutures) along the entire circumference. These features emerge directly on the surface except where they are overlapped by young sedimentary troughs on front of mountains (fore-Taimyr, fore-Verkhoyansk). The boundary suture at the periphery of the Baikal–Patom fold system was regenerated as boundary overthrust along which formations of Baikalides overlapped the southeastern edge of the Angara–Lena pericratonal downwarp.

The southeastern projection of the craton which includes the Aldan Shield, especially its southern part (the Stanovoi Belt), underwent rather intense Mesozoic–Late Cimmerian (J_3–K_1), and recent (N–Q) orogenic reactivation[1], which is displayed in its composition and structure.

However, this region did not lose its contiguity with the remainder of the craton body. There is therefore no sense in excluding it from the latter, as was done, e. g., by SAVINSKY et al. (1971). It is for the same reason inopportune to exclude from the craton the fore-Sayan and fore-Baikalian Late Precambrian and Jurassic troughs, which are linked with a gradual and shallow subsidence of its southwestern and southeastern periphery.

It is also hardly possible to agree with SAVINSKY et al. (1971), who refer to the craton the miogeosynclinal zones of the Baikalides of its western and southern surroundings, and who locate the craton boundaries along deep faults separating the mio- and eugeosynclinal domains. These faults, on the other hand, appear to be very important from many aspects and, in particular, are most evident in geophysical fields, being accompanied by basic and ultrabasic magmatites. There is no doubt that the miogeosynclinal zones developed on the same old basement. In this case, however, it is necessary to include in

---

[1] Earlier and, moreover, repeated reactivation, characteristic of the Stanovoi Belt (see below), began in the Early Proterozoic.

the Siberian Craton almost the entire Verkhoyansk–Kolyma system in the east, as well as a large part of the Taimyr in the north, i. e., to revert to the contours of the North Asian Craton (see below).

The largest salient of the basement of the Siberian Craton is the Aldan Shield which occupies an eccentric position in its structure and forms its southeastern corner. The northern slope of the shield is formed on the surface by a thin, flat, monoclinal cover of Vendian–Lower–Middle Cambrian sediments. This caused some scientists to refer to the Aldan Anteclise instead of the Aldan Shield. This, however, is incorrect since practically all other shields of old cratons, including the Baltic and Canadian shields were periodically, even if not for long, covered by sediments of marine transgressions.

The term "anteclise" is fully justified for the other large Anabar salient of the basement of the Siberian Craton in the north. Facies and thickness of the Riphean–Ordovician sediments surrounding the Precambrian Anabar Salient (Anabar Massif) prove that the present Anabar Massif was covered by the sea almost continuously for a very long geological time (1,600 to 450 m.y. B.P.); only after the end of the Ordovician did it emerge as a positive tectonic structure subject to erosion.

The entire area of the Siberian Craton outside the Aldan Shield, including the Anabar Anteclise, can be separated from it as the Central Siberian Platform. Its northeastern part comprises the Anabar Anteclise, while the remaining area contains the largest syneclises (Tunguska and Vilyui), as well as the comparatively smaller structural elements described in Ch. 4.3. A characteristic of the Central Siberian Platform is the repeated reworking of its structural pattern under the obvious influence of movements in adjacent geosynclinal belts. This is why its present structure presents a complex combination of subsidence regions and of newly formed or buried relict uplifts, of different ages. The thickness of the sedimentary platform cover is 10 to 12 km. The thickness of the Earth's crust is 30 to 40 km. It tends to reduce beneath deep syneclises, especially the Tunguska (see below), at the expense of the granite layer. Magnetotelluric soundings indicate that the surface of the asthenosphere lies under the Siberian Craton at a depth of about 120 to 150 km.

The Siberian Craton, in contrast to the Eastern European and North American cratons, is characterized by more "pronounced" deep faults not only in the basement but also in the cover. This, obviously, is also linked with its much more intense and prolonged magmatic activity. The most important fault zone, which divides the platform in the meridional direction into two halves, runs in the north along the western slope of the Anabar Anteclise and its junction with the Tunguska Syneclise (Kotuy Fault). In the south it bisects the Irkutsk Amphitheater, as Suess (1885) called the southern corner of the platform, which is bordered by the mountainous terrain of Eastern Sayan and cis-Baikalia. The importance of this Taimyr–Baikalian lineament in the structure and history of the Siberian Craton (and its southern folded framing) resembles that of the Nemaha–Boothia line in the structure of the North American Craton. In particular, the western boundary of the zone of influence of the Pacific mobile belt runs along it. Its continuation can be traced far to the south outside Siberia, i. e., to the Shillong Spur of the Hindustan Craton and even to the Ninety East Ridge in the Indian Ocean.

The lineament next in importance has a northeasterly trend and includes the Primorye boundary suture. It is then continued by the axial faults of the Vilyui Syneclise and the aulacogen in its basement. This lineament connects two re-entrant corners of the craton, i. e., the Patom re-entrant in the southwest and the Verkhoyansk re-entrant in the northeast. Its continuation can be traced further in this direction within the Verkhoyansk fold system itself.

The third lineament appears in the south as the Zhuya Fault which forms the western boundary of the Aldan Shield (including the Stanovoi Belt). It then runs along the western edge of the Urin intracratonal fold zone, cuts across the Vilyui Syneclise, and probably forms two branches in the north. The western branch is the Udzha Fault, with abundant and various magmatic manifestations, between the Anabar and Olenek uplifts; the eastern branch runs along the lower Lena, where it coincides with the craton boundary.

Two further latitudinal lineaments should also be mentioned. The first runs from the mouth of the Podkamennaya Tunguska to the large bend of the Lower Tunguska and further to the latitudinal reach of the Lower Aldan. The second runs along the lower reaches of the Angara River to the vertex of the Patom Salient of the Baikalides and to the northern boundary of the Aldan Shield.

It is obvious that all these lineaments cutting across the craton, parallel to its boundary sutures and together with the latter, correspond to the main directions of the planetary rhegmatic net: latitudinal (Pyasina–Northern Tukuringra–Angara–Central Tunguska), meridional (Yenisei–Taimyr–Baikal–Zhuya–Udzha–Nel'kan), northwestern (Biryusa), and northeastern (Primorye–Vilyui).

4.2 Aldan Shield, Anabar Massif, and general structure of craton basement

Aldan Shield. It has already been stated that the largest and most stable salient of the basement of the Siberian Craton is the Aldan Shield (Fig. 23). It has the overall shape of a quadrilateral (or rather, a trapezium) with long sides in the east-west direction. It is bounded on three sides by faults, i. e., in the east and west by the straight Nel' kan and Zhuya faults (running NNW and NNE respectively). In the south it is bounded by the Northern Tukuringra Fault which is slightly convex to the south. In the north the basement quite gradually descends in the direction of the Vilyui Syneclise and the southern branch of the Verkhoyansk Foredeep. The Aldan Shield extends for almost 1,200 km in the latitudinal direction, while its meridional cross-section is up to 800 km long.

Like all similar Early Precambrian shields, the Aldan Shield has a pronounced block structure (GLUKHOVSKY & STAVTSEV, 1973; and others). Its two largest blocks (megablocks) are the Aldan Block proper and the Stanovoi Block. The latter is better called Stanovoi Belt in analogy with the Grenville Belt of the Canadian Shield, the Mozambique Belt in Africa, and the Eastern Ghats Belt in India, which resemble it very closely. The Stanovoi Belt is separated from the Aldan Megablock by a major fault zone, the Stanovoi Fault zone.

The Aldan Megablock is by a system of a minor faults running north–south (in the west) and NNW–SSE (in the east) divided into five smaller blocks (Chara, Olekma, Aldan–Timpton, Aldan–Uchur, Maimakan–Batomga). Practically all these blocks, and thus the Aldan Megablock as a whole, are formed by rocks of Archaean age, 4,500 (??) to 2,500 m. y., only by places radiometric age being rejuvenated. This rock complex, which has a thickness of 20 to 25 km, is called Aldan Complex ("Aldanian") and forms one of the most complete sequences of the ancient continental crust of the Earth. The Aldanian Complex was first explored in the classic work of KORZHINSKY (1936). However, this was difficult in the "preradiogeochronological" era, since composition, extent of metamorphism, and structure of the various blocks differ considerably. Different litho-stratigraphic units were therefore distinguished in them; it is only now possible to compare them with some reliability (cf. table in the paper by GLUKHOVSKY & STAVTSEV, 1973; this author follows their exposition in principle). However, certain discrepancies still remain (SALOP, 1973). A stratum of basic crystalline schists (Nesmurin, Sutam, etc. series), metamorphosed under the conditions of sillimanite-hypersthene subfacies of the granulite facies and dated as 4,500 m.y. old[2], is now assumed to be older than the Aldan gneissic complex proper. The density of these schists is 3.5 to 3.9 g/cm^3. PAVLOVSKY (1970), followed by GLUKHOVSKY and MORALEV, considers this stratum to be the oldest, possibly the original basaltic crust of the Earth and dates it of Katarchaean (GLUKHOVSKY et al., 1977). These rocks are found under particular conditions, i.e., in horsts along tectonic sutures at the boundary between the western blocks of the Aldan Megablock; they are also found along the Stanovoi Fault zone at the southern boundary of this megablock. It seems that this is the result of later (at the verge of Archaean and Proterozoic?) tectonic processes, and not some primary system of orthogonal greenstone arches, as suggested by GLUKHOVSY & STAVTSEV (1973).

The lower part of the Aldan Complex proper, i.e. the Lower Archaean Yengra series and its analogues, is formed by highly aluminous gneisses and quartzites. These may respectively represent the crust of weathering of an older basement and the product of its redeposition[3]. The remaining sequence of the Yengra series and its analogues consists of basic crystalline schists and gneisses with intercalations of amphibolites and quartzites. The Yengra series is developed in the Aldan–Timpton Block which occupies a central position in the Central Aldan Megablock. Its age has been determined radiometrically as 3,500 to 3,200 m.y. The age of the probable analogue of the Yengra series in the Chara Block (Chara series) is more than 3,100 m.y.[4]. Slightly lower, probably still too low values, obtained for the Olekma series of the Olekma Block, are 2,750 m.y. All data confirm that the lower half of the Aldan Complex belongs to the Early Archaean. This part of the Aldan Complex is not found in the eastern blocks of the shield, where rocks of the upper half of the Aldan Complex predominate. These are the Timpton and Dzheltula series, which are now usually combined (after FRUMKIN, 1968) into a single Timpton – Dzheltula series, and its analogues. These are predominantly basic (and even ultrabasic) crystalline schists and gneisses with marble intercalations. Quartzites are found relatively rarely and occur in thin strata. FRUMKIN (1968) found that the Timpton–Dzheltula series lies slightly unconformably on the Yengra series. The unconformity is evident in changes of the structural pattern. He apparently also succeeded in establishing a certain zonality

---

[2] This dating of the rocks considered, by the K/Ar method, about the age of the Earth (!) raises obvious doubts.

[3] N.V. FROLOVA (1951) and several researchers after her consider the quartzites of the lower Yengra series to be of chemical and not of clastic origin. This hypothesis is disproved by the textures found in the quartzites, which are characteristic of terrigenous sediments (cross bedding, etc.).

[4] All these radiometric datings are now considered as unreliable (E. V. BIBIKOVA).

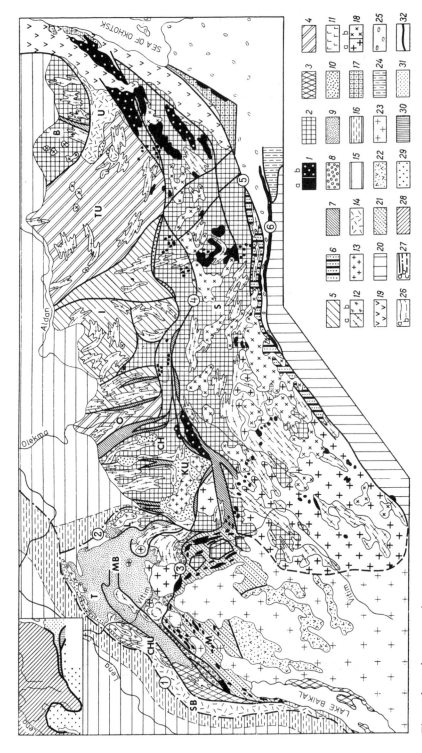

Fig. 23 legend see opposite page.

in changes of the lithological composition of the Timpton–Dzheltula series. Primary-sedimentary rocks predominate in the west (west of the Tarkanda Fault, obviously a region of earlier consolidation). An increasingly greater role is played in the east by primary-volcanic formations. This recalls the division of later geosynclines into mio- and eugeosynclinal zones.

The age of the Timpton–Dzehltula series has been radiometrically established at 2,500 m.y.; the age of its analogues in the most easterly block, the Maimakan–Batomga (Batomga series), was found to be 2,300 to 2,100 m.y. The latter values, possibly also the

---

Fig. 23.   Tectonic map of the Aldan Shield and N part of the Baikalian fold region (after LEITES & FEDOROVSKY, 1978). 1 = complexes of melanocratic basement: a) massifs and bodies of ultrabasics, metagabbro and metagabbro-norites, gabbro-amphibolites, melanocratic amphibolites and crystalline schists, b) massifs of anorthosites and gabbro-anorthosites; 2–19 = continental crust formed by the beginning of Riphean (1.7–1.6 b.y. B.P.): Aldan Shield: complexes of protometamorphic layer, originated in Archaean (3.5–3.0 b.y. B.P.): 2 = complexes of melanocratic basic-ultrabasic basement (primary Earth crust) of Chara–Aldan and Stanovoy Range lithosphere plates and equivalents of rocks of the oceanic stage, undivided, fragments of allochthonous slices in the region of Early Paleozoic continental crust; 3 = ibid., in Bodaibo and Baikal–Amur plates in Peribaikalian region; 4–5 = complexes of deeply metamorphosed sedimentary-volcanic cover of the primary Earth's crust (4.0–3.5 b.y. B.P.); 4 = equivalents of the rocks of the oceanic stage with extensive development of basic volcanics, 5 = equivalents of rocks of the oceanic and transitional stages, undivided; 6–14 = complexes and lithologic assemblages of the time of formation and evolution of the granite-metamorphic layer of the first continental crust: 6 = of the oceanic stage (2.6 b.y. B.P.), 7 = sedimentary-volcanic and ferruginous-siliceous rocks of rift-like suture troughs (greenstone belts, V. Kh.) (3.0–2.6 b.y. B.P.); 8–13 = of the transitional stage: 8 = island-arc sedimentary-volcanic calc-alkaline rocks with carbonate formation in the upper part of the sequence, 9 = sandstone-shale and carbonate sediments of the protocontinental shelf, slope and rise (2.6–2.0 b.y. B.P.), 10 = variegated Cu-bearing sandstones (2.6–2.0 b.y. B.P.), 11 = layered gabbro, 12 = plagiogranite-gneiss and gneiss-migmatite granitoids (2.4–1.7 b.y. B.P.): a) gneiss-like mainly autochthonous, more rarely allochthonous, b) massive allochthonous; 13 = granodiorite-granite allochthonous granitoids (2.4–1.7 b.y. B.P.); 14 = volcano-plutonic belt – indicative of formation of continental crust (1.8–1,6 b.y. B.P.); 15–19 = complexes of continental stage: 15–16 = Siberian Platform cover: 15 = in the region of shallow basement and small thickness of the sedimentary cover, 16 = in suture zones with deep basement, great thickness and intense deformation of the platform cover; 17 = Paleozoic terrigenous sediments of the continental slope; 18 = granitoid and alkaline plutonic rocks of adjacent regions of young crust: a) Early Paleozoic, b) Late Paleozoic and Late Mesozoic; 19 = Late Mesozoic volcano-plutonic belt; 20–23 = continental crust formed in Early Paleozoic: 20 = undivided, 21 = complexes of the oceanic stage (2.0 b.y. B.P.), 22 = complexes of the transitional stage (Late Precambrian and Cambrian), 23 = Early Paleozoic granitoids – indicative of the formation of the continental crust; 24–25 = young continental crust: 24 = Late Paleozoic, 25 = Late Mesozoic; 26 = boundaries of complexes: a) on the surface, b) under the platform cover; 27 = tectonic sutures, boundaries of nappe units, faults: a) on the surface, b) under the platform cover, c) hypothetical.
Legend to the schematic insert map of distribution of the continental crust of different age: 28 = pre-Riphean crust; 29 = Early Paleozoic crust; 30 = Late Paleozoic crust; 31 = Late Mesozoic crust; 32 = boundaries of different age crust.
Main structural units: blocks of the Aldan Shield: CH = Chara, O = Olekma, I = Yengra (Aldan–Timpton), TU = Timpton–Uchur, B = Batomga, KU = Kodar–Udokan, U = Ulkan volcanic belt, S = Stanovoi Belt; Baikalian fold region: SB = North Baikalian (Akitkan) volcanic belt; CHU = Chuya, T = Tonod, L = Longdor uplifts (anticlinoria), MB = Moma–Bodaibo Trough (Synclinorium), M = Muya zone; faults (encircled figures): 1 = Baikalian marginal suture, 2 = Nichat (Zhuya), 3 = Mama–Vitim, 4 = Stanovoi, 5 = Okhotsk, 6 = Tukuringra.

former, are too low. They are probably due to the formation, at the beginning of the Proterozoic, of the widespread (except in the two central blocks) palingenetic-metasomatic granite-gneisses and plagiomigmatites (2,420 m.y. old), and augen-metasomatic granitoids (2,130 m.y.). The rocks of the Yengra series in the central blocks enclose older granite-gneisses, charnockites, and migmatites dating from the Archaean or even Early Archaean. They are considered now as pre-Yengra rocks – equivalents of "Gray gneisses" of other shields. SALOP (1973) believes that the entire Aldan Complex (which he calls series), including the Timpton and Dzheltula series, are older than 3,500 years and therefore belong to the Archaean in his sense; all radiometric datings indicating a lesser age are considered by him to be rejuvenated.

Most peculiar from the structural aspect is the Aldan–Timpton Block, where large, oval (with 80 to 350 km long major axes) structures (gneiss fold ovals, identified and described in detail by SALOP, 1971) predominate in the terrain of the Yengra series. Gneiss domes are usually located in the centers of these ovals. These often have granite cores and arc-shaped folds at the periphery. They are often isoclinal, complicated by smaller folds, down to microplications. These folds exhibit centripetal vergency. The plans of these ovals are often very complex ("amöeboidal" according to SALOP). SUDOVIKOV et al. (1965) and GLUKHOVSKY (1971) believe that the original structural pattern here was linear, and that domes and ovals were formed during later (Early Proterozoic?) granitization and ultrametamorphism. The importance of dome-shaped and oval structures is less to the east and west of the Aldan–Timpton Block, whereas that of linear structures is greater. The predominant orientation of the folds is NNE in the Chara Block, north-south in the Olekma Block, northwesterly in the Aldan–Uchur Block (arc-shaped-sublatitudinal in the north), and northeasterly in the Maimakan –Batomga Block.

In concluding the survey of the Aldan Megablock we note that the five blocks forming it may be combined into three larger blocks as regards age and composition of the rocks forming them, and their structural features. These are the Chara–Olekma Block in the west, the Central–Aldan Block in the center, and the Maimakan–Batomga Block in the east. The Stanovoi Belt (Megablock) can also be divided into blocks. The boundaries between these blocks and the blocks of the Aldan Megablock to the north are often common, as is the predominant trend of the folds (see Fig. 23). However, this division into blocks is less evident here, being suppressed by the general sublatitudinal orientation of the belt. The Stanovoi Belt was for a long time, in contrast to the Aldan Shield, considered to be a region formed during the Early Proterozoic (or Late Archaean), where folding occurred at the end of the Early Proterozoic. All or nearly all researchers nowadays admit that the Stanovoi rock complex has the same age as the Aldan Complex and a generally similar sequence. Rocks of the lower part of the Aldan (Aldano–Stanovoi) Complex are common in the western part of the Stanovoi Belt, as in the Aldan Megablock, while rocks of the upper part of the said complex predominate in the eastern part. The principal difference between the Stanovoi Belt and the Aldan Shield proper, however, is that at the end of the Early Proterozoic the Archaean rocks of the Stanovoi Complex underwent intense reworking, in particular granitization and regressive metamorphism of the amphibolite facies superposed on the primary granulite facies.

As regards the structure: "A characteristic feature of the Stanovoi Megablock is the association of primary, large, wide, linear folds with oval brachyform domes and bowls, formed as a result of Early Proterozoic granitization. The linear structures as a whole have a northwesterly orientation, becoming sublatitudinal only near the Stanovoi and Northern Tukuringra faults" (GLUKHOVSKY & STAVTSEV, 1973: p. 69).

An important feature of the Stanovoi Belt is the abundance, mainly along the Stanovoi suture, of large plutons of gabbro-anorthosite composition (Olekma, Dzhugdzhur). It is quite obvious that formation of the Stanovoi suture must have been a precondition for the isolation of the Stanovoi Belt, and therefore occurred at the beginning of the Proterozoic. Separation of the megablocks into blocks must have preceded this, and thus occurred at the end of the Archaean.

The structure of the Aldan Shield was complicated during the Early Proterozoic or more probably Late Archaean by the formation of a series of narrow troughs or paleoaulacogens according to LEITES, MURATOV & FEDOROVSKY (1970). They were later regenerated as wider and flatter troughs. These troughs were filled with volcano-terrigenous and iron-silicious formation having a thickness of 2 to 7 km, metamorphosed in the amphibolite and greenschist facies. They are deformed into narrow linear folds and intruded by interstitial bodies of gabbro, gabbro-amphibolites, and ultrabasics. These structures extend for hundreds of km. Their width is a few km or tens of km. They are considered now as deeply eroded greenstone belts.

The next generation of Early Precambrian structures of the Aldan Shield includes the Kodar–Udokan Trough (250 × 100 km) in the extreme west and the Ulkan Trough in the southeast of the Central Aldan Megablock. These clearly protocratonic structures are filled with terrigenous and volcanic rocks. The former predominate in the Udokan series

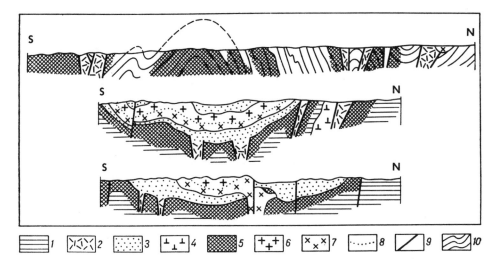

Fig. 24.   Cross-section through the Precambrian of the central part of the Olekma–Vitim mountain region (after LEITES & FEDOROVSKY, 1972, simplified). 1 = Archaean salients, 2 = Early Proterozoic sutural troughs (Late Archaean greenstone belts, V. KH.); 3 = Early Proterozoic Kodar–Udokan protoplatform downwarp; 4 = Archaean anorthosite massif; 5 = Early Proterozoic granite-gneisses; 6–7 = Kodar–Kemen lopolith of Early Proterozoic granitoids; 6 = granites of the main intrusive facies, 7 = granodiorites and quartzdiorites of the marginal basal facies; 8 = boundaries of the lopolith facies; 9 = faults; 10 = structural surfaces in Archaean and Lower Proterozoic.

of the Udokan Trough, forming a red, continental, clastic formation of enormous thickness (up to 12 or 13 km). Its lower part was metamorphosed in the greenschist, partly even the amphibolite facies. Their attitude at the periphery is complicated by mantled gneiss domes. These are the product of remobilization of the Archaean basement. Palingenesis of the basement also created the large allochthonous Kodar–Kamenka granite lopolith located in the central part of the trough in the middle of its sequence (Fig. 24). The structure of this trough part is complicated further only by brachyform folds. Structure and development of the Kodar–Udokan Trough were described in detail by LEITES & FEDOROVSKY (1972) who noted its synsedimentary nature.

The Ulkan Trough is smaller and filled mainly with volcanics of predominantly acid composition. The overall thickness of the strata is up to 4 km. They have a flat attitude except near faults. The Lower Proterozoic age of the Udokan and Ulkan series is proved by their being intruded by granitoids having an age of 1,900 ± 100 m.y. Subsidence continued also during the Middle Proterozoic in the Ulkan Trough and the even smaller Bilyakchan Trough in the east of the shield. It was accompanied by accumulation of the red molassoid Uyan series. The Ulkan laccolith of alkaline granitoids in the east of the Batomga Block has been dated as having an age of 1,660 m.y. The main part of the shield underwent uplift in the Middle Proterozoic, and the whole shield also in the Late Proterozoic. Granite stocks intruded in the Chara–Olekma zone at the end of the Proterozoic (1,000 ± 100 m.y. B.P.). Ultrabasic alkaline intrusions of the central type occurred during the Baikalian epoch (650 m.y. B.P.); kimberlites intruded in the Batomga zone at that time.

Mobility became less at the beginning of the Vendian. Uplift in the Aldan Megablock gave way to gentle subsidence with transgression of the sea and the accumulation of thin carbonaceous shelf sediments of the Yudoma series and of the Lower and Middle Cambrian. The Stanovoi Belt retained its uplifted position. The entire Aldan Shield again underwent prolonged uplift from the Late Cambrian. This was most pronounced in the Stanovoi Belt where it was accompanied by the intrusion of gabbro and granitoids in the Devonian. Plutons of alkaline syenites of the same and younger age (Permian) have also been found in the Chara–Olekma zone.

Intense orogenic reactivation involved the Stanovoi Belt at the beginning of the Mesozoic and especially in the Jurassic, when overall arc-like uplift occurred. The southern part of the Aldan Shield, immediately to the north of the Stanovoi Fault, at that time underwent compensatory subsidence with the accumulation of up to 1.5 km thick molassoid coal-bearing formations. Deposits of these formations remain at present in isolated depressions, the largest of which is the Chul'man Depression. The Stanovoi Swell was later overthrust at a small angle on the system of Jurassic troughs in the south of the Aldan Shield. This overthrust most likely dates from the Late Jurassic or Early Cretaceous. Individual depressions, possibly of the rift type, also formed in the Stanovoi Belt proper, mainly in the west and south. The axial part of the belt became a region of pronounced granitization at the same time or slightly later (in the Early and Middle Cretaceous this process migrated from west to east). Granite plutons were associated with faults and fissures, and together with them are mainly distributed within the fields of Lower Proterozoic granite-gneisses, being probably formed due to palingenesis of the latter. Important skarn and hydrothermal mineralization is linked with the Mesozoic granitoids of the Stanovoi Belt.

Uplift and fragmentation of the Stanovoi Belt continued during the Cretaceous. The southern part of the Aldan Megablock was also drawn into this process; a fairly large number of plutons of alkaline granitoids were formed there. Cretaceous magmatism manifested itself also in effusive form in a strip gravitating from the north to the Stanovoi Fault, as well as inside the Stanovoi Belt. It caused eruption of andesites and rhyolites, and the formation of subvolcanic bodies of similar composition. The Cretaceous lavas are preserved in the depressions together with coarse molasses of the same age, thus indicating an intensive uplift of the Stanovoi Belt.

It is quite obvious that all this tectonomagmatic activity in the south of the Aldan Shield during the Mesozoic, and much less also during the Paleozoic, was linked with the geosynclinal-orogenic evolution of the Mongol–Okhotsk system. Tectonic activity decreased after this at the end of the Cretaceous and Paleogene. A period of peneplaination began. However, reactivation of upward movements resumed at the beginning of the Neogene. The Stanovoi and Dzhugdzhur ranges were formed in the Stanovoi Belt, while the Kodar and Udokan ranges were formed in the Chara–Olekma zone of the Aldan Megablock. A rift zone penetrated from the Baikal into the shield and in particular formed the Chara Depression; it runs parallel to the Stanovoi Fault, but slightly north of it. Centers of basaltoid volcanism gravitated towards this zone and also towards the fault itself. Inheritance manifests itself in very recent movements along a Mesozoic structural pattern (GLUKHOVSKY et al., 1972).

The Aldan Shield, situated on the periphery of the Siberian Craton and at the junction of two geosynclinal belts (the Ural–Okhotsk Belt and the Pacific Belt) thus exhibited much more prolonged and greater mobility than most other shields of old cratons.

The **Anabar Massif** is the second large salient of the basement of the Siberian Craton. It is generally assumed that the Precambrian formations (basic and in part highly aluminous crystalline schists and amphibolites, interlaid by quartzites in the lower part and by carbonaceous rocks in the upper) exposed by erosion, are analogues of the Aldan Complex and were also metamorphosed in the granulite facies. The similarity extends also to manifestation of intrusive magmatism represented by charnockites, granodiorites, alaskitic granites, and migmatites. Radiometric dating (by the K/Ar method) gives an age varying between 2,900 and 2,350 m.y. for the hypersthenes. This obviously corresponds most likely to the real age of the rocks. Radiometric dating gives also an age of 2,500 to 2,300 m.y. for the amphiboles and 2,000 to 1,830 m.y. for the whole rocks samples. This clearly reflects later reactivation processes. The fold structure of the Anabar Archaean is quite complicated (up to isoclinal), has not yet been finally deciphered, and is characterized by a general predominance of northwesterly trends. SALOP (1973) believes, on the basis of a study of aerial photographs, that fold ovals similar to those of the Aldan Shield formed here.

Archaean rocks analogous to those of the Anabar and Aldan salients also form the Sharyzhalgai salient of the basement in the extreme southwestern corner of the craton, west of Irkutsk, as well as the Kan Block in the Yenisei Ridge and the Biryusa Block in Eastern Sayan. Both the latter already lie outside, but almost directly at the border of the craton proper and are undoubtedly its separated fragments. A certain, obviously secondary difference between the rocks forming all these blocks and the Aldan Complex is the more pronounced regressive metamorphism and correspondingly the radiometric rejuvenation. Rather different rocks are preserved in the Onot Graben in the western part of the

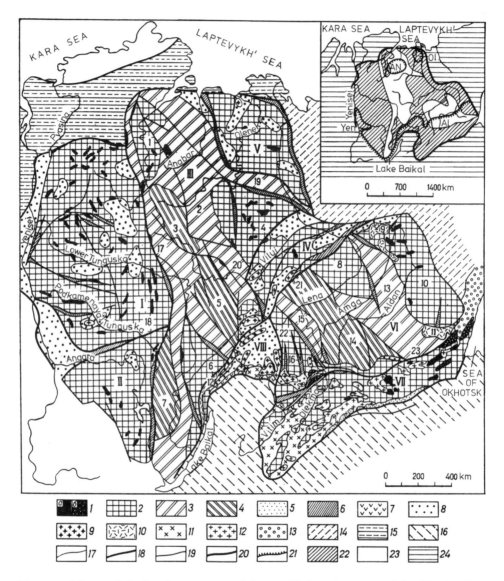

Fig. 25.   Scheme of the interior structure of the pre-Riphean basement of the Siberian Craton
(after GAFAROV, LEITES, FEDOROVSKY, PROZOROV, SAVINSKAYA & SAVINSKY, 1978). 1–10 = continental crust formed by the beginning of Riphean (1.8–1.6 b.y. B.P.): 1–4 = complex of protometamorphic granulite-basite layer originated in Archaean (3.5–3.0 b.y. B.P.): 1 = melanocratic
basement: a) massifs and bodies of ultrabasics, metagabbro, gabbro-amphibolites, melanocratic
amphibolites and crystalline schists; b) anorthosite and gabbro-anorthosite massifs; 2 =
melanocratic basement and Archaean equivalents of rocks of the oceanic stage of evolution, subject
to regional Early Proterozoic granitization and recurrent regressive high-temperature metamorphism (2.4–1.7 b.y. B.P.); 3–4 = deeply metamorphosed Archaean volcano-sedimentary cover of
the primary Earth's crust: 3 = equivalents of the rocks of the oceanic stage with extensive
development of basic volcanics; 4 = equivalents of rocks of the oceanic and transitional stages,
undivided, with extensive development of primary sedimentary components; 5–13 = complexes of

Sharyzhalgai salient. They include jaspilites, primarily metamorphosed in the amphibolite, but not granulite facies. These rocks obviously date from the Upper Archaean or Lower Proterozoic.

A quite different composition of the basement formations manifests itself over a small area in the opposite corner of the craton, in the center of the Olenek Uplift. These terrigenous rocks are metamorphosed in the greenschist facies (sandstones, siltstones, schists in the flyschoid sequence), intruded by gabbro-diabases, quartz-diorites, and granites. Dating (by the K/Ar method) of mica from the granites and metamorphites gave an age of the order of 2,000 m.y. This shows that these formations probably belong to the Lower Proterozoic.

These are "reference data" on the basement of the Siberian Craton. Together with drilling data and aeromagnetic surveys, they made it possible to begin deciphering the inner structure of the basement for the whole platform (Fig. 25). These constructions, however, are far less verified and much more controversial than those for the Eastern European Craton. There are two reasons for this. Firstly, the number of boreholes drilled in order to explore the basement is much smaller; moreover, they were almost wholly located in the southern part of the craton. Secondly, interpretation of aeromagnetic data for large areas of the Tunguska Syneclise is greatly hampered by the extensive development of traps.

The age of the basement of the eastern craton half (east of the Taimyr–Baikal Lineament) has been determined fairly reliably. Interpretation of the data for this part of the

---

the time of formation and evolution of the granitic-metamorphic layer: 5 = sedimentary-volcanic rocks of the oceanic stage (2.6 b.y. B.P.); 6 = sedimentary-volcanic and ferruginous-siliceous rocks of the rift-like suture troughs (greenstone belts, V. Kн.); 7 = island-arc sedimentary-volcanic calc-alkaline rocks with carbonate formation in the upper part of the sequence (2.6–2.0 b.y. B.P.); 8 = flyschoid and variegated sandstone, shale and carbonate rocks of the protocontinental rise, slope, shelf and inner troughs (2.6–2.0 b.y. B.P.); 9 = allochthonous granitoids of the plagiogranite-gneiss and granodiorite-granite formations (2.0–1.7 b.y. B.P.); 10–13 = rocks of the continental stage: 10 = volcano-plutonic, associated with the late molasse – indicators of the continental crust formation (1.8–1.6 b.y. B.P.); 11–12 = granitic and alkaline bodies , 11 = Early Paleozoic, 12 = Late Paleozoic and Late Mesozoic; 13 = Late Mesozoic volcano-plutonic belt; 14–16 = young continental crust (framing of the Siberian Craton): 14 = Late Precambrian–Paleozoic, precise age unestablished, 15 = Early Paleozoic, 16 = Mesozoic; 17 = boundaries of complexes; 18 = tectonic sutures – boundaries of the lithosphere plates; 19 = main faults, tectonic sutures.
Lithosphere plates: I = Tungussian; II = Lower Angarian; III = Anabaro–Baikalian; IV = Viluyan; V = Olenekian; VI = Chara–Aldanian; VII = Stanovoi Range; VIII = Bodaibo.
Fold regions and zones of lithosphere plates: 1 = Lower Kotuy; 2 = Anabar–Mirny; 3 = Verkhoyansk; 4 = Tung; 5 = Botuoba; 6 = Kirensk; 7 = Ust–Kut; 8 = Sin; 9 = Lower Amga; 10 = Khandyga; 11 = Ulkan; 12 = North Baikalian; 13 = Timpton–Uchur; 14 = Yengra; 15 = Olekma; 16 = Chara.
Inter-plate tectonic sutures: 17 = Sayan–Anabar; 18 = Irkineevo; 19 = Udzha–Zhigansk; 20 = Linda; 21 = Baikal–Vilyui; 22 = Nichat (Zhuya); 23 = Stanovoi Range.
Legend to the insert scheme of distribution of different types of continental crust of eroded salients and buried pre-Riphean basement of the Siberian Craton: 20 = boundaries of the craton; 21 = contours of salients (hatchured toward the salients); 22 = complexes and structures of the peripheral ring of mature continental crust – nucleus of the Siberian Craton; 23 = complexes and structures of the region of the protometamorphic layer; 24 = young crust of the craton's framing.
An = Anabar Massif; Ol = Olenek Salient; Al = Aldan Shield, Yen = Yenisei Range salient.

platform is greatly facilitated by the fact that two large basement salients exist in this area. These are the Anabar and the Aldan salients respectively, which are practically identical as regards the complex of rocks forming them (Archaean). They have a common predominantly NNW–SSE trend. There are thus hardly any doubts (except in the view of Bulina & Spizharsky, 1970) on the original unity of the Anabar-Aldan region (belt) of epi-Archaean consolidation. The situation is, however, complicated by the fact that this belt is in its axial part interrupted by the Vilyui Syneclise, where anomalies of northwesterly trend are accompanied by anomalies of near-latitudinal and northeasterly trend. These run parallel to the axis of this syneclise and of the underlying aulacogen. This strip is bounded by faults having the same northeasterly trend and is characterized by an overall lower strength of the magnetic field. Gafarov (1971) and Savinsky (1972) identified it as the separate Vilyui zone or basement megablock with an Archaean basement reworked in the Proterozoic. This basement was explored by means of a stratigraphic borehole on the median Suntar Uplift and its age in fact determined as Proterozoic. The Vilyui zone separates two older megablocks, i. e., the Anabar and the Aldan megablocks. These have been identified by practically all researchers, although with different boundaries[5]. All researchers also distinguish the Olenek Megablock in the northeastern corner of the platform. Its basement, at least in its upper part, must be formed by a Lower Proterozoic complex found in the center of the Olenek Uplift. This is in agreement with the uniform magnetic field with small positive anomalies. However, the gravitational maximum observed here also indicates that an Archaean complex saturated with basic magmatites lies here at no great depth.

A strip with northeasterly trend stands out in the basement of the eastern craton half. This strip runs in the basement of the Angara–Lena Trough parallel to the front of the Baikal–Patom fold system, and is called Baikalian Megablock by Savinsky (1972). The existence of a northeasterly trend on the background of a generally mosaic field obviously indicates that the Archaean basement was reworked here by Early Proterozoic tectogenesis.

The Angara Block exhibits the most definite characteristics in the western half of the craton. This megablock extends along the upper south-north reach of the Angara. Its features are the more pronounced, since in the south it is in direct contact with the Sharyzhalgai Salient of the Archaean basement.

Linear, almost exactly north-south anomalies are observed here on the background of a weak negative magnetic field. This permits the conclusion that the Angara Archaean is the direct continuation of the Anabar Archaean. It also confirms the view of Bulina & Spizharsky (1970), who included the Angara Block in the Anabar Megablock (the northern part of the Angara Block is by Gafarov (1972) included in his "Tunguska Massif").

The Sayan Megablock more to the west, identified by Savinsky (1972) (Taseev Megablock according to Bulina & Spizharsky, 1970), differs in its predominantly northwesterly magnetic anomalies which clearly run parallel to Eastern Sayan. This leads to the

---

[5] Gafarov (1971) has detailed the structure of the Aldan Megablock in his scheme. In the southeastern part of the Anabar Megablock he especially distinguishes the Early Archaean Tyung Massif which, in contrast to the whole Anabar Megablock, has a mosaic magnetic field instead of a linear one.

conclusion, firstly, that the Archaean basement, which further to the west cross out into the Kan and Biryusa blocks was reworked during the Early Proterozoic (as were the blocks mentioned); secondly, that separate protogeosynclinal zones of the Lower Proterozoic may exist, like those of the Onot Graben in the Sharyzhalgai Block.

Least evident is the nature of the basement of the Tunguska Megablock, which in general spatially coincides with the syneclise of the same name. The high intensity of the magnetic field, noticed in this megablock, the existence of wide and long magnetic anomalies of nearly latitudinal and often arc-shaped trend, and the many local maxima and minima of the magnetic field cause a marked difference between this megablock and adjacent ones. These phenomena are mainly due to the trap magmatism which is very pronounced here. However, part of these anomalies remain also when the observed field is smoothed out and when it is converted into the upper half-space. This shows that it is partly linked with the peculiar composition and structure of the basement (GAFAROV, 1971). Several researchers, using this and indirect evidence, identify in the Tunguska Syneclise a series of isometric Archaean blocks, separated and also joined by zones of Proterozoic (Early or Middle) folding (DASHKEVICH et al., 1968). These constructions remain very hypothetical, but continuation of the Archaean into the southeastern part of the Tunguska Megablock in a direction from the Angara Megablock cannot be doubted.

It is in general quite obvious that the basement of the Siberian craton is formed predominantly by an Archaean complex. This basement is either unaltered (Aldan, Anabar, Angara megablocks) or was reworked in the Early, partly in the Middle Proterozoic. This causes the Siberian Craton to stand out from other cratons on the Earth as the oldest and largest region of epi-Archaean consolidation[6]. Moreover, there is no doubt that its basement includes also Early Proterozoic, and possibly Middle Proterozoic fold systems which most likely developed from narrow "protogeosynclines" of the "Saksagan" type, analogues of similar fold zones in the Aldan Shield. An only exception may be the peripheral Taimyr–Tunguska protogeosynclinal system.

The most recent version of the basement structure of the Siberian Craton is due to GAFAROV, LEITES, FEDOROVSKY et al. (1978). (Fig. 26). It is based mainly on extrapolation of data on the Aldan–Stanovoi Shield. The decisive predominance of Archaean formations is even more obvious in this version. However, their distribution over the area and their sequence differ in this representation considerably from those given in earlier versions. The oldest, melanocratic formations of protooceanic type (according to these authors) include the Stanovoi, Chara, and Batomga blocks as well as the Tunguska, Angara, Tyung, and Olenek blocks (reworked in the Proterozoic – PR_{1+2}), the basement of the Vilyui Syneclise, and some smaller blocks. On the other hand, the Chara–Aldan and Anabar blocks, as well as the continuations of the latter in the Baikal direction, are considered to be younger. Fairly numerous, but narrow and pinching out strikewise Late Archaean protogeosynclines (greenstone belts?) are located at the boundaries between the large blocks, and partly within them. Finally, a whole series of depressions of the "Udokan" type occur within the oldest blocks. These depressions are filled with protoplatform sediments of the Lower Proterozoic.

---

[6] One should, however, remember that in size the Siberian Craton approximately corresponds to individual blocks of consolidated Archaean of the considerably larger North American and African cratons.

4.3 Structure of the sedimentary cover of the Mid-Siberian Platform

The accumulation of protoplatform formations in the Siberian Craton began, as just mentioned, already in the Early Proterozoic. An even wider, but by no means continuous accumulation of platform deposits occurred in the Middle and Late Riphean. On the other hand, a continuous cover began to be formed, as on the Russian Platform, only in the Vendian.

The main features of the cover structure become apparent through a cursory glance at any, even small-scale geologic map of the Siberian Craton. They are the Anabar Uplift and its Olenek appendix in the northeast, the Aldan Shield in the southeast, the Mesozoic Vilyui Syneclise between them, the huge Triassic trap-filled Tunguska Syneclise with the Upper Paleozoic framing in the northeast, and finally the monoclinal trail of Lower Paleozoic sediments, sloping gently from the Baikalian fold surroundings to the southwest and south, and from the Aldan Shield to the southeast, towards one another and in the direction of the Tunguska and Vilyui syneclises. However, the structural pattern of the basement surface, represented as smooth and with no faults by SPIZHARSKY (1958) and TUGOLESOV (1970), and as formed of blocks by SAVINSKY (1971 b), differs considerably from the structure exposed at the present erosional surface, reflecting the algebraic sum of vertical movements during the long period of platform development. The main difference lies in the existence of an almost completely buried uplift strip of northeasterly trend, which cuts the platform in a diagonal direction from the middle reaches of the Angara to the Lena, and in the depth separates the Tunguska Syneclise from the Vilyui Syneclise. This strip has been identified under various names, e. g., Mid-Siberian Anteclise (SAVINSKY, 1972; TUGOLESOV, 1970), Angara–Lena Anteclise (or Uplift, by SPIZHARSKY, 1958), Nepa–Botuoba Anteclise by petroleum geologists (Geology ..., 1981) and many other names referring to its separate elements. All the names mentioned are justified but not practical: the former, because the structure considered occupies only part of the Mid-Siberian Plateau; the adjective concerned is therefore better applied to the entire platform, as is done here. The second designation is impractical because the Angara–Lena pericratonal downwarp was identified much earlier (PAVLOVSKY, 1959). It is inconvenient to give the same name to both downwarp and uplift. The third name does not cover all the extent of this structure. It is therefore recommended to name this structure which is rather a ridge than an anteclise, the Central Siberian Ridge or Anteclise; (cf. also "Directions of ...", 1977).

The principal positive structural elements of the Mid-Siberian Platform are thus the Anabar and Central Siberian anteclises, while the principal negative elements are the Tunguska and Vilyui syneclises. Furthermore, the Turukhansk–Noril'sk dislocation zone is identified in the northwestern peripheral strip of the platform, the narrow Vel'mo Trough along the eastern slope of the Yenisei Ridge, and the Baikit Anteclise to the east of it in the western part of the platform, the fore-Sayan Trough in the southwest, and the fore-Baikalian Trough in the south along the basement and Upper Riphean and Lower Cambrian strata; in higher Paleozoic strata this trough is part of the wider Angara–Lena Trough located also on the southeastern limb of the Central Siberian Anteclise.

The **Anabar (Anabar–Olenek) Anteclise** is the largest positive structure of the Mid-Siberian Platform, being about 1,000 km across. Its culmination is the flat **Anabar Uplift** which has a rounded triangular shape and is about 400 km across. In its center there is an

Archaean basement salient of the same name, surrounded by outcrops of a lower, Riphean structural complex of the cover. The Anabar Uplift is slightly oblong in the southeastern direction, in accordance with the predominant trend of the basement rocks. A very interesting detail of its structure is the rounded Popigai "Graben" superposed on the corner of the anteclise. It has a diameter of 70 to 80 km and is filled with Mesozoic and Paleogene sediments accompanied by magmatites. MASAITIS et al. (1976) have shown that the Popigai "Graben" belongs to the category of astroblems, but this opinion is disputed by some others.

The **Muna Swell** is located in the southeastern continuation of the axis of the Anabar uplift in the Lena Basin. The basement of the Muna Swell is located at a depth of less than 1 km. The Anabar and Muna uplifts are separated by the quite deep **Sukhan Trough,** of the same trend, from the second summit of the anteclise, the **Olenek Uplift** having outcrops of the Lower Proterozoic (or reworked Archaean?) in its highest part. The peripheral position of the Olenek Uplift in the protruding corner of the craton also caused its pronounced fragmentation by faults running northwest to southeast. These faults in particular border the superposed **Kyutyungdy Trough** in which Upper Paleozoic sediments, extending there from the Verkhoyansk Range, have been preserved from erosion.

The Anabar–Olenek Anteclise is covered mainly by Riphean sediments beginning by its lower strata, and by the Lower Paleozoic. It first became an erosion zone probably at the end of the Ordovician; however, it obtained its sharp southwestern boundary only during the Late Paleozoic when the Tunguska Syneclise became isolated. The anteclise underwent steady uplift during the Mesozoic and Cenozoic. Small superposed depressions formed only in some parts. Annular bodies of ultrabasic-alkaline rocks intruded along the western slope of the anteclise, in the region of the Taimyr–Baikal Lineament, in the basins of the Maimecha and Kotui rivers.

The Baikit anteclise trends east of the Yenisei Ridge, roughly parallel to the latter and separated from it by the narrow Velmo and Teria troughs. The basement lies at a depth of 2–3 km and more, the cover is made up of predominantly Riphean and Vendian–Lower Paleozoic sediments. In the southern part of the anteclise there is a very interesting Chadobets dome with an alkaline intrusive body exposed in the central part.

The **Central Siberian (or Nepa–Botuoba) Anteclise (Ridge)** is a linear, almost completely buried (except at the headwaters of the Lower Tunguska) uplift having a northeasterly trend. It extends from the middle reaches of the Angara to the middle reaches of the Vilyui. Continuation of this uplift is here interrupted by the transverse Upper Muna Trough of northwesterly trend. This trough originates on the southern slope of the Anabar Anteclise. The Muna Swell further to the northeast may be considered to be an element of the same uplift zone.

The highest part of the anteclise considered is the **Katanga Swell** (after which it was first identified as Katanga Anteclise by TUGOLESOV, 1970 and OFFMAN, 1959). The basement lies at a depth of less than 1.5 km beneath the surface of this swell (also called Nepa Swell). This has been confirmed by drilling and seismic investigations (using the exchanged-wave method) along the headwaters of the Lower Tunguska (PRITULA et al., 1973). Cambrian sediments, directly overlain by coal-bearing Carboniferous are here exposed on the surface. The southwestern plunge of the anteclise in the region of the Irkutsk Amphitheater is frequently described as the Angara–Lena structural terrace.

Various swells have been identified in it (Bratsk, Ust–Kut, etc.). The basement lies here at a depth of 2 to 3 km. The Mirny Swell is identified in the northeastern part of the anteclise where the basement lies at a depth of less than 1.7 km.

The Central Siberian Anteclise formed actively during the earliest stages of platform development, up to the Caledonian or rather including the Ordovician. No Riphean (and Lower Vendian?) sediments are found in its highest part. The thickness of the Upper Vendian, Cambrian, and Ordovician is much less than in the adjacent troughs. The anteclise participated in the general uplift of the southern part of the platform from the end of the Silurian till the Middle Carboniferous. Its northwestern slope became the southeastern edge of the Tunguska Syneclise during the Late Paleozoic.

The Central Siberian Anteclise greatly resembles the Transcontinental Arch of the North American Craton in its history and its role in the structure of the Siberian Craton. The Transcontinental Arch also cuts the North American craton diagonally and disappeared in the Late Paleozoic. A considerable historical similarity with the Volga–Urals Anteclise of the Eastern European Craton is also evident.

Fore-Baikalian and Angara–Lena troughs. The Central Siberian Anteclise is in the southeast separated from the Baikalian mountainous fold structure by the narrow fore-Baikalian Trough whose basement lies at a depth of up to 4 or 5 km. The trough is filled mainly with molassoid sediments of the Upper Riphean and Vendian (Ushakovka and Moty suites). In fact it represents a buried Baikalian Foredeep (KLITIN, PAVLOVA & POSTEL'NIKOV, 1970, and others). This foredeep is overlain by a younger, wider, and flatter subsidence zone joined to the epi-platform Caledonian orogen of the Baikal –Patom Upland and accordingly formed of Cambrian, Ordovician, and Silurian sediments (Fig. 26). This subsidence zone encompasses the southeastern slope of the Central Siberian Anteclise and is usually called Angara–Lena Trough (or fore-Patom Trough, see "Geology …", 1981). It was cited by PAVLOVSKY (1959) as an example of pericratonal downwarp.

The Angara–Lena Trough, which is not so much a pericratonal as a posthumous foredeep, underwent folding at the end of the Silurian. A peculiar zone of linear folds was formed. These are typically ejective anticlines and wide, flat synclines. The anticlinal

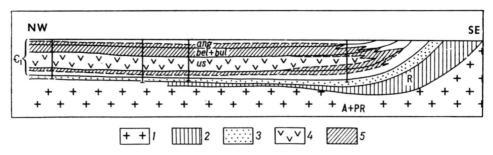

Fig. 26.   Schematic cross-section through the SE limb of the Angara–Lena Trough from Kurtun River to Severny Kutulik River.
1 = Early Precambrian crystalline basement; 2 = Riphean (Baikalian "three-member" complex); 3 = Ushakovka series; 4–5 = Early Cambrian; 4 = evaporite series (us = Usol series), 5 = sub- and supra-evaporite strata (bel + bul = Belskaya and Bulayskay series undivided, ang = Angara series).

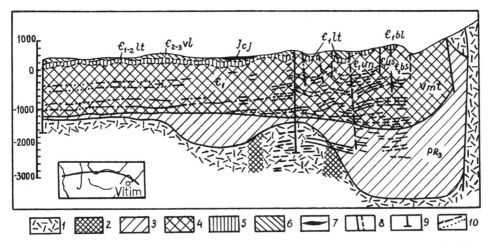

Fig. 27.　Seismic cross-section through the Angara–Lena zone of dislocations (along the line in the insert). 1 = Early Precambrian basement; 2 = fault zones in the basement; 3 = Riphean; 4 = Vendian (Ushakovka and Moty series) and Lower Cambrian, Aldanian stage (Vendian boundary conventional); 5 = Lower Cambrian, Lena stage; 6 = Middle and Upper Cambrian, Verkholensk series; 7 = Jurassic; 8 = faults; 9 = boreholes; 10 = conventional seismic horizons and their correlation.

zones are 200 to 300, even 350 km long, and are frequently complicated by upthrusts-overthrusts (Fig. 27). The folding does not manifest itself at all in the almost horizontal attitude of the basement and of the lowest cover strata. It clearly owes its existence to the presence, in the Lower Cambrian, of an evaporite formation. However, it seems that the manifestation of salt diapirism in the form found here was caused not only by a low salt density in this region, but also by a lateral pressure acting from the direction of the Baikalian orogen. The latter manifested itself clearly in the orogen being overthrust on the southeastern limb of the Angara–Lena Trough (NAUMOV, 1974). The suggestion that this dislocation zone has a gravitational origin is less likely.

The Baikalian Trough is in the northeast separated by a saddle from the deep **North Patom** or **Nyuia Depression** whose basement lies at a depth of more than 9 km, and which extends in a nearly east-west direction. This depression is bordered by the Baikalides of the Patom Upland in the south and by the Urin branch of the Baikalides in the west. The sequence of the depression contains mainly Riphean and Vendian sediments accompanied by Cambrian, Ordovician and Lower Silurian. The southern edge of the depression is complicated by linear folds and faults; its remainder has a smoother structure.

The **Berezov Trough** running north to south is symmetrically located on the other, eastern side of the Urin Anticlinorium, and separates it from the northern slope of the Aldan Shield. It is much more narrow and slightly less deep. It has a rather strongly deformed western limb, a gently folded axial zone, and a monoclinally stepped eastern limb. The trough becomes wider in the north, flattens out, becomes deeper, and vanishes under the Vilyui Syneclise.

The **Irkutsk Jurassic Foredeep** is superimposed on the southwestern end of the fore-Baikalian Trough. Its formation is linked with the Jurassic reactivation of the Baikalian orogen. It is filled with a molassoid continental formation and continues from the southernmost corner of the platform further to the northwest along the foot of Eastern Sayan. The Precambrian of the Baikalian Upland is overthrust at a small angle on the southern edge of the Irkutsk Trough at the source of the Angara. This famous **Angara Overthrust** was at the time the subject of sharp arguments between TETYAEV (1932) and V. A. OBRUCHEV. The former had tried to show, on the basis of this overthrust and overthrusts found in Transbaikalia, that the whole Precambrian of the Baikal has an allochthonous attitude. OBRUCHEV disputed its existence. Drillings already carried out have confirmed the existence of a flat overthrust of Precambrian on the Jurassic, but TETYAEV's "nappistic" view of the Baikalian and Transbaikalian tectonics was not confirmed in general.

The **Tunguska Syneclise** occupies an area of more than 1.2 million km^2. Its maximum width is about 1,500 km. It is thus not only the biggest, but also the most peculiar structure of the Siberian Craton. This peculiarity is due to a large trap (plateau-basalt) formation of the Upper Permian–Lower Triassic, which extends over practically the whole area of said syneclise[7]. The formation consists of a tuffogenous subformation in the lower part (up to 600 or 800 m thick) and a subformation of lava in the upper part (up to 2.5 or 3 km thick). The Permian–Triassic tuff and lava, as well as the older sedimentary strata beneath them, are pierced by a large number of dykes and sills. A cluster of dykes is located mainly along the periphery of the syneclise, where it attains a length of 400 to 500 km (Kotui zone). Individual dykes are up to 100 m thick; the sills are up to 350 m thick. Most interesting among the intrusive bodies are the relatively large differentiated plutons of the Noril'sk type. The differentiation manifests itself in the formation of picritic dolerites and troctolite dolerites in the lower parts of the plutons, often with sulfides of copper and nickel; the central zone contains olivine and non-olivine dolerites, while the upper parts contains leukocratic and quartzose gabbro-dolerites and even granodiorites, partly of hybrid origin. With this complex of intrusions, common in the northwestern part of the syneclise north of the basin of the Lower Tunguska, the known copper-nickel sulfide ores of the Noril'sk type are linked. The predominant distribution of intrusive bodies in the present erosion cut is typical of the peripheral parts of the syneclise. This caused zones of pronounced permeability associated with deep faults, to be identified early here. The specific formations are volcanic pipes filled with basaltic material, common along the southeastern periphery of the syneclise, in a strip where it joins the Central Siberian Anteclise (OFFMAN, 1959; and others). However, geologic mapping during recent years (AVDALOVICH, 1974; and others) proved the existence of a large number of volcanic-tectonic structures also in the central part of the syneclise. These are, obviously, associated with ancient volcanic activity[8], they exhibit a certain gravitation towards linear pervious zones, i. e., deep faults cutting across the syneclise instead of bordering it. The distribution of dolerite intrusions inside the syneclise, as first noted by OFFMAN (1959), can be represented by a complex mesh pattern similar to the location of

---

[7] The closest analogue of the Tunguska Syneclise is the Paraná Syneclise in South America.

[8] The existence of paleovolcanoes of the central type in the Tunguska Syneclise is disputed by some researchers who believe that the eruptions were here of areal character.

salt intrusions in regions of salt-dome tectonics. This analogy, observed by A. M. SYCHEVA-MIKHAILOVA (oral commun.) in data obtained by AVDALOVICH (1974), suggests that basaltic magma saturated with gas floated up beneath the entire surface of the Tunguska Syneclise in accordance with the law of density inversion. It follows that a gigantic magma basin existed at that time deep below.

The extensive development of a trap formation with its specific magmatogenic dislocations makes determination of the internal structure of the syneclise very difficult. The depth of the basement in its central part is estimated as 8 to 10 (possibly up to 12) km. For the region east of Tura settlement on the Lower Tunguska this was confirmed by seismic investigations with equipment of the "Zemlya" type. The depth of the basement roof was established as 9 km. The same researchers (PRITULA et al., 1973) also found that the thickness of the crust is only 29 to 30 km in the Tura region, as against 37 to 38 km at the southern edge of the syneclise. This general thinning-out of the Earth's crust, accompanied by a considerable increase in the thickness of the sedimentary layer (from 2 to 9 km), is compensated by a reduced thickness of the granitic layer. This had also been suggested already by earlier researchers (some even assumed its complete disappearance).

There is no doubt that the Tunguska Syneclise is not a single depression, but consists of several more partial depressions varyingly identified by different researchers. The largest is the Kureika Depression which occupies the northwestern part of the syneclise. The second largest is the Kochechum Depression in the center of the syneclise. The Noril'sk and Kharayelakh depressions lie north of the Kureika Depression, while the Vanavara and Mura depressions are located south of the Kochechum Depression.

The Tunguska Syneclise acquired its present configuration only in the "post-trap" time, i. e., since the Middle Triassic, when the Turukhansk–Noril'sk dislocation zone obtained its definite shape. Traps took part in these dislocations. However, there is a noticeable reduction in the abundance of traps both in this zone and more to the west, on the left bank of the Yenisei, to the south, in the Yenisei Ridge, and practically everywhere along the periphery of the syneclise outside its borders. The type of magmatism is also different (e. g., in the east). This shows that isolation of the syneclise took place already during the period of basaltic volcanism, i. e., during the Permian and Triassic. There is also reason to assume that the entire syneclise is a huge volcano-tectonic structure (AVDALOVICH, 1974). Formation of this structure is assumed to have begun in the Middle to Late Carboniferous and Early Permian, at the time when a coal-bearing formation accumulated, since this crops out from beneath the traps everywhere along the perriphery of the syneclise. The Tunguska Syneclise as such did not exist earlier (Middle and Early Paleozoic). Fairly pronounced subsidence occurred only in the west and northwest, in a trough formed in the north in the direction of the Taimyr, and partly bounded in the west by the system of synsedimentary formed embryonic arches of the Turukhansk–Noril'sk zone. Conversion of this trough into a syneclise in the Middle and Late Carboniferous was due to the growth of the Sayan–Yenisei Uplift zone (PAVLOV, 1974) as well as to pronounced subsidence on the more distant slopes of the Anabar and Central Siberian anteclises. Previously, during the earliest stage of the development of the platform, in the Riphean time, the strata probably became gradually thicker from the Anabar Anteclise towards the Turukhansk–Noril'sk zone and farther, to the northern continuation (beyond the Yenisei), of the Yenisei Ridge Geosyncline. The western part of the future syneclise thus formed a region of pericratonal subsidence. This is to some extent also true for the Early and Middle Paleozoic. However, the regional slope was at that time complicated by linear uplifts.

The comparatively small but quite deep (7 to 8 km) **Taseevo Syneclise** is located to the southwest of the southern end of the Tunguska Syneclise. Its western and northwestern part overlies unconformably the Baikalian structures of the Yenisei Ridge. It conceals the southern continuation of its foredeep which is filled with a quite thick molasse layer of the uppermost Riphean and Vendian. Cambrian, particularly Lower Cambrian, and Lower Paleozoic sediments in general have a considerable thickness in the sequence of the syneclise. No Devonian was deposited here. Subsidence was small during the Carboniferous and Permian. However, a region of accumulation of Upper Paleozoic coal-bearing and tuffaceous formations developed here nevertheless. Subsidence ceased in the Middle Triassic but partly resumed in the Jurassic, when the continuation of the Angara–Vilyui Trough (see below) extended there.

The narrow **Sayan Foredeep** forms the southeastern continuation of the Taseevo Syneclise. It corresponds to the edge of the craton which was twice involved in compensatory subsidence caused by the uplift of Eastern Sayan. The first time this occurred at the end of the Riphean and in Vendian; the second time this happened, to a much smaller extent, during the Jurassic. The Taseevo Syneclise, together with the Sayan and Baikalian troughs and the buried southwestern part of the Central Siberian Anteclise, forms the present structure which is frequently called Angara or Irkutsk Syneclise; it corresponds in general to the Irkutsk Amphitheater as defined by SUESS (1885).

As already mentioned, the **Turukhansk–Noril'sk dislocation zone** borders the Tunguska Syneclise in the west. It extends for 600 km along the right bank of the Yenisei and is bounded in the west by a major fault (boundary suture) of the craton, which clearly manifests itself in the magnetic field. The width of the zone varies from 50 km in the south to 140 km in the north. Its trend varies from NNW to NE in the same direction. This zone consists of a series of anticlines whose limbs are inclined at angles of 50 to 70°, and which are complicated by flexures and faults. The entire fold system is in the north overlain unconformably by a cover of Jurassic and Cretaceous rocks of the Yenisei–Khatanga Trough. A saddle between its two main elements (the Ust'–Yenisei and the Khatanga depressions) overlies its continuation here.

The Turukhansk–Noril'sk zone seems to merge with the Baikalides of the Yenisei Ridge in the south. This fact caused said zone to be included in the Baikalides in earlier tectonic schemes. This is contradicted by the facial character of its Upper Riphean–Vendian sequence and the mainly germanotype style of the deformations, but with some eastern vergence. It is now known that the Baikalides continue beyond the Yenisei directly to the west of the zone considered, so it would obviously be correct to assume that the Turukhansk–Noril'sk dislocations evolved at the edge of the old craton in its pericratonal downwarp, as reflection of deformations in orogens extending to the west. It is known that folds in the zone evolved during a long time and synsedimentary throughout the Paleozoic, terminating their formation by block movements at the end of the Triassic and the beginning of the Jurassic, i. e., at the time of the Early Cimmerian deformations of the Taimyr.

The Middle Riphean sediments of the Igarka region in the north of the zone considered are a flyschoid terrigenous formation. Schists accompanied by siliceous and basic volcanics of the spillitic type predominate in them. They may therefore on the basis of their composition be considered to be geosynclinal, also because of the thickness of their strata (more than 2 km) and the intensity of their deformations. They are disconformably overlain by the Upper Riphean whose lower strata contain coarse clastic sediments resembling molasses. Lower Cambrian rocks directly overlie the Middle Riphean in the Turukhansk region in the south, also with a sharp angular unconformity. This permits the conclusion that tectogenesis occurred here in the Grenvillian (Dalslandian) epoch, similar to the Taimyr (POGREBITSKY, 1971). This Pre-Late Riphean fold zone may thus extend here. The Grenvillian unconformity manifests itself clearly also south of the Noril'sk–Turukhansk zone in the Yenisei Ridge (VOLOBUYEV, 1976; VOROBYEV et al., 1969).

The **Vilyui Syneclise** is the second largest depression of the Mid-Siberian Platform. It is also the youngest as regards its definite formation. It is from the surface formed by Mesozoic sediments and has the shape of a wide gulf whose axis runs in a northeasterly direction. This gulf is open to the NE in the direction of the Verkhoyansk Foredeep where the trend of the latter changes from meridional to latitudinal. The presence of erosional remnants of thin Jurassic sediments on the western continuation of the syneclise marks the flat and very shallow **Angara–Vilyui Trough** which represents its continuation. The thickness of the Mesozoic sediments in the syneclise proper attains 6 to 8 km. The thickness is maximum in the east, near the junction with the Verkhoyansk Foredeep, where the depth of the basement in the syneclise is 10 to 12 km. The Mesozoic deposits have in general a very gentle attitude except near the Kempendyai dislocations (see below). The structure of the Vilyui Depression becomes very complicated below the Mesozoic. In fact, it assumes there the character of a complicated aulacogen (Fig. 28). The **Suntar Swell** is located in the central zone of the southwestern part of the syneclise. A large horst of the crystalline basement is located here, directly beneath the Jurassic sediments, at a depth of less than 400 m. This uplift is paralleled in the northwest by the deep **Ygyatty (Markha) Trough,** and in the southeast by the **Kempendyai Trough.** Their depths are 7 and 9 km respectively. They are mainly filled with Paleozoic and

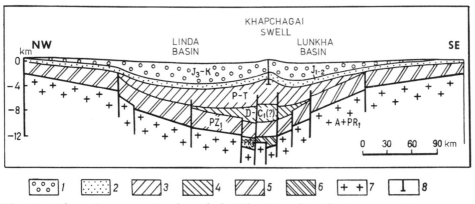

Fig. 28. Schematic cross-section through the Vilyui Syneclise (after SOKOLOV & LARCHENKOV, 1978). 1 = Cretaceous and Upper Jurassic; 2 = Middle and Lower Jurassic; 3 = Triassic and Permian; 4 = Lower Carboniferous and Devonian (?); 5 = Lower Paleozoic; 6 = Riphean (?); 7 = crystalline basement; 8 = boreholes.

Upper Proterozoic rocks. The Kempendyai Trough contains an evaporite Devonian formation in its sequence. It underwent inversion during its evolution; two zones of salt-dome uplifts, known as the Kempendyai dislocations, were formed.

An analogue of the Suntar Swell is the more deeply buried latitudinal **Khapchagai** (Central Vilyui) **Swell** in the eastern part of the syneclise. Both anticlines and the troughs paralleling them are complicated by local uplifts. The latter, however, are very shallow and nowhere as pronounced as the Kempendyai folds. Formation of the complex Paleo–Vilyui Aulacogen is clearly linked with the activity of the Patom–Vilyui Lineament which connects the two re-entrants of the craton; this lineament is undoubtedly the weakest zone in the body of the craton. The movements were accompanied by magmatic activity which manifested itself in the eruption of basalts in the Middle and Late Devonian (their flows are preserved in the sequences of the Ygyatty and Kempendyai troughs). More alkaline lavas (including trachytes) were less important. Kimberlite pipes intruded in the northwestern surroundings of the trough at the end of the Devonian and beginning of the Carboniferous. These pipes contain more diamonds than the Mesozoic ones (MASAITIS, 1969; MASAITIS et al., 1977).

The ancient Siberian Craton formally also includes the southern limb of the Yenisei–Khatanga Trough, as stated at the beginning of this chapter. This, however, is better discussed after analysing the structure of the Taimyr (cf. Ch. 5.6).

4.4 Main stages of craton evolution

4.4.1 Archaean stage – 3,500 to 2,500 m.y. B.P.

The sequence of events in the pre-Proterozoic history of the Siberian Craton is still rather vague. This is due to the complexity of the geological relationships of the oldest rocks, as well as to insufficient, contradictory, and unreliable radiogeochronometric data. The supracrustal melanocratic formations of the Stanovoi Belt are quite probably the oldest in the exposed sequence of the craton. Their absolute age, however, has not been established, and their relationship to the Yengra series and its analogues is unclear. The presence in the corresponding series (Kurul'ta–Gonam, Sutam, and their analogues) of quartzites in the upper parts of the sequence may indicate the existence of even older granite-gneisses. This is even more likely for the Yengra series, since the origin of the quartzites of this series is probably clastic. The sequence hyperbasites-metagabbro-anorthosites-melanocratic strata of the Stanovoi Belt–Yengra series (LEITES & FEDOROVSKY, 1977, 1978) cannot be considered to be rigorously proven. It may be that the granite-gneisses of tonalitic composition must be included in it as a basal member. These are only recently found in Siberia, but are known to be present in most other shields on the Earth. The Archaean age of the anorthosites must also still be definitely proved. Most geologists continue to consider them, together with the accompanying gabbroids, to be Early Proterozoic. Comparison of these anorthosites with lunar ones, as well as relating the Kurul'ta–Gonam and Sutam rocks and the supposed annular structures in their terrain, observed on space photographs, to a "lunar stage" of development of the Earth's crust with its impact craters (PAVLOVSKY, 1970; GLUKHOVSKY et al., 1977) is only a nice theory insufficiently supported by proofs and little likely on the basis of general consideration.

One thing is not in doubt: the continental crust in the region of the future Siberian Craton grew intensely during the Archaean, via metamorphism and granitization of volcano-sedimentary strata, comprising those of the greenstone belts, as in other regions of the future continents.

The region of epi-Archaean consolidation, about 2,500 m.y. B.P.[9], encompassed almost the entire area of the future craton and extended far beyond its boundaries. This follows from the presence of Archaean blocks in the structure of the Baikalides of the entire southwestern and southern surroundings of the platform, including the Verkhoyansk–Chukotka region (Okhotsk, Omolon, Taigonos massifs). This not only adequately justifies identifying the so-called North Asiatic Craton (KOSYGIN, 1964), but also causes one to suspect that this craton was connected with other Eurasian cratons, primarily the Sinian Craton. The thickness of the crust must have attained 25 to 30 km already at the end of this stage, judging from the universally manifested metarmorphism of granulite type. Most of the greenstone belts, examples of which are found on the Aldan Shield, became closed in the beginning or in the middle of the Early Proterozoic. They underwent folding, regional metamorphism in the amphibolite, more rarely in the greenschist facies, and granitization. The older consolidated blocks of the continental crust were again welded into one whole. The blocks were intensely heated; palingenetic-anatectic and metasomatic granitoids were formed, accompanied by regressive metamorphism of the amphibolite facies.

4.4.2 Early Proterozoic stage – 2,500 to 2,000 (1,900) m.y. B.P.

Some of the protogeosynclines continued their development during the Early and even the Middle Proterozoic. However, the protoplatform cover began to be formed already during the Early Proterozoic in some parts of the Aldan Megablock. These were the Udokan, Ulkan, and other smaller troughs and grabens. This platform cover has the features transitional from orogenic to platform-like proper. It underwent metamorphism to the lowest stages of amphibolite facies in its lower strata. In the Udokan region it encloses a large sheet-like body of granites which, like a series of peripheral granite-gneiss domes, evolved as a result of basement remobilization. This indicates that an intensive heat flux still occurred then. This is also confirmed by the pronounced manifestation of acid volcanism while the Ulkan series accumulated in the trough of this name in the east of the Aldan Shield.

Formation of the very large, nearly latitudinal Stanovoi Fault zone also occurrred in the Late Archaean and Early Proterozoic. The Stanovoi Belt running in the same direction became separated. Its Archaean (Lower Archaean) rocks were altered markedly during the Early Proterozoic by regressive metamorphism and granitization. The Stanovoi Belt, formed directly to the north of the Ural–Okhotsk geosynclinal belt underwent repeated activation. Very large plutons of a gabbro-anorthosite formation intruded along the Stanovoi Fault at the end of the Early Proterozoic. Relatively small bodies of anorthosites of the same age are also known in the Anabar Massif. They may therefore also be common beneath the cover of the Mid-Siberian Platform.

---

[9] SALOP (1973) claims that this consolidation took place even earlier, at the end of the Archaean according to his nomenclature, i.e., about 3,000 m.y. B.P., but it is less probable.

4.4.3 Middle Proterozoic stage – 2,000 (1,900) to 1,800 (1,700) m.y. B.P.

Most of the present Siberian Craton became stabilized already at the beginning of this stage. Its continental crust was completely consolidated and fairly thick. The Middle Proterozoic stage is documented only in the extreme southeast of the craton, i. e., in the Batomga Block of the Aldan Shield. Red molassoid formations of the Uyan series evolved on its periphery. They were accompanied by sills of diabases and flows of acid volcanics associated with graben-troughs (paleoaulacogens). Accumulation of this series ended with the intrusion of alkaline granites of the Ulkan complex (1,660 m.y. B.P.).

4.4.4 Late Proterozoic stage including Baikalian – 1,700 to 1,650 m.y. B.P.

Fully cratonic conditions prevailed at the beginning of the Late Proterozoic over practically the entire Siberian Craton, except (see below) the Turukhansk–Noril'sk peripheral zone. The craton boundaries were established at the same time, with the formation in the west and south, of the geosynclinal systems of the Turukhansk–Noril'sk Zone (closed by the Late Riphean, the boundary receding to the west), the Yenisei Range, Eastern Sayan, the Baikal–Patom Upland, and the Mongolo–Amur system. In the southeast the Yudoma–Maia meridional trough of the intracratonal type was formed; it separated the Okhotsk Massif from the Aldan Shield. In the north, the platform extended to the southern border of the Taimyr. The situation remains unclear in the northeast and east. The similarity between the Riphean, Vendian and Lower Cambrian deposits of the Okhotsk, Kolyma, and Omolon median massifs, and the analogous formations of the Siberian Craton might confirm the continuing existence of a single North Asiatic Craton. However, metamorphics are found in various regions, e. g., in the Polousny Range, on the Novosibirsk Islands (Greater Lyakhovsky Island), and on Wrangel Island. These metamorphics may be Riphean or even older according to radiometric dating of sporomorphs. This necessitates the assumption that geosynclinal conditions prevailed in the axial zones of both the Verkhoyansk–Kolyma and the Novosibirsk–Chukotka fold systems, in accordance with REZANOV's suggestions (1968). An indirect proof of the probable existence of a Riphean geosyncline to the east of the present Siberian craton is the Yudoma–Maia Trough, and in particular the fact that this trough becomes wider and deeper towards the north, until it becomes buried beneath younger formations. If the Early to Middle Paleozoic structural patterns of the Verkhoyansk–Kolyma system have been inherited from the Late Proterozoic, then it may be assumed that the Riphean geosyncline extended far to the north; it may have turned northwest, toward the Taimyr in the region of the Laptev Sea (bypassing the Laptev Massif from the north?).

The Upper Precambrian formations can be divided into series both in the Siberian Craton proper and, particularly clearly, in its immediate surroundings. These series are separated by unconformities and by manifestation of magmatic activity. In the Yudoma–Maia Trough the sequence of the Upper Precambrian is quite complete and thick; it can be quite easily differentiated since metamorphism is practically absent. Four series are distinguished here: the Uchur series of the Lower Riphean (1,520 to 1,400 m.y.

old[10], most likely 1,600 to 1,350 m.y.), the Maia series of the Middle Riphean (1,190 to 890, but most likely 1,350 to 1,000 m.y. old), the Uia series of the Upper Riphean (down to 680 m.y. old), and the Yudoma suite of Vendian, linked with the Lower Cambrian by a gradual transition. All these series consist of a cyclic sequence of sedimentary rocks: terrigenous rocks, coarser rocks below, and carbonaceous (predominantly dolomitic)

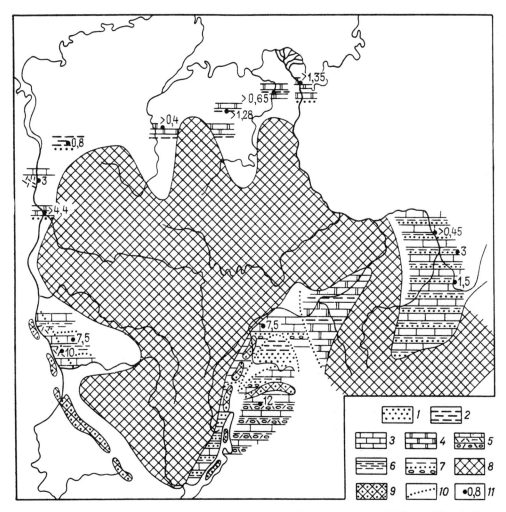

Fig. 29. Lithologic-paleogeographical scheme of the Siberian Craton, Middle and first half of Late Riphean (after Trunov, 1975, simplified). 1 = sandstones; 2 = shales, siltstones; 3 = limestones and marls; 4 = dolostones; 5 = silt-shales and coarse clastic sediments with intercalations of acid volcanics; 6 = shaly-sandy sediments; 7 = ibid., and coarse clastic sediments; 8 = dry land; 9 = islands, temporarily covered by sea; 10 = lithofacial boundaries; 11 = thickness of sediments in km.

---

[10] Here and below, dating is by the K/Ar method for glauconites.

rocks on top. Each series in fact corresponds to a separate tectonic stage. The first is
Gothian, the second Grenvillian; the third to last series are Baikalian. The independence
of the Grenvillian stage was most clearly established in the Yenisei Range where it was
completed by intrusions of granitoids of the Teia complex (VOLOBUYEV, 1976; VORO-
BYEV, 1969). Considerable movements and intrusions of granitoids are assumed to have
occurred at that time in the Taimyr (POGREBITSKY, 1971). This has now been confirmed
from data obtained by VOLOBUYEV (oral commun.).

The Grenvillian deformation probably continued from there to the Noril'sk–
Turukhansk zone which lies between the Taimyr and the Yenisei Range. Lastly, stocks of

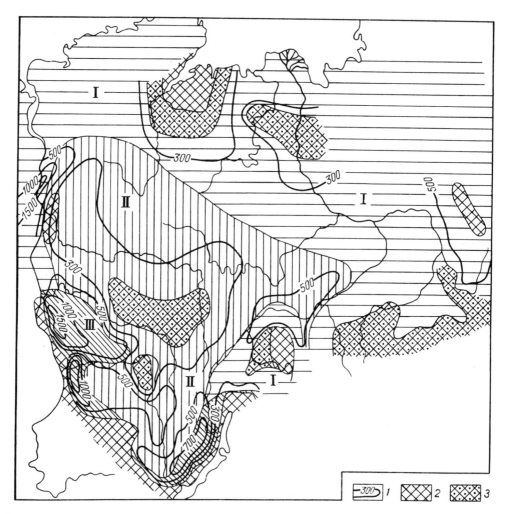

Fig. 30.   Areal distribution of Vendian thicknesses in the Siberian Craton (after KIRKINSKAYA &
POLIAKOVA, 1978, simplified). 1 = isopachs in hundreds of meters; 2 = dry land; 3 = ibid., only in
Early Vendian. I = open sea, II = sea with high salinity, III = desalted shallow zones of the marine
area.

granites, having an age of 1,000 ± 100 m.y., are known to exist on the northwestern slopes of the Aldan Shield in the Chara–Olekma zone (GLUKHOVSKY et al., 1972).

However, the structural pattern of the main part of the Siberian Craton proper remained comparatively stable during the entire Riphean (Fig. 29). Most of it evolved in an environment of gentle and moderate subsidence. This was accompanied by the accumulation of predominantly carbonate rocks (mainly dolomites), partly quartzose sandstones (in the basal part of the sedimentary cycles), and siltstone-clayey sediments. It is most likely that the main region of uplift and erosion within the platform was the Central Siberian Anteclise which runs in a northeasterly direction. The absence of Riphean in it has been proved at some points by drilling. It is less likely that uplifts (shown in some paleogeographic reconstructions) existed in the zone of the Anabar Uplift, and especially in the center of the Tunguska Syneclise.

General inversion occurred in the Yenisei–Sayan Geosyncline during the second half of the Late Riphean. A foredeep formed along the southwestern border of the craton. This foredeep was from then on till the beginning of the Cambrian filled with red lagoonal-continental molasse of the Chingasan–Taseevo (Yenisei Range) and Karagas–Oselok (Sayan) series (Fig. 30). The same happened in the Baikalian Riphean geosyncline; the Baikalian Foredeep was formed and the Ushakovka and lower Moty series accumulated. However, Baikalian movements in the northeast, towards the summit of the Patom Upland, grew weaker; the molasse formation was correspondingly substituted by red continental formation of the platform type, belonging to the Aldan stage of the Lower Cambrian.

4.4.5 Caledonian stage (Cambrian to Middle Devonian)

Formation of the Baikalian mountains at the periphery of the present Irkutsk Amphitheater, which were relatively low and to a large extent levelled before the beginning of the Lena age, and intensification of uplift along the ancient Anabar–Aldan direction inherited from the Early Precambrian, which manifested itself in the formation of barrier reefs, created the preconditions for the isolation of a huge saline basin in the southwestern and central parts of the platform. This was accompanied by accumulation of a thick evaporite formation in the **Lena and Amga** age (Fig. 31). Conditions of an open but shallow sea still prevailed in the northern and eastern parts of the platform. This led to deposition of a carbonaceous formation. The depth increased to the east; thin bituminous lime and clay sediments ("Siberian domanik") were deposited at the periphery of the Aldan Anteclise and in the Yudoma–Maia Trough during the Early and at the beginning of the Middle Cambrian. Subsidence in this eastern pericratonal trough probably exceeded the accumulation of sediments. Analogous conditions, accompanied by the deposition of similar sediments between arches with reef edifices, prevailed also on the northwestern slope of the craton. Inheritance from the quite old structural pattern also manifested itself in the Turukhansk–Noril'sk Belt which continued in the direction of the Taimyr.

Uplift occurred **during the second half of the Middle Cambrian (Maia age)** (Fig. 32) in the southwestern and southern mountainous framing of the craton. This was an echo of the powerful Salairian tectogenesis of more interior zones of the Central Asiatic segment of Mongol–Okhotsk Belt. This uplift extended also to the southern half of the craton

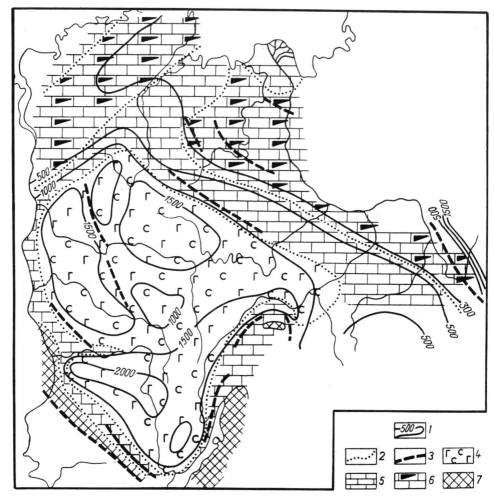

Fig. 31. Map of thicknesses and lithologic assemblages of the Siberian Craton, Early Cambrian and Amga age of Middle Cambrian (after KIRKINSKAYA & POLIAKOVA, 1978). 1 = isopachs; 2 = boundaries of assemblages; 3 = faults influencing sedimentation: 4–6 = assemblages; 4 = sulphate-evaporite, 5 = carbonate, 6 = shaly-carbonate (bituminous); 7 = dry land.

where it interrupted the accumulation of sediments until the beginning of the **Late Cambrian.** The uplifts are reflected in the composition of the sediments of the Late Cambrian. These are a variegated molassoid, slightly saline, Verkholensk (more to the northwest, Evenky) rocks. These lagoonal-marine sediments were in the northern half of the craton replaced by more normal marine sediments of carbonaceous composition, conformably overlying the Middle Cambrian. They contain traces of deposition under extremely shallow-water conditions and increasing regression at the end of that period.

This regression was followed by a large-scale transgression at the **beginning of the Ordovician.** The sediments of this period are therefore very common in the Siberian

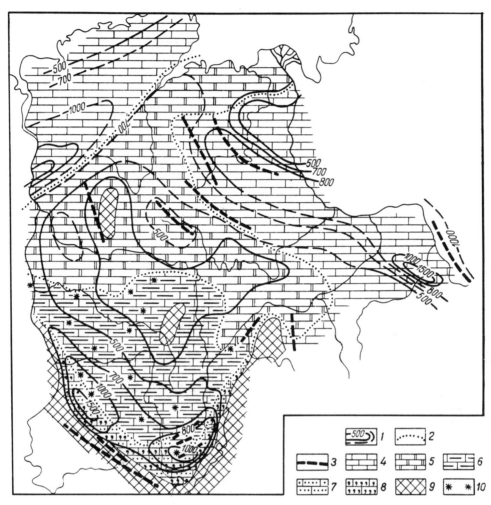

Fig. 32. Map of thicknesses and lithologic assemblages of the Siberian Craton, Maia age of Middle Cambrian and Late Cambrian (after KIRKINSKAYA & POLIAKOVA, 1978). 1 = isopachs; 2 = boundaries of lithologic assemblages; 3 = faults influencing sedimentation; 4–8 = assemblages: 4 = carbonate, 5 = dolostone, 6 = carbonate-terrigeneous, 7 = arkose sandstone, 8 = sandy-carbonate, 9 = dry land; 10 = red beds.

Craton, practically everywhere except the Anabar Uplift and the Aldan Shield. These are a shelf carbonaceous-terrigenous formation of moderate thickness (a few hundred m, Fig. 33). The change of its composition in the vertical direction and over the area shows that new, gradually increasing regression occurred at the end of this period. It was most pronounced in the south and east of the craton. Barren red beds were formed. Subsidence (up to 900 m) lasted longest in the western and northwestern part of the Tunguska Syneclise in the continuation of the Taimyr Trough.

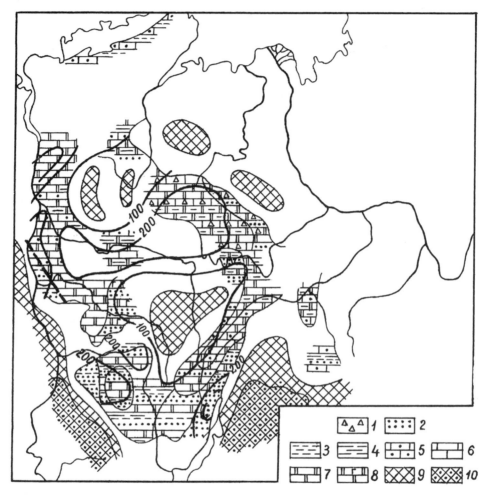

Fig. 33.   Lithological-paleogeographical scheme of the Siberian Craton, Early Ordovician, Ust–Kut time (after BGATOV, BONDAREV, KNIAZEV, LEBED, E. P. & L. G. MARKOV, MATIUKHINA, 1969, simplified). 1 = breccia; 2 = sandstones; 3 = siltstones; 4 = shales; 5 = sandy limestones; 6 = limestones; 7 = dolostones; 8 = anhydrites; 9 = low land; 10 = high land.

Uplift **at the end of the Ordovician** was a reflection of the Early Caledonian movements along the craton periphery. When the sea returned in Central Siberia at the **beginning of the Silurian,** it occupied a much smaller area than during the Ordovician. The entire Anabar–Aldan strip became exposed as a vast erosion zone. The sea covered only the more western regions. The Silurian thus practically overlies everywhere the Ordovician with traces of disconformity. A new regression becomes noticeable from the Wenlockian onward. It caused the basin to become much smaller at the end of the period considered, and led to an accumulation of variegated, gypsum-bearing strata in residual troughs. The region of most stable and pronounced subsidence (up to 700 m) was the northwestern part of the Tunguska Syneclise. Graptolitic shales are present in the Lower Silurian also further south, up to the Podkamennaya Tunguska. This underlines the features common with the Taimyr.

The mobile frame of the craton in the west and south experienced a powerful impact by the Caledonian tectogenesis **at the end of the Silurian and beginning of the Devonian.** This caused emergence of the Irkutsk Amphitheater, folding of the Cambrian–Silurian strata of the Angara–Lena Trough, final formation of the Urin anticlinorium and folding in the troughs bordering it, and increased growth of the anticlines of the Turukhansk–Noril'sk dislocation zone. The Vilyui Aulacogen was formed or possibly regenerated (provided it originated during the Riphean, which is quite likely) at the same time in the body of the craton proper. Subsidence of this aulacogen was accompanied in the Middle and Late Devonian by quite intensive basaltic, partly more acid subalkaline volcanism. Dykes, sills, and at some points stocks of dolerites and gabbro-dolerites were formed at the periphery of the aulacogen and near its southwestern closure. Intrusive bodies of alkaline-basaltoid magma were also formed. Pipes of diamond-bearing kimberlites tend to the Vilyui–Markha Fault zone (the northwestern boundary of the Vilyui Aulacogen). Several plutons of alkaline-ultrabasic rocks are known along the eastern edge of the platform, more precisely of the Anabar–Aldan Uplift zone. They extend from the Udzha River basin in the north to the western slope of the Sette–Daban Range. Similar intrusions are found at the opposite periphery of the craton, i. e., on the northeastern slope of the Eastern Sayan. To this should be added the presence of local tuffogenic material of basaltic composition in the Turukhansk–Noril'sk zone, and of 370 m.y. old gabbro and granite massifs in the extreme southeastern corner of the craton, near the Okhotsk coast. The important Late Caledonian tectonic-magmatic reactivation of a large part of the craton thus becomes clear.

This reactivation was directly reflected in the facial character of the Devonian sediments. These are represented by a red carbonaceous-terrigenous formation of continental-littoral origin, containing large amounts of dolomites and gypsum (anhydrites). The evaporite formation in the Kempendyai Graben of the Vilyui Aulacogen and in the Nordvik region in the extreme north of the craton is significant by itself. It contains large deposits of rock salt.

The main region of accumulation of Devonian sediments, apart from the Vilyui Aulacogen, is the Tunguska Syneclise. A continuous sequence of Devonian (lagoonal variegated sediments at the bottom, shallow-water marine variegated sediments at the top, accompanied by sulfates but predominantly in carbonaceous facies), having a thickness of up to 1000 m or more, is found only in the extreme northwest of the platform, north of the Kureika River[11], as well as in the extreme northeast, near the mouth of the Lena (in the latter region there are purely carbonaceous open-sea sediments). The Devonian here conformably overlies the Silurian; no Caledonian movement is thus apparent. Neither was such movement observed in the Taimyr, despite earlier assumptions.

---

[11] A strip of saline lagoons extended in the east, in the Igarka-Noril'sk region, and more to the north in the Central Taimyr and Severnaya Zemlya. In the west it gave way, via a barrier reef, to a carbonaceous formation.

4.4.6 Hercynian stage (Late Devonian to Middle Triassic)

Uplift decreased already at the end of the Middle Devonian; transgression with the deposition of pure biogenic limestone increased. The upper part of this carbonaceous formation belongs to the Lower Carboniferous. Transition from the Devonian to the Carboniferous is marked in the east and south by the disappearance of the Famennian stage (partly or completely) and of the Lower Tournaisian from the sequence. A gradual (?) transition from the Devonian to the Carboniferous, already in the continental facies, is noticed in the Vilyui Aulacogen. Marine sediments accumulated in the extreme north of the platform till the end of the **Dinantian** and possibly also in the **Namurian**. However, this process was interrupted in the remaining platform area at that time. A radical change occurred in the nature of sedimentation. This later involved also the northern region. Marine sediments of the arid zone were replaced by continental sediments of a limnic coal-bearing formation of the Middle and Upper Carboniferous and Permian. The main accumulation zone of this formation was the Tunguska Syneclise. It acquired its present size at that time, due to involvement of the slopes of the Anabar and Central Siberian anteclises in the subsidence (Figs. 34, 35). Moreover, subsidence also involved the Taseevo Syneclise already during the **Early Carboniferous.** The marine Upper Paleozoic of the Vilyui Aulacogen and of the newly formed Kyutyungdy Graben of the Olenek Uplift is closely linked with the lower part of the Verkhoyansk complex of the geosynclinal system bearing the same name. A connection with the Verkhoyansk Geosyncline is evident already for the Devonian. Formation of the Tunguska Syneclise must similarly be linked with the beginning of pronounced subsidence of the Southern Taimyr Trough. The marine carbonaceous formation of the Cambrian to Namurian was there first replaced by a flyschoid terrigenous marine formation of the Middle Carboniferous to Lower Permian. This was followed, from the Kungurian age onward (earlier in Western Taimyr) by a paralic formation, and from the Tatarian age, by a limnic coal-bearing formation. The Taimyr Trough forms the northwestern branch of the Verkhoyansk-Kolyma geosynclinal system (POGREBITSKY, 1971). Evolution of the Siberian Craton from the **Middle Carboniferous** onward was therefore largely determined by the influence of the Taimyr–Verkhoyansk Geosyncline, and not by the Yenisei–Baikalian orogen, as happened during the Early and Middle Paleozoic. This orogen, however, and especially its Baikalian part, remained an important source of clastic material, and may even have undergone glaciation.

A new important phase in the evolution of the Siberian Craton started at the **end of the Permian and the beginning of the Triassic.** The Tunguska Syneclise became the site of very intensive trap magmatism. It began with an outburst of explosive activity which was superseded by an eruption of voluminous nappes of lava. As already stated in the chapter on craton structure, the centers of this eruption are mainly associated with zones of high permeability, i. e., zones of deep faults. These are either peripheral to the Tunguska Syneclise or intersect it, mainly in a latitudinal direction. Intrusive bodies of dolerites and gabbro-dolerites are also concentrated in these zones. The region of magmatic activity extended far beyond the boundaries of the present Tunguska Syneclise, especially in the north, where it included the present Yenisei–Khatanga Trough and the Taimyr fold system. Trap formation (more precisely, association) partly coexisted with generally younger ultrabasic, alkaline, and kimberlite intrusive formations in the east

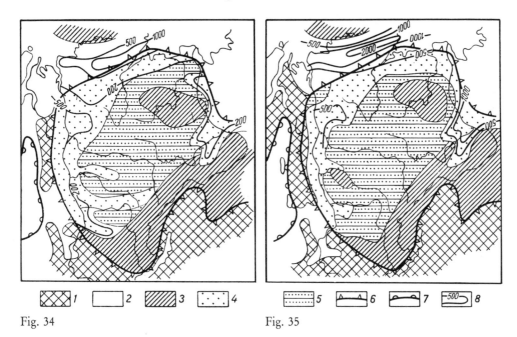

Fig. 34   Fig. 35

Fig. 34. Paleotectonic scheme of the Siberian Craton, Early and Middle Carboniferous (after GOLUBEVA, 1975, simplified). 1 = areas of intense uplift outside the craton; 2 = ibid., of subsidence; 3 = areas of gentle and moderate uplift within the craton; 4 = ibid., of subsidence; 5 = alternating uplift and subsidence; 6 = boundaries of the craton; 7 = boundaries of the geosynclines (Ob'–Zaysan Geosyncline); 8 = isopachs.

Fig. 35. Paleotectonic scheme of the Siberian Craton, Late Carboniferous–Permian (after GOLUBEVA, 1975, simplified). Legend see Fig. 34.

(Maimecha–Kotui region) and southwest (Chadobets Uplift) of the syneclise surroundings. Kimberlite pipes and dykes are known to exist also in the Anabar–Olenek Anteclise at the middle reaches of the Olenek.

This magmatic activity reached its highest point during the second half of the **Early Triassic;** it ended in general in the **Middle and Late Triassic** and the **Early Liassic.** The structure of the Tunguska Syneclise and of its western boundary, the Turukhansk –Noril'sk dislocation zone, acquired at that time its definitive form. The evolution of the Taimyr fold system in fact also ended at that epoch. Both events concluded an important stage in the evolution of the north of Central Siberia as a whole.

4.4.7 Late Cimmerian stage (Late Triassic to Early Cretaceous)

The principal events of this stage occurred in the north and east of the craton. The wide Yenisei–Khatanga (cis-Taimyr) Trough was formed in the north during the Middle Liassic. This trough continued eastward to the lower reaches of the Anabar, Olenek, and Lena (Lena–Anabar Trough). It evolved steadily throughout the remaining Mesozoic and Cenozoic. The marine environment was gradually replaced by a continental one.

Generally running in a nearly latitudinal direction, the Yenisei–Khatanga Trough forms a superposed structure in relation both to the structures on the north of the craton, and to the system of arches and troughs of the Turukhansk–Noril'sk dislocation zone which link the two. The Yenisei–Khatanga Trough undoubtedly forms at the same time a structure related to the Taimyr–Severnaya Zamlya Uplift which evolved quite actively at that epoch. This justifies calling that trough cis-Taimyr Trough or Taimyr Foredeep. In relation to the Siberian Craton its position is pericratonal. The Early Jurassic transgression, which initiated the evolution of the trough, at the same time advanced from both east and west, reaching its maximum in the **Late Jurassic.** It was followed by regression at the end of this epoch. A new cycle of trough evolution began by transgression in the **Berriassian–Valanginian,** which was soon followed by regression with a maximum in the **Aptian–Albian.** Conditions were instable during the **Late Cretaceous.** Rare ingressions from the west alternated with prolonged periods of emergence and predominance of limnic and fluvial-plain environment. The Cenozoic history of the trough will be considered later.

Formation of the Verkhoyansk pericratonal trough and the Vilyui Syneclise also occurred at the beginning of the stage considered, in the Late Triassic and Early Liassic. Formation of these subsidence zones is linked with the second main phase of evolution of the Verkhoyansk Geosyncline. The latter, in contrast to its Taimyr apophysis, again subsided after Middle Triassic upheavals. The broad Vilyui Syneclise, which opens into the Verkhoyansk Trough, is superposed unconformably (but with the same overall trend) on the more narrow Paleovilyui Aulacogen which has a complicated structure and, thus, is a classic example of degeneration of an aulacogen into a syneclise. From the Vilyui Syneclise the Jurassic subsidence zone continued in the west to the lower reaches of the Angara. It formed the shallow Angara–Vilyui Trough which existed only from the end of the Middle Liassic till the beginning of the Dogger.

All these subsidence zones were filled with terrigenous, sandy-clayey, neritic, paralic and continental formations. These were the products of erosion from an extensive dry land which included the whole northern half of the craton and a large area in the south.

The Verkhoyansk pericratonal trough became a foredeep at the beginning of the Cretaceous, due to orogeny in the Verkhoyansk Range. Accumulation of marine sediments was from the end of the **Hauterivian** or the beginning of the **Barremian** followed by the accumulation of continental sediments, since subsidence lagged behind the supply of clastic material. It led, as in the Yenisei–Khatanga and Lena–Angara troughs, to establishment of a swampy lake-alluvial plain environment, and caused the accumulation of a thick coal-bearing formation. This formation constitutes the entire Cretaceous sequence in the Vilyui Syneclise.

Conditions in the southern half of the craton were during the Jurassic largely determined by considerable reactivation of uplifts in the Sayan–Baikalian framing and the Stanovoi Belt. Foredeeps were formed here as a reaction to these uplifts. These were the Sayan Foredeep (Kansk and Irkutsk depressions) and the Stanovoi Foredeep (Chul'man, Toka, and other depressions). They were filled with quite coarse (molassoid) limnic coal-bearing formations, of moderate thickness in the former foredeep and up to 1,500 m thick in the latter. The Stanovoi Belt, partly also the southern part of the Aldan Shield, became regions of very intense magmatic activity which continued also in the Cretaceous. It is remarkable that magmatic reactivation spread from west to east in the course of time. This paralleled the development of the orogenic process in the adjacent Mongol– Okhotsk geosynclinal system, which manifested itself in the west during the Jurassic, in the central part of the Stanovoi Belt during the Jurassic and Early Cretaceous, and in the east at the verge of Early and Late Cretaceous (GLUKHOVSKY et al., 1972).

Tectonic activity in general decreased during the second half and especially at the end of the Cretaceous. Denudational leveling of the surface of the entire Siberian Craton, including the Stanovoi Belt, took place. From the beginning of the Cenozoic the craton underwent general slow uplift. Thin layers of continental sediments accumulated only at its periphery, in particular along the Verkhoyansk and Sayan ranges, here and there in the foredeeps. Transgression of the sea occurred during the **Paleogene** only in the Yenisei–Khatanga Trough. This continued until the Middle Oligocene. Renewed transgression occurred in this trough at the **end of the Pliocene** and the **Quaternary.** Small portions inside the craton, in particular along the boundary between the Anabar–Olenek Anteclise and the Vilyui Syneclise, underwent insignificant relative subsidence accompanied by accumulation of thin layers of limno-fluvial sediments. Maximum uplift during recent times occurred in the northern part of the Tunguska Syneclise, which previously had experienced the deepest subsidence. The Putorana Plateau was formed there. It has an altitude attaining 1,500 to 1,700 m, and exhibits erosional dissection of the centrifugal type. Such a relief inversion is characteristic of all trap syneclises on Earth. It must obviously be due to density reduction of the upper mantle taking into consideration the thinning of the crust, a thick pile of basaltic lavas on top and the absence of isostatic anomaly, or possible basaltic underplating.

4.5. Some conclusions on the history and structure of the Siberian Craton

We state once more that the Siberian Craton is the largest relict of the regions in which the continental crust was formed mainly at the end of the Archaean and definitely 2,000 m.y. B.P. Its long history demonstrates convincingly the interrelation between evolution of the craton and development of the adjacent mobile belts (geosynclines, orogens).

Subsidence dominated in the craton surroundings at the earliest stage, i. e., during most of the Riphean (between 1,600 and 850 m.y. B.P.). At that time the craton extended beyond its present boundaries in the north and east. The craton proper also underwent subsidence, increasing towards the periphery. It formed a zone in which cyclically alternating terrigenous, quartzose-sandy and carbonaceous, limestone-dolomite shelf formations accumulated.

The situation changed radically at the beginning of the Baikalian orogeny in the west and southwest of the craton framing. A system of foredeeps was formed there, which became filled with red continental molasse of the Upper Riphean and Vendian. Uplift in the northeasterly Baikalian direction occurred at the same time in the central part of the craton. Elsewhere, far from the orogens, accumulation of terrigenous-carbonaceous sediments of the same type as before continued.

The southwestern and northwestern parts of the craton subsided mainly in the Cambrian, Ordovician and Silurian, during the Caledonian stage. The former part became a zone of downwarping, connected with recurrent Baikalide orogen, while the other part became a pericratonal trough linked with the Taimyr branch of the Verkhoyansk

Geosyncline and possibly with the northern part of the Ural–Siberian geosynclinal region. The broad Anabar–Aldan zone of, at first relative (Cambrian) and then absolute, uplift of northwesterly orientation appeared at that time in the platform east. The relatively quiet evolution of events during this stage was twice interrupted by general upheaval and (or) regression occurring at first in the Middle Cambrian and then at the end of the Ordovician. It reflected respectively the Salairian and Early Caledonian reactivation of uplifts in the western and southern surroundings of the craton.

Reactivation during the Devonian was pronounced also inside the craton proper, except for its northern periphery. It caused reworking of its structural pattern and cessation of subsidence in the region of the Angara Syneclise. Subsidence became relatively stronger in the extreme northwest. The Vilyui Aulacogen was formed; littoral-marine and lagoonal-continental red molassoid and evaporite formations accumulated, basaltic, trachybasaltic, and kimberlite magmatism and folding occurred in the Angara–Lena Trough in the northwestern frame of the Vilyui Aulacogen. Deformations of a northeasterly direction resumed at that time at least in the southern half of the craton.

This Devonian reactivation marks the end of the Caledonian stage of craton evolution and the beginning of the Hercynian stage. The latter was characterized by concentrated subsidence in the west and northwest, in the region of the Tunguska Syneclise. The latter was closely linked in its evolution with the peculiar downwarping zone of Southern Taimyr and probably with the northern continuation of the Ob'–Zaisan Geosyncline. The Hercynian stage ended in the second half of the Triassic. In fact it "absorbed" also the Early Cimmerian stage of global tectogenesis.

The sequence of evolution of this "double" Hercynian–Early Cimmerian stage includes transitional (from the Caledonian stage) Devonian continental-lagoonal-marine red beds and evaporites, a marine carbonaceous-terrigenous formation of the Late Devonian and Early Carboniferous, a limnic coal-bearing formation of the Middle and Upper Carboniferous and Permian, and a trap formation of the Upper Permian and Lower and Middle Triassic. The Siberian Craton was situated in an arid belt also during the first half of the Hercynian–Early Cimmerian stage, as during the Baikalian and Caledonian stages. It was shifted in temperate humide conditions after the Middle Carboniferous, and remained in this position during most of the Mesozoic and Cenozoic.

The last stage of craton evolution includes the Jurassic–Quaternary period and corresponds to the Late Cimmerian and Alpine stages of global tectochronology. This stage is characterized by a high position of the inner regions of the craton, except for the Angara–Vilyui zone of weak and moderate (near the periphery) downwarping, which cuts diagonally across the craton. The principal subsidence during this period occurred at the craton edges. It was due to epi-geosynclinal orogeny of the Verkhoyansk and Mongol–Okhotsk systems. It was also connected with recurrent orogeny at the western and southern craton framing and, to a lesser extent, in the Taimyr. Orogeny in the Stanovoi Belt and partly in the Aldan Shield was particularly important in the Late Mesozoic. This was linked with the primary orogeny of the Mongol–Okhotsk system, but by itself represented recurrent orogeny.

The general, final uplift of the Siberian Craton, contrary to its initial almost complete subsidence in the Riphean, is correctly linked by SPIZHARSKY (1958) with complete decay of the surrounding geosynclines. These geosynclines at first behaved "aggressively" toward the Siberian Craton and separated considerable areas from it in the southwest, and then in the north and east. Then the old craton grew through the addition of young platforms after the end of the Paleozoic, first in the west, later in the north, and finally in the east. Least quiet was the southern framing of the craton, which was repeatedly involved in orogenic reactivation.

One of the most important features of the Siberian Craton is also the repeated manifestation of magmatism (trapean, trachybasalt, ultrabasic-alkaline, and kimberlite). This was first demonstrated in its full extent by V. L. MASAITIS (1969, 1977). Outbursts of this magmatism were regularly associated with the boundaries between tectonic stages. This began with the Grenvillian and continued in the Baikalian, Caledonian, Hercynian–Early Cimmerian, and finally the Late Cimmerian and Alpine stages. The regions of ultrabasic-alkaline and kimberlite magmatism were generally associated with uplift zones; regions of trapean and partly of trachybasalt magmatism were associated with subsidence zones. Moreover, they migrated upon the craton area from stage to stage, naturally gravitating to the surrounding zones of maximum tectonic activity. The actual location of volcanic centers and intrusive bodies was controlled by zones of deep faults, characterized by high permeability possibly linked with high degree of consolidation and thus high brittleness of its basement[12].

---

[12] Not considering the period of establishment of the cratonic regime in the Middle and beginning of the Late Proterozoic.

5. Taimyr–Severnaya Zemlya fold region

5.1 General features of the relief and main views on tectonics of the region

The Taimyr–Severnaya Zemlya region lies in the extreme north of Central Asia. Until now it did not occupy a definite place in the tectonic schemes of the Soviet Union. This is partly due to its still inadequate state of knowledge; partly, however, this is due to the particular features of the evolution of this region. Its principal structural zones can be identified quite clearly (Fig. 36). There are, however, disputes on assessing the tectonic nature of its individual parts and of the whole structural ensemble which is distinguished by certain complications.

The **Kara Massif** is located in the center of the ensemble. This is a large uplift in the recent structure, and is composed mainly of crystalline Precambrian rocks. The massif includes the northern part of the Taimyr Peninsula and southeastern part of the Severnaya Zemlya Archipelago (Bolshevik Island and the eastern half of October Revolution Island). It also includes the southeastern corner of the Kara Sea area which it borders. The massif has a crescent-shaped form, being convex towards southeast. The inner structure is characterized by a predominance of northeasterly trends which are nonconformal in relation to the ENE trend of the Paleozoic deformations of the southern limb of the Kara Uplift.

The northwestern part of the Severnaya Zemlya Archipelago (Pioneer and Komsomolets islands, as well as the western part of October Revolution Island) form the second structural zone of the region. This is the **Northern Severnaya Zemlya zone** which consists of moderately folded sediments of the Upper Proterozoic, and of the Lower and Middle Paleozoic up to and including the Devonian. The trend of the folds varies from northwesterly to NNE; they envelop the northern extremity of the Kara Massif.

The Kara Massif is in the south directly framed by the **Central Taimyr zone** in which Vendian, Lower and Middle Paleozoic sediments are developed. This zone is located symmetrically with respect to the Northern Severnaya Zemlya zone. It differs from the latter in the composition of the sediments and in the greater extent and regularity of its deformations. In the south this zone gives way to a wider strip of thick, also folded Upper Paleozoic and Triassic formations. This is the **Southern Taimyr zone** whose relief manifests itself in the Byrranga Ridge which attains a height of more than 1,500 m. A wide swampy lowland extends to the south of the Byrranga Mountains. In the west it includes the basins of the Pyasina and Taimyra rivers which flow into the Kara Sea. In the east it includes the Khatanga River whose estuary forms the Khatanga Inlet of the Laptev Sea. This lowland separates the Taimyr Mountains from the Central Siberian Tableland. From

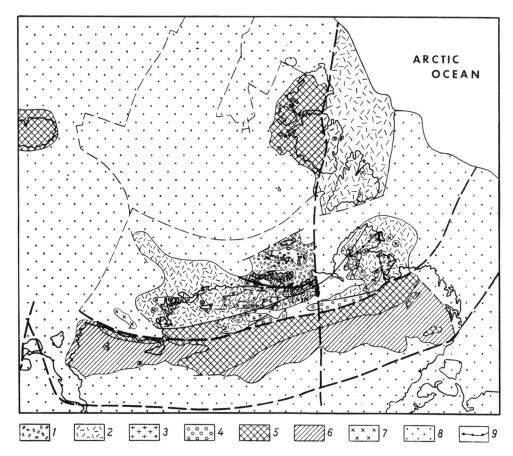

Fig. 36.   Tectonic scheme of the Taimyr–Severnaya Zemlya region (after PoGREBITSKY, 1971, modified according to the data of M. I. VOLOBUYEV, personal commun.). 1 = Early Precambrian metamorphic complex; 2 = Late Precambrian metamorphic complex; 3 = Late Precambrian granitoids; 4 = Late Riphean–Vendian (?) volcanic molasse; 5 = Early and Middle Paleozoic deformed cover of Taimyr, Severnaya Zemlya and Novaya Zemlya; 6 = Late Paleozoic–Triassic fold complex of Taimyr; 7 = Late Paleozoic–Early Mesozoic granitoids; 8 = Post–Triassic sedimentary cover; 9 = ophiolite sutures.

the tectonic aspect it separates the Taimyr fold structure (anticlinal uplift in the neotectonic plan) from the ancient Siberian Craton and forms the deep Meso-Cenozoic **Yenisei–Khatanga Trough.** This is the northeastern branch of the West Siberian Megasyneclise.

These are the principal structural zones of the region. The views on their tectonic nature and their relationships underwent the following evolution. N. N. URVANTSEV, who in fact was the first to investigate the Taimyr, at the beginning of the thirties expressed the opinion that the Taimyr represents a fragment of the central zone and southern limb of a Hercynian geosynclinal fold structure. He suggested that old formations of a central metamorphic zone (Northern Taimyr, see below) had to a considerable extent been thrust over a peripheral zone which evolved in the Paleozoic. He also pointed to overthrusts in the latter. These, however, were not confirmed by later researchers (this question, however, should not be considered to be solved definitely–see below). URVANTSEV's predictions regarding the Late Paleozoic age of the bimicaceous granites of the Taimyr were partially confirmed through radiometric dating.

A basically different concept of the structural evolution of the Taimyr and Severnaya Zemlya formed in the fifties. It was suggested by VAKAR & EGIAZAROV (1965), that the structure of the Taimyr evolved as a result of Late Proterozoic, Caledonian, and Hercynian geosynclinal tectogenesis. The fold structure expanded off the central zone composed of metamorphic rocks and granitoids.

This structural pattern was, however, refuted later by POGREBITSKY (1971). He drew attention to the non-correspondence of the inner structure of the Precambrian massif to that of the Paleozoic (and Upper Riphean (?)–Vendian) framing, to the non-geosynclinal nature of the Lower and Middle Paleozoic sedimentary complex, and to the absence of any Caledonian unconformity. POGREBITSKY (1971) concluded that the Taimyr Trough originated in the Carboniferous on a cratonic basement like that of the ancient Siberian Craton. He assumed that the formation of a fold system on its place has been the result of deformations ending only at the beginning of the Jurassic. He believes that reactivation of the Taimyr region was linked with evolution of the Verkhoyansk Geosyncline and of the Pacific Belt as a whole. The tectonic regime in the Taimyr during the Late Paleozoic and Early Triassic is by him defined as parageosynclinal.

Tectonic maps of the USSR show the Taimyr and Severnaya Zemlya sometimes as Caledonides and Hercynides, sometimes only as Hercynides, sometimes as Baikalides and Hercynides. This reflects the differences in existing views.

Let us now consider briefly the features of the separate zones. We shall then, at the end of the chapter, deal with the tectonic nature of the region as a whole.

5.2 The Kara Massif

The Kara Massif comprises the northern Taimyr Coast and the southeastern part of the Severnaya Zemlya Archipelago. It is 1,200 km long and up to 450 km wide; it consists mainly of a complex of rocks regionally metamorphosed in the amphibolite and greenschist facies. The Precambrian age of these rocks has been reliably determined by the unconformity with which Vendian–Lower Cambrian (the latter being characterized by fauna) sediments overlie them.

There exist two quite contrasting opinions about the possible subdivision and more precise determination of the age of the Precambrian complex. One view, expressed by Krasnoyarsk geologists (ZABIYAKA, 1971; MAKHLAEV & KOROBOVA, 1972), is that this complex represents a single and continuous sequence of sedimentary-volcanic formations. These range from flyschoid, mainly silty-lutitic, via flysch, sometimes substituted by basic and intermediate volcanics, to volcano-carbonaceous with acid effusives. The thickness of the complex is estimated at 12 to 13 km. Metaterrigenous rocks predominate in it. Differences in the degree of metamorphism are due to metamorphic zonality. The granitoids enclosed by this complex are also considered to be Precambrian, since their pebbles are found in basal conglomerates of the Vendian. They, as well as the whole complex, are assumed to be of Upper Proterozoic age. The deformations, metamorphism, and granitization therefore must be referred to the Baikalian epoch of tectogenesis. The Kara Massif (anticlinorium according to the authors mentioned) is considered to be the analogue of the other Baikalian structures surrounding the Siberian Craton (Yenisei Range, Eastern Sayan, etc.).

Another group of researchers (RAVICH & POGREBITSKY, 1965; POGREBITSKY, 1971) identify in the Precambrian complex of the Taimyr several independent stratigraphic and structural subdivisions separated by unconformities and discontinuities in metamorphism. These researchers state that the unconformities, especially between the oldest subdivisions, appear only locally; rocks of the amphibolite facies of metamorphism have at certain points been converted into greenschist facies, due to diaphthoresis. Different strata therefore contain different structural elements belonging to different stages of tectonic evolution.

The oldest series identified by M. G. RAVICH & YU. E. POGREBITSKY is the **Kara series** of plagiogneisses (almandine-biotite and almandine-hornblende) and crystalline schists (almandine-biotite-quartzose), with members of amphibolites. They form the northwestern Taimyr and the northwestern part of the Severnaya Zemlya Archipelago from October Revolution Island onward, as well as a small independent block near Cape Chelyuskin. The structure of the series is characterized by an abundance of gneiss domes. The foliation of the gneisses is oriented latitudinally or in the northeasterly direction. The age of the series is conventionally considered to be Archaean. Direct data on the age of the Taimyr metamorphics were not available until recent time. Radiometric determination by means of the K/Ar method pointed merely to the end of the Paleozoic and beginning of the Mesozoic (275 to 225 m.y. B.P.).

The next structural-stratigraphic subdivision of the Precambrian is considered to be a series (Lenivaya series) of basic volcanics metamorphosed in the greenschist facies, terrigenous rocks (lutites, arenites), and some carbonaceous rocks, having a thickness of up to 5 or 7 km. POGREBITSKY (1971) believes that this series fills a geosynclinal trough of NNE trend, extending through Central Taimyr to Bolshevik Island. It is deformed into a system of isoclinal folds which as a whole represent a structure of the type of a synclinorium. POGREBITSKY assumes that this geosynclinal system (bounded in the west by the Archaean "Kara Platform" and in the east by the Chelyuskin median massif) extends to the southeast, cross-cutting the present Yenisei–Khatanga Trough up to the region of the present Turukhansk–Noril'sk dislocations. This geosynclinal system also includes a higher complex of slightly less metamorphosed terrigenous (sandy-clayey) and volcanic rocks. It is significant that the latter form a bimodal formation consisting of basic (spilites) to intermediate (porphyrites, albitophyres) and acid (felsites) volcanics.

The two stratigraphic subdivisions are defined by YU. E. POGREBITSKY (1971) as lower and middle complexes of the Proterozoic and are conventionally related to the Middle Proterozoic and the Lower to Middle Riphean. Their analogues exist also in the surroundings of the Central Taimyr Geosyncline, but in much lesser thicknesses.

The uppermost part of the Precambrian complex of the Taimyr differs considerably from the remaining sequence. It represents a slightly metamorphosed and relatively little deformed terrigenous-carbonaceous formation. It begins with quite thick layers of conglomerates which are then superseded by variegated schists, gritstones, and sandstones, followed by carbonaceous rocks containing stromatolites, oncholites, and catagraphies of the (Middle?) Riphean[1]. This part of the sequence is 3.7 km thick. Sediments of this age are absent in the east. POGREBITSKY (1971) relates this uppermost series to the Upper Riphean.

The metamorphic complex of the Taimyr encloses many quite large bodies of granitoids. Two principal generations can be distinguished in them. The older one comprises synkinematic palingenetic-anatectic porphyroblastic granitoids. The younger formation consists of bimicaceous or, according to RAVICH (1958), muscovitized granites. The Precambrian age of the porphyroblastic granitoids is not in doubt. Most researchers also relate the bimicaceous granites to the Precambrian. However, POGREBITSKY (1971), following N. N. URVANTSEV and DAMINOVA (1960), considers them to belong to the Late Paleozoic, in full agreement with radiometric dating.

Formation of the granitoids was preceded by gabbro intrusions. These were later converted into drusitic orthoamphibolites. Sills and dykes of diabases and gabbro-diabases intruded after the granites. These were also altered into amphibolites.

Recent investigations by a group of investigators of Moscow University, led by M. I. VOLOBUYEV, partly solved the contradictions, cited above, as regards the estimated age of the Precambrian Taimyr metamorphics. These investigations confirmed the Precambrian age of the crystalline schists and gneisses of the northwestern Taimyr Coast, as well as of the granites and granitogneisses intruding them. The age of the zircons from the crystalline schists at the mouth of the Taimyra River was found to be 2,200 ± 100 m.y. (U-Pb-Th method). The age of the granites was determined as 2,400 ± 350 m.y. (on orthite, Pb/Pb and U/Pb ratios). This permits the assumption that the complex considered belongs to the Lowest Proterozoic or Upper Archaean. Also confirmed was the Early Precambrian age (approx. 1,600 m.y. according to an oral communication by YU. E. POGREBITSKY) of the rocks of the Chelyuskin Block.

The remainder of the Kara Massif in fact belongs to the Riphean. This is confirmed by the Baikalian age of the whole structure. M. I. VOLOBUYEV found that the region of development of Riphean rocks may be subdivided into eu- and miogeosynclinal zones separated by a fault. The miogeosyncline surrounds the Kara Massif, while the eugeosyncline extends from it to the southeast. The eugeosynclinal zone is associated not only with the already mentioned basic volcanics, but also, according to ZALALAYEV & BEZZUBTSEV (1975), with the Chelyuskin ultrabasic belt discovered by them. The latter consists of highly deformed serpentinized harzburgites. Another belt formed by serpentinite melange was found by M. I. VOLOBUYEV (oral commun.) at the western periphery of the Chelyuskin Block.

The miogeosynclinal zone consists of a schistose and flyschoid formations, less deformed and metamorphosed than the eugeosynclinal strata. The metamorphism exhibits zonal character and is linked with bodies of allochthonous granitoids having an age of 1,100 ± 50 m.y. (U-Pb method). Granitoids of the same age outcrop in the cores of granitogneiss domes in the eugeosynclinal zone. This shows that the enclosing rocks mainly belong to the Middle Riphean. It should be remembered that the Middle Riphean of the Yenisei Ridge is similarly represented and also underwent Grenvillian deformation. However, the final epoch of the geosynclinal stage proper of evolution of the Kara

---

[1] It has recently been found that the Middle Riphean remains of vegetation belong to a lower stratum and not to the one considered.

Massif was Early Baikalian. This follows from the presence, in Eastern Taimyr, of plagiogneissogranites having an age of 850 ± 50 m.y. (M. I. VOLOBUYEV).

Unconformably overlying the metamorphics and conformably underlying the Cambrian strata coarse detrital as well as shelf-carbonaceous formations are developed mainly at the southern periphery of the Kara Massif. According to recent information they represent Baikalian molasse and belong, like those of the Yenisei Ridge, to the Vendian[2]. It is possible that they belong also to the upper part of the Upper Riphean. Acid volcanics are associated with them in the Chelyuskin Peninsula on the eastern coast of Taimyr. This molasse is thus volcanic there. It fills a depression superposed on the eugeosynclinal zone.

Various, still fragmentary data indicate that the inner structure of the Precambrian complex of the Taimyr is very complicated. Nappes are possibly present there. Noteworthy in particular is an inlier of highly metamorphosed rocks, bounded by tectonic contacts, in the middle of a Riphean field south of the Northern Taimyr crystalline massif in the Shrenk–Mamontov region.

5.3 The Northern Severnaya Zemlya zone

A quite sharp transition from the metamorphosed complex forming the Kara Massif to the region of development of the Upper Riphean–Devonian sediments forming this zone occurs in a strip extending in a near-longitudinal direction through October Revolution Island and turning into the latitudinal direction in the southeastern part of Komsomolets Island. The type of transition indicates that a boundary flexure-fault zone exists there. This is underlined by the intensity of fold deformations and the presence of small hypabyssal intrusions of acid composition.

The base of the sequence of the Northern Severnaya Zemlya zone is a slightly metamorphosed, tuffogenous-terrigenous-carbonaceous formation identical with that in the Kara Massif. It is unconformably overlain by a conglomeratic-carbonaceous formation of Vendian age (0.5 km) and a flyschoid terrigenous formation of the Cambrian (about 3 km). The Ordovician also overlies the Cambrian unconformably at certain points. It is represented by a lagoonal gypsum-bearing, variegated, carbonaceous formation having a thickness of up to 2.5 km. A larger unconformity separates this formation from a shelf carbonaceous Silurian formation having a thickness of up to 1.7 km. The unconformity between the Devonian and the Silurian is much inferior to the pre-Silurian one. Transition at some points is continuous. The Devonian represents a lagoonal-marine gypsum-bearing, red, clastic-carbonaceous formation having a thickness of 1.6 km.

The Upper Riphean–Ordovician sediments are deformed into steep linear folds. Their trend is meridional in the south, with a virgation further north. The western branch turns to the northwest, while the eastern branch extends in the NNE direction. Faults are fairly rare; there are, however, some overthrusts. The Ordovician sediments with gypsum interlayers exhibit intensive disharmonic folding. The Silurian and Devonian sediments are much less deformed than the Ordovician–Upper Riphean ones. Folding is of the

---

[2] This was confirmed by K/Ar dating of the phyllites of the Stanovaya series (650 m.y. B.P.).

ejective type: narrow and steep anticlines are separated by wide and rounded brachysynclines. The folds have no definite orientation. However, a northeasterly and latitudinal
trend predominates in the whole. The faults are of normal type and also have a predominantly latitudinal trend. Deformation becomes generally weaker toward the west.

The facial character of the sediments, their thickness, and the style of deformation
indicate a gradual reduction of the "geosynclinality" toward the top of the sequence. The
geosynclinal (miogeosynclinal) complex can by convention be bounded on top by the
Ordovician. This permits us to consider the Upper Riphean–Ordovician fold complex as
Caledonian. However, the overlying Silurian and Devonian sediments do not exhibit the
characteristics of being orogenic, but rather form a platform cover. The Caledonian
geosynclinal evolution cannot therefore be considered to be normally complete.

5.4 The Central Taimyr zone

This zone is composed of the same range of strata as the Northern Severnaya Zemlya
zone. However, it extends on the other, southern side of the Kara Massif in a generally
ENE direction. The junction of the Precambrian massif with its Riphean–Paleozoic
framing resembles that noted for Severnaya Zemlya. It manifests itself as a strip of
flexures and faults. Predominant among the latter are upthrusts–overthrusts directed
toward the south. It is quite possible that overthrusts may occur here in a considerable
scale, as assumed by N. N. URVANTSEV.

The sequence of Cambrian–Devonian sediments of the Central Taimyr zone differs
from the sequence of Severnaya Zemlya in being continuous and exclusively marine. The
facies are not only of neritic, but also of comparatively bathyal type. The zone considered
is now 30 to 50 km wide; with a surprising constancy it was from the Ordovician onward
(i. e., for a period of about 150 m.y.) divided into two facial regions (zones) running
conformal with its trend. The northern zone represents a deep trough having a minimum
depth of about 750 m (as estimated by POGREBITSKY, 1971). It contains accumulations of
black graptolite schists with interlayers of dark bituminous limestones with a fauna of
pteropods. Its overall thickness is 1.5 to 2.0 km. The southern zone represents a shelf
with accumulation predominantly of limestones and dolomites, with an abundant fauna
of brachiopods, gastropods, bryozoa, corals, and other shallow-water bottom organ-
isms. The thickness of this carbonaceous formation is 4.2 to 4.7 km.

The Central Taimyr zone underwent uplift and partial emergence at the boundary
between the Devonian and the Carboniferous. This led to the formation of a thick
weathering crust with bauxites along a salient of meridional trend. It underwent uplift in
the south, toward the Anabar Massif, and subsidence in the north, in the direction of
Severnaya Zemlya. Uninterrupted accumulation of sediments continued in the west and
east. The Lower Carboniferous represents a shelf carbonaceous formation with minor
amounts of shales and gritstone-sandstones (the latter in the Namurian, only in the west).

The Central Taimyr zone is structurally characterized by an intense linear folding
complicated by overthrusts. This zone is overthrust on the Southern Taimyr zone, as the
Kara Massif is overthrust on the Central Taimyr zone. The amplitude of this overthrusting is not known but may be considerable. The Kara Massif was hardly part of a bathyal

depression during the Early and Middle Paleozoic. However, it is at present in direct contact with sediments of this depression.

5.5 The Southern Taimyr zone

This zone is much wider than the Central Taimyr zone. It extends for 100 to 150 km in its exposed part. In the south it is overlain by the Yenisei–Khatanga Depression and possibly underlies its entire northern limb. Its overall width is thus of the order of 200 km. The general structure of this zone is a deep trough. This trough is filled with a thick pile (about 5 km in the west and 8 km in the east) of sediments from the Middle Carboniferous up to and including the Lower Triassic. The Middle and Upper Carboniferous and the Lower Permian (up to and including the Artinskian stage) represents a marine-terrigenous flyschoid formation, mainly of gritstones and argillites with minor amounts of sandstone. The type of fauna (brachiopods, pelecypods) and the frequent presence of remains of terrestrial flora indicate its shallow-water origin; it is more likely to be a lower marine molasse than flysch. It is generally overlain conformably by a paralic coal-bearing formation. The latter by volume represents the main part of the filling of the Southern Taimyr Trough; it forms almost the entire Upper Permian. The sequence of the trough ends with a trap formation of the Upper Permian and Lower Triassic. Its thickness is up to 1 km in the west and up to 3 km in the center and east. In the east the traps are overlain by Middle and Upper Triassic molasse which is continental in the center, and paralic, coal-bearing in the east.

The sediment complex considered is in the present structure localized in two depressions. These are the smaller Western Taimyr and the larger Eastern Taimyr depressions. They are separated by the diagonal Taimyr Arch having a southwesterly trend. Lower and Middle Paleozoic sediments occur in this arch (more precisely, anticlinorium). These sediments belong to the southern shelf subzone of the Central Taimyr zone. The northern edges of the two depressions, which border this zone, exhibit linear folds. Ejective folds occur along the axes of the depressions, and brachymorph folds along their southern edges. The structure of the eastern depression is complicated by a system of straight latitudinal normal and wrench faults which probably join the fault paralleling the Kara Massif. The parts where traps are developed represent graben-synclines.

5.6 The Yenisei–Khatanga Trough

The Taimyr–Severnaya Zemlya fold region, which had been formed by the Middle Triassic, is separated from the ancient Siberian Craton by the Yenisei–Khatanga Trough. The latter extends from the lower reaches of the Yenisei to the Khatanga Inlet over a distance of some 900 km, and has a width of the order of 300 km. In the west, the trough wedges into the West Siberian Megasyneclise[3]; in the east, it is separated by the Anabar Saddle from the less deep Lena–Anabar Trough. The latter in its turn goes over into the

---

[3] The very flat Tanama–Yenisei Saddle lies on their boundary (SAPIR, 1971).

Verkhoyansk Foredeep. The Yenisei–Khatanga Trough is thus a link in the system of Mesocenozoic troughs bordering the Siberian Craton in the north and east.

The main part of the Yenisei–Khatanga Trough is superposed on the northern slope of the Siberian Craton. Only its northern edge (the Southern Taimyr Monocline) overlies the southern periphery of the Taimyr fold system. Evolution of the trough as a single structure began apparently already with the deposition of sediments of the Middle and Late Triassic, which cover practically the entire bottom of the trough and part of its southern limb. Subsidence increased sharply in the Jurassic and continued till the Middle Oligocene. It resumed thereafter only at the end of the Pliocene and in the Quaternary. The sediments consist of sand and clay. Lagoonal-continental, more sandy facies alternate with marine, more clayey ones. The thickness of this complex, which begins with the Jurassic, and thus the depth of the trough down to the base of the Jurassic sediments, is 6 to 9 km. The maximum depth occurs in the west, in the Yenisei Valley. It generally decreases in the direction of the Khatanga Inlet (down to 2 km). A zone of large arches runs along the axis of the trough (**Tanama, Rassokha, Balakhna**). In the west it approaches the southern edge, and in the east the northern edge. This bend is complicated by a fault. The arches are apparently of inversion origin: they were troughs in the Permian and Triassic, but evolved in their present form from the Jurassic onward. They divide the axial zone of the entire trough into two partial zones of subsidence consisting of separate depressions along the trend. The axial zone of the trough is separated from the edges by buried basement faults. The basement is also cut across by transverse faults, e. g., along the present Pyasina and Taimyra valleys. Many local uplifts are known in the trough; their amplitudes are tens or some hundreds of meters.

It has already been mentioned that the eastern continuation of the Yenisei–Khatanga Trough is the sublatitudinal **Lena–Anabar Trough.** In fact, however, this trough prolongs only the southern strip of partial troughs of the Yenisei–Khatanga zone. It is bounded in the north by the **Olenek branch** of the Verkhoyansk fold system, which becomes less pronounced in the west. This fold system extends from the Lena Delta to the Khatanga Inlet along the coast of the Laptev Sea, and includes the southeastern part of this sea. It has been suggested ("Tectonics of Yakutia", 1975), that the Olenek branch in its turn splits into two opposite the mouth of the Khatanga. One branch runs north to a junction with the Taimyr Mountains, bypassing the Kara Massif. The other branch runs west, ending on the Anabar–Khatanga Saddle. It is, however, possible that the zone of central arches of the Yenisei–Khatanga Trough just forms the continuation of this most westerly apophysis of the Verkhoyansk system.

5.7 Main stages of evolution

5.7.1 Early Precambrian stage

The presence of a complex of plagiogneisses and crystalline schists in the Precambrian massif of the Taimyr and Severnaya Zemlya, which belongs to the Lowest Proterozoic or the Upper Archaean, proves that this region may initially have formed part of the Siberian (Northern Asiatic) Craton. This union with the Siberian Craton probably continued up to and including the Middle Proterozoic, i. e., until some 1,700 to 1,600 m.y. B.P.

5.7.2 Late Precambrian stage

However, the region north of the axis of the present Yenisei–Khatanga Trough became fragmented and underwent distension at the beginning of the Late Proterozoic, in the Early or Middle Riphean. This caused isolation of the Kara microcontinent (median massif). A region with a crust of oceanic type appeared between it and the Siberian Craton. This has been recently documented by ultrabasics discovered in the Taimyr, as well as by basic volcanics of the lower Lenivaya series. This geosynclinal basin was later filled with a carbonaceous-terrigenous series (Middle Riphean) which had undergone deformations, regional metamorphism, and granite intrusion during the Grenvillian tectonomagmatic epoch. Subsidence resumed at the beginning of the Late Riphean; however, the volcanics now acquired a bimodal nature. This, together with the granitoids, may indicate the emergence of a volcanic arc.

Evolution of the Taimyr Geosyncline ended in the middle of the Late Riphean. Final deformations, metamorphism, and formation of plutons of plagiogneissogranites having an age of 850 m.y. took place. The Riphean evolution of the Taimyr (and of Severnaya Zemlya) as a whole greatly resembled the evolution of the Yenisei Ridge. It shows that the Taimyr and Severnaya Zemlya form the northern continuation of the Baikalian western framing of the Siberian Craton.

As in the Yenisei Ridge, the top of the Upper Riphean (?) and Vendian form a molasse (volcanic in the northeast). The general impression is that evolution of the Taimyr Geosyncline in the Riphean and Vendian was linked with the presence of a Benioff zone dipping northward underneath the Kara Massif. Calc-alkaline volcanism and plutonism was also connected with this zone.

The Vendian molasse in the Taimyr and Severnaya Zemlya is conformably overlain by the Cambrian.

5.7.3 Caledonian and Early Hercynian stage (Cambrian to Early Carboniferous)

The northern part of the region (Northern Severnaya Zemlya) evolved more actively during the Cambrian than the southern part. Sediments accumulated here in a thickness of 3.5 km on a continental margin or inside a deep intracontinental trough (i. e. in a pericratonal-miogeosynclinal or intracratonal environment). A more quiet, clearly cratonic environment prevailed in the south (Taimyr). The thickness of the sediments was only 600 to 800 m.

Evolution of the two zones on either side of the Kara Massif differed even more during the Ordovician. Subsidence in Severnaya Zemlya began to lag behind the accumulation of sediments, and normal marine conditions were superseded by shallow-water, lagoonal ones. Considerable folding took place here at the end of the Ordovician. It was accompanied by small intrusions of hypabyssal bodies of acid composition. Short-term upheaval at the **verge between the Ordovician and the Silurian** was followed by re-establishment of a shelf environment with the deposition of typically platform, shallow-water, carbonaceous formations. Renewed uplifting and smaller deformations occurred at the **boundary between the Silurian and the Devonian.** The environment of a coastal plain was then established here during the Devonian. This plain was at times flooded by the sea, and bordered in the south by dry land (Kara Massif).

An epicontinental sea extended southward to Central and Southern Taimyr in the Cambrian when carbonates accumulated. Differentiation took place in the Ordovician. A wide shelf gradually suffered upheaval to the south, towards the Siberian Craton and a deep depression formed to the north of it; clayey or limestone-clayey sediments accumulated in the latter. This deep-water basin existed up to and including the end of the Devonian. It probably was connected in the east with an analogous basin of the inner zones of the Verkhoyansk–Kolyma system, and most likely underwent shoaling in the west. Its junction with the Kara Massif is still very problematic. The latter, at least its northern part, must have supplied detrital material to the basin of the northwestern part of Severnaya Zemlya. This refers in particular to the quartzose sandstones in the Ordovician–Devonian sequence. However, no signs of erosion from the Kara Massif exist in the Central Taimyr bathyal trough. It is possible that the transitional zone between the region under erosion and the bathyal trough is now overlain by the overthrust of the Kara Massif on the Central Taimyr zone.

The sea was for a long time after the Devonian absent from the Severnaya Zemlya region. The Central Taimyr bathyal trough had ended its independent existence by the **beginning of the Carboniferous.** Considerable parts of this region were emergent for a short time at the verge between the Devonian and the Carboniferous; a bauxite weathering crust formed. Accumulation of carbonates on the Taimyr Shelf gradually gave way to an accumulation of terrigenous sediments during the Early Carboniferous. This indicates extension and uplifting of the Kara dry land (including Severnaya Zemlya).

5.7.4 Late Hercynian–Early Cimmerian stage (Middle Carboniferous to Late Triassic)

The zone of maximum subsidence moved to Southern Taimyr after the **Middle Carboniferous.** Subsidence itself intensified sharply. This was obviously linked with increased uplifting of the Kara Massif. Large amounts of detrital material were supplied by this massif to the Southern Taimyr Trough during the Late Paleozoic and Early Triassic. The massif itself underwent intense tectonic reactivation. The Precambrian basement was remobilized and laccoliths of subalkaline granitoids were formed.

The general similarity between the Upper Paleozoic sequence of Southern Taimyr and the sequence of the Tunguska Syneclise of the Siberian Craton is quite obvious (despite the larger role of marine facies and more important thickness of sediments). It leads to the conclusion that both regions belonged to the same basin in the **Late Paleozoic** and at the beginning of the Mesozoic. The depth of this basin increased toward the Taimyr, and from west to east. The structure of this basin was apparently complicated by a synsedimentary evolving uplift at the site of the Tareya Arch and the southern boundary of the present Yenisei–Khatanga Trough. These uplifts may be responsible for some differences in the fauna of Taimyr and of the Tunguska Syneclise.

Ingressions of the sea, the most important of which were the Late Carboniferous, Kazanian and Indian (Seissian) ones, invariably invaded the Southern Taimyr Basin from the east, i. e., from the Verkhoyansk Geosyncline during this stage. Each following ingression extended less than the preceding one: the first encompassed all Taimyr and reached the western edge of the Tunguska Syneclise, the second stopped in front of the Tareya Arch but encompassed the northern part of the Tunguska Syneclise, and the third was practically restricted to the Khatanga Basin.

Taimyr, like the Tunguska Syneclise, was involved in important trap magmatism at the **end of the Permian** and especially during the Indian time. This magmatism also occurred in Severnaya Zemlya to a lesser degree and exclusively in intrusive form. The thickness of the basic volcanics attained 3 km in southwestern Taimyr.

Taimyr underwent rather strong folding and overthrust deformation during the second half of the **Early Triassic.** This followed a prolonged period of quiescent evolution which began after the Vendian. The Taimyr fold system was also formed mainly by these deformations, and accreted the Kara Massif. It is possible that the much weaker deformations of the Paleozoic sedimentary cover of Severnaya Zemlya, which extends up to the Devonian, also occurred at that time or even earlier, already in the Late Paleozoic.

The final evolutionary stage of the Taimyr fold system belongs to the **Middle and Late Triassic.** Sediments of this age sometimes overlie formations (from up to and including Indian strata) quite unconformably, sometimes supersede them almost conformably in residual troughs. They represent a quite typical upper molasse. Its accumulation ended in tangential fold-fault deformations. These, however, were much less intense than at the end of the Early Triassic.

Taimyr and Severnaya Zemlya stood out as a single arch-like uplift by the **end of the Triassic,** when formation of the Yenisei–Khatanga Trough began south of them. Indirect data indicate that the initiation of the trough may have been preceded by formation of a rift. These data refer to the shape and depth of the trough, as well as the nature of the magmatic activity, i. e., the presence of alkaline and ultrabasic-alkaline ring intrusions both at the southern edge of the trough (Maimecha–Kotui complex) and north of it, in Taimyr (subalkaline and alkaline granites, syenites, nepheline syenites). The rift may have been originated along the joint between the ancient basement of the Siberian Craton and the younger basement of Taimyr. In the west it may have joined (triple junction) with the principal rift system of the basement of the West Siberian Megasyneclise, which has a meridional orientation.

5.7.5 Alpine stage (Jurassic to Quaternary)

This stage is characterized by a predominant tendency of upheaval of the Taimyr–Severnaya Zemlya Welt and, until the Middle Oligocene, uninterrupted subsidence of the Yenisei–Khatanga Trough. The trough became filled with strata of sandy-clayey sediments. These strata were thicker, predominantly sandy and continental, at the beginning of the Jurassic and Cretaceous. They gave way to lagoonal-continental ones in the Middle Jurassic and Cretaceous, and then to purely marine, predominantly clayey sediments in the Late Jurassic, Late Cretaceous, and Paleogene. The reduced thickness of the strata of clayey sediments may indicate that the rate of their accumulation was less than the rate of trough subsidence. The growth of partial uplifts, which had appeared already at the beginning of the Jurassic, decreased during these periods. On the other hand, intense growth of these structures coincided with an increased supply of detrital material and with shoaling of the trough (SAPIR, 1971).

The sea left the Yenisei–Khatanga Trough for a long period in the Middle Oligocene. This trough was during the neotectonic stage involved in the general uplift of Central Siberia; it was subject to renewed ingression of the sea only from the end of the Pliocene to the Middle of the Pleistocene. This was accompanied by the deposition of sandy-clayey sediments in strata of moderate thickness (up to 250 m). Ingression took place from the Laptev Sea which had formed by that time. The superposed depression of the Kara Sea formed approximately at the same time in the west. The Shokal'sky Strait graben (rift) between the October Revolution and Bolshevik islands also formed in the Neogene. The subsidence of these regions had its opposite in the uplift of the central parts of the islands of the archipelago, and of the Byrranga Range in Southern Taimyr. The inversional Putorana Plateau rose at the same time on the other side of the Yenisei–Khatanga Trough, in the northern part of the Siberian Craton. The neotectonic stage was also the time of most intense growth of the internal structures (arches, local uplifts) which complicate the Yenisei–Khatanga Trough.

5.8 Some deductions

We must now analyse the structure and evolution of the Taimyr–Severnaya Zemlya region as a whole, explain its tectonics, and consider possible lateral connections.

The Late Precambrian history of this region, despite being not yet completely clear, is obviously quite different from that of the old Siberian Craton. They differ mainly in the fact that cratonic conditions were established in Taimyr and Severnaya Zemlya not earlier than about 1,000–800 m.y. B.P. (but not later than during the Vendian in Taimyr, and during the Ordovician in Severnaya Zemlya). Geosynclinal conditions in Severnaya Zemlya successively degraded from the Riphean to the Vendian, and from the Cambrian to the Ordovician, through folding, but without any noticeable orogeny.

Taimyr evolved under cratonic conditions (epi-Baikalian Craton) during the Early and Middle Paleozoic, till the Middle Carboniferous, as did Severnaya Zemlya from the Ordovician onward. Only the Central Taimyr bathyal trough slightly disturbs the general pattern, which represented a structure of the aulacogen type, opening out into a geosyncline in the east (triple junction!). Pronounced activation of movements began at the end of the Early and beginning of the Middle Carboniferous. This coincided precisely in time with the turning point in the evolution of the Verkhoyansk–Chukotka region and undoubtedly was an echo of events there. Without a geosynclinal preparatory stage the epi-Baikalian Taimyr–Severnaya Zemlya Platform entered the orogenic stage of evolution. Thick strata of Upper Paleozoic molasse and Lower Triassic traps accumulated in the South Taimyr Trough.

The South Taimyr Trough and the Kara–Novaya Zemlya Uplift occupy, in relation to the ancient Siberian Craton and adjacent geosynclines, a position which to some extent resembles that of the Donets Basin and the Ukrainian Shield in the structure of the Eastern European Craton, and also of the Wichita zone and the Bend Anteclise in the structure of the North American Platform. The South Taimyr Trough contains formations which are very similar to the coal-bearing formations of the Donets Basin. The main difference lies in the fact that the Ukrainian Shield and the basement of the Bend Anteclise were formed in the Early Precambrian and were not involved in any noticeable Hercynian reactivation.

As stated, Southern Taimyr underwent its main deformations in the Triassic; they were more pronounced at the end of the Early Triassic and ended at the verge between the Triassic and the Jurassic. Southern Taimyr should be regarded as a Late Hercynian structure in relation to the first epoch, but as an Early Cimmerian one in relation to the second epoch. Present information indicates that the latter is more likely (POGREBITSKY, 1971).

The Taimyr–Severnaya Zemlya region returned to the cratonic regime of evolution at the beginning of the Jurassic. The Yenisei–Khatanga Trough is the link uniting the West Siberian Megasyneclise with the Verkhoyansk Foredeep, but is closer to the former in its evolution and structure. Most of the clastic material came to the Yenisei–Khatanga Trough from the Siberian Craton and not from the Kara Swell. The latter was at times, especially in the Early Cretaceous, itself a region of sedimentation. The Kara Swell underwent maximum upheaval in the Jurassic and Cenozoic, till the Late Pliocene.

Geophysical data indicate that the Taimyr fold zone dies out in the west between the Yenisei Gulf and the Gydan Inlet. It must, however, be separated only by a comparatively narrow bridge from the northern continuation of the Ob'–Zaisan Late Hercynian fold system whose trend is in the direction of Yamal. The latter either retained the character of a true geosynclinal fold system (SURKOV & ZHERO, 1981), or degenerated into an aulacogen within the epi-Baikalian Platform (SHABLINSKAYA, 1979). Taimyr joins the northwestern branch of the Verkhoyansk geosynclinal system in the east. However, the Upper Paleozoic and Triassic sequence is here of miogeosynclinal type; it retains this character also in the Jurassic. The age of the main folding becomes younger – Late Jurassic and Early Cretaceous.

The Paleozoic fold zone of northern Severnaya Zemlya is in the west probably linked with the northern end of the Novaya Zemlya system. The facial character of the Silurian and younger strata here differs from that in Severnaya Zemlya, but their attitude near Cape Zhelanya becomes nearly horizontal. The meridional branch of the Northern Severnaya Zemlya zone is obviously cut by the rim of the Eurasian Basin of the Arctic Ocean. It may originally have been connected with the Innuitian geosynclinal system of the northern edge of North America, which however, exhibits much greater and longer-lasting mobility.

6. Hercynian geosynclinal fold system of the Urals

6.1 Principal features of the relief, state of geological knowledge and general structural zonality

The Uralian Mountainous system extends for 2,500 km from north to south, from the Kara Sea Baidaratsky Inlet to the southern end of the Mugodzhary Range. The relatively low Urals Range was shaped by recent tectonic movements in place of the western and central zones of the Late Hercynian fold system; the eastern zones of this system are expressed orographically over small areas only: the Voikar–Syn'insky Massif, the Pai Er and Rai Iz massifs in the south of the Polar Urals, the Vishnevye and Il'menskie mountains at the contact between the Middle and Southern Urals, and the Mugodzhary Range at the southernmost end of the range. The eastern zones of the Hercynian Urals on the remaining territory are a slightly sloping plain with hummocky topography descending from 500 to 300–200 m and covered by semiconsolidated Mesozoic and Cenozoic rocks. The Paleozoic and Precambrian deposits here are exposed only in deep valleys; their outcrops extend particularly far east between the latitudes of Chelyabinsk and Kustanai; however, even here they do not exhibit a complete cross-section of the Uralian fold system which is about 500 km wide, the exposed part ranging in width from 100 to 250 km.

In the west, the Hercynian fold system of the Urals is separated from the Eastern European and Timan–Pechora platforms by the cis-Uralian Foredeep; it is only south of Aktyubinsk that this zone is overlapped by the eastern margin of the North Caspian Syneclise which partly overlaps the western and central zones of the Urals proper. Geophysical evidence and some data of deep drilling indicated that a major deep fault – a tectonic suture traceable east of Tyumen', Kurgan, and Kustanai – marks the eastern boundary of the Uralian system. This fault separates the younger Uralian system, which occupies the most outward position in the Uralo–Siberian Belt, from the region of earlier consolidation: the Caledonian–Precambrian massif in the western part of Central Kazakhstan and its northern continuation within the Western Siberian Meso–Cenozoic megasyneclise.

The Pai Khoi–Novaya Zemlya Late Hercynian–Early Cimmerian fold system forms the northern continuation of the Urals, more specifically, of its western zones. In the south, the buried continuation of the Uralian western zones turns southwestwards and can be followed along the southeastern margin of the North Caspian Syneclise to the Caspian Sea, whereas eastern zones retaining their meridional trend reach the southern shore of the Aral Sea, even though they may be buried under the cover. With these data taken into account, the Uralian system (broadly speaking) reaches 4,000 km in extent; it should, however, be stressed that its structure is sufficiently typical only in its main part. However, the Uralian Hercynian system is much more uniform along the strike than many other geosynclinal fold systems, Mesozoic and Alpine in particular. Yet, there is a distinct segmentation in its structure, as well, which was especially pronounced in the last evolutional stage and is therefore well expressed in the topography. The following parts have been identified: **Southern Urals,** approximately south of the upper reaches of the rivers Ufa and Iset' or 55.5° N; this is the widest part of Mountainous Urals with an orographic culmination in Bashkiria (the Yamantau Mountain: 1,640 m). **Middle Urals,** to the contact of the Timan and Urals; this is the lowermost part of the range. **Northern Urals,** to the northeastern turn of the range; the range is narrow here, yet the tallest, about 1,500 m high over a considerable distance. **Polar Urals,** which, unlike the rest of the range, has a northeastern strike, and which gradually plunges and lowers toward the Baidaratsky Inlet. The highest point of the range, the Narodnaya Mountain (1,894 m), is situated at the junction of the Polar and Northern Urals.

The Kustanai Saddle separating the Turgai Trough from the Central Siberian Megasyneclise and forming a bridge between the Urals and Western Kazakhstan Massif is situated east of the Southern Urals. The Paleozoic folded basement within the Kustanai Saddle occurs several tens of meters above ocean level and has been traversed by numerous boreholes, which has made it possible to compile a geological and a tectonic map of the surface of the Hercynian complex. One can, therefore, get a very clear idea of the Uralian system structure throughout its width on this particular cross-section. More to the North, the Paleozoic folded basement of the Uralian eastern zones plunges under an increasingly thick Mesozoic and Cenozoic cover and has been reached by a limited quantity of boreholes only.

As for the exposed part of the Urals, it has been covered by sufficiently extense geological and geophysical studies, especially the Southern and Middle Urals, since this region is one of the main and oldest ore bases of the Soviet Union. In addition to the fundamental works by KARPINSKY (1884), CHERNYSHEV (1889), ZAVARITSKY (1941), NALIVKIN (1943), and KUZNETSOV (1933), the principles of modern concepts of the Urals tectonics have been advanced by GORSKY (1958) and developed further by I. D. SOBOLEV et al. (see "Geology of the USSR", Vol. 12, 1969), and those of the Uralian tectonic history by KHERASKOV (KHERASKOV & PERFILIEV, 1963), and PRONIN (1965, 1971).

Studies of the Urals tectonics and tectonic history have now entered a critical stage: the period of sharp confrontation of two opposed concepts. In the 1940's through 1960's there developed a "verticalistic" approach to interpretations of the structure and history of the Urals pictured as a system of narrow blocks corresponding to individual structural-facial zones and separated by long-lived deep faults along which basic, ultrabasic, and then also granitic magma intruded. That nappes or even any significant and flat overthrusts exist in the Urals has been totally denied, despite the fact that their individual manifestations were noted as far back as the 1930's by KUZNETSOV (1937), FREDERIKS (FREDERIKS & EMELYANTSEV, 1932), BLOKHIN (1932), YANSHIN (1932), NEIMAN-PERMYAKOVA (1940), and VARSANOF'EVA (1940). ARKHANGEL'SKY (1932) and O. S. VIALOV & R. I. VIALOVA (1939) even tried to prove a general nappe structure of the Urals: each zone was regarded by them as an independent nappe. This "verticalistic" concept underlay the materials on the Urals tectonics discussed in the relevant volumes of the "Geology of the USSR" (Vol. 2, 1963; Vol. 12, 1969; Vol. 13, 1969; and Vol. 21, 1970). In the 1960's geophysics and drilling confirmed the former conclusions on the existence of overthrusts and nappes at the western slope of the Urals (KAMALETDINOV, 1974), and in the beginning of the 1970's PEIVE, SHTREIS (PEIVE et al., 1971, 1977), as well as PERFIL'EV (1979) and

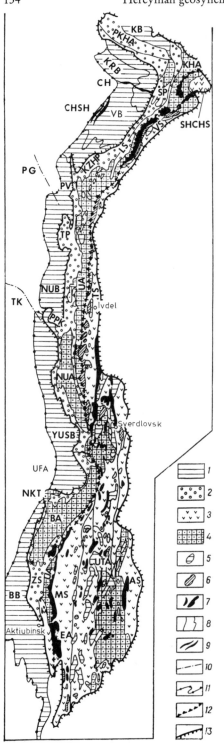

Fig. 37. Main tectonic elements of the Urals (after KAMALETDINOV, 1974).

1 = Upper Paleozoic deposits of the Uralian Foredeep;
2 = Paleozoic deposits of miogeosynclinal zone;
3 = the same, of eugeosynclinal zone;
4 = Precambrian formations;
5 = granitoids;
6 = gabbroids (autochthon);
7 = ultrabasic rocks (allochthon);
8 = boundaries of structures;
9 = anticlinal structures;
10 = Timan Range;
11 = western boundary of Uralian Foredeep;
12 = base of the Main Uralian nappe;
13 = boundaries of outcropping folded Urals.

Anticlinoria: BA = Bashkirian Anticlinorium with Karatau nappe (NKT), UA = Uraltau Anticlinorium, UTA = Uralo–Tobol Anticlinorium, NUA = Central Uralian Anticlinorium; PKHA = Pai Khoi Anticlinorium, KHA = Kharbei Anticlinorium, EA = Ebeta Anticlinorium;

Uplifts: PPK = Polyudov Range, KZHP = Kozhim, SP = Sob';

Ridges: PG = Pechora Ridge, CHSH = Chernyshev Ridge, CH = Chernov Ridge, TK = Timan Ridge;

Synclinoria: ZS = Zilair Synclinorium; MS = Magnitogorsk Synclinorium; AS = Ayat Synclinorium; UFA = Ufa Amphitheater; TS = Tagil Synclinorium; LS = Lemva Synclinorium; VS = Voikar Synclinorium; SHCHS = Shchuch'ia Synclinorium;

Depressions of Uralian Foredeep: BB = Belaya Depression; YUSB = Yuryuzan–Sylva Depression; NUB = Northern Uralian Depression; PV = Pechora Depression; VB = Vorkuta Depression; KRB = Karataikha Depression; KB = Kara Depression.

RUZHENTSEV (1971) of the Geological Institute of the USSR Academy of Sciences not only breathed new life into the nappe concept of the Urals structure, but also supplemented it, together with the Sverdlovsk geologists IVANOV, SMIRNOV (IVANOV et al., 1972, 1975), and others, with the hypothesis that the Uralian Geosyncline had originated in the Early Paleozoic over an oceanic crust.

This novel concept, which has so far been discussed in a few papers only, seems to us much more attractive than the conventional "verticalistic" ideas of the Urals tectonics. Much time and effort are, however, necessary to reinterpret the entire vast material on the Urals tectonics. It is, therefore, quite understandable that the new concept is not yet sufficiently well substantiated in every detail (this concerns particularly the interpretation of the Urals tectonic history) and that supporters of the traditional concept, who persists in their opinions, find weak points and contradictions in the "innovators'" constructions.

The Uralian geosynclinal fold system is generally quite well subdivided into several longitudinal zones which can be followed over a considerable part of its extent. These zones had traditionally been regarded as "through going" ones, i.e., as zones which had developed steadily at the same site as well as the intervening faults at least over the entire Paleozoic. From the novel positions, the present tectonic zonality of the Urals represents a structural pattern of the Late Paleozoic only and formed in its present aspect after the major horizontal movements had ceased. The vertical and subvertical faults traceable at the surface and followed down to the mantle by the deep seismic sounding profiles (DSS) must be Late Hercynian, which formed after the nappes and which may coincide with older faults only partly.

Hence, the following main zones can be identified in the present structure of the Paleozoic and Late Precambrian geosynclinal complex of the Urals and Transurals (Fig. 37).

6.1.1 Zone of Uralian foredeeps

which originated at the beginning of the Permian at the margin of the Eastern European and Timan–Pechora plates in the form of a trough with an uncompensated subsidence and accumulation of deep-water sediments, accompanied by the development of a barrier reef at the western limb. This trough was later filled in with thick molasses beginning by the upper part Permian and up to Lower Triassic and at the level of the Kungurian stage with salt-bearing molasses (with the exception of the extreme north where the salt-bearing molasses are replaced by coal-bearing ones). The major folding took place at the limit of the Early and Middle Triassic, at the extreme north at the limit of the Triassic and Jurassic. The zone consists of four segments which have been termed (preferred nomenclature) **Polar Uralian** Foredeep (in a different nomenclature, Vorkuta or Kos'yu –Rogov), **Northern Uralian** (or Upper Pechora) Foredeep, **Middle Uralian** (Sylva–Solikamsk), and **Southern Uralian** (Belaya) Foredeep. These foredeeps are separated by saddles: Kolva (Middle Pechora), Ksenofontovo, and Karatau, which are regarded as transverse uplifts of the trough basement by some authors and as its tectonic overlaps by others. The Early Precambrian basement, identical to the basement of the Russian Platform, plunges to a depth of 11–12 km at the contact of the cis-Middle- and cis-Southern Uralian foredeeps.

The Uralian foredeeps exhibit a transverse asymmetry typical of such structures; their western platform limbs did not experience any appreciable internal deformations and the eastern limbs have a folded-overthrust structure which becomes more complicated toward the Urals.

6.1.2 The Frontal Bashkirian Uplift (Anticlinorium or Meganticlinorium)

This uplift is traceable only in the northern part of the Southern Urals and is composed of a thick (10 to 12 km) complex of Middle and Upper Proterozoic (Riphean) unmetamorphized or weakly metamorphized sediments as well as Paleozoic deposits crumpled into broad folds grouped in bands which are separated by regional faults. These faults have been regarded (KAMALETDINOV, 1974) as major west-trending overthrusts with an amplitude of several dozens of kilometers and the bands bounded by them as allochthonous nappes. Nevertheless, the total amplitude of the Bashkirian Anticlinorium displacement is assumed to be moderate, so that it must be a parautochthonous rather than a fully allochthonous unit.

6.1.3 The Western Uralian zone

This zone borders on the zone of cis-Uralian foredeeps in the east, with the exception of a stretch corresponding to the Bashkirian Uplift which may be looked on as an uplifted basement of this zone. It is represented in its typical aspect at the western slope of the Northern and Middle Urals, where it is composed of a thick carbonate formation of Ordovician to Lower Permian age with individual sandstone horizons at the level of the base of the Ordovician, the lower part of the Middle Devonian (the Takata sandstones), the base of the Upper Devonian, the Lower and Lower-Middle Viséan, and sand-shale alternation members. These sediments are similar in the structural-facial respect to the deposits of the same age at the eastern margin of the Russian Platform (some investigators believe them to be of platform type), but are characterized by a much more complete sequence due to the presence of the Ordovician–Middle Devonian sediments, which are missing on the platform, by a greater total thickness and linear distribution of facies and thicknesses, that follows the Uralian meridional trend. The zone exhibits a sufficiently complicated, also distinctly linear fold structure of western vergence, complicated by overthrusts; the fold-system mirror plunges in the same direction, and the zone as a whole is overthrust on the eastern limb of the foredeep.

A stretch of the Western Uralian zone south of the latitudinal bend of the Belaya River, south of the Bashkirian Anticlinorium, which wedges in between this anticlinorium and the Uraltau Anticlinorium farther northeast, is known as the **Zilair Synclinorium.** It occupies a position a little more internal with respect to the main part of the Western Uralian zone and is composed generally, instead of the carbonate formation, of a thick terrigenous flyschoid Upper Devonian–Lower Carboniferous Zilair series underlain by relatively thin carbonate-terrigenous Ordovician–Lower Carboniferous formations. The western limb of the Zilair Synclinorium is overthrust at a small angle on the cis-Southern Uralian Foredeep; the Zilair series overlaps the subplatform carbonate Middle Paleozoic along this overthrust revealed by drilling.

The typical sequence of the Western Uralian zone is amagmatic and has an obvious miogeosynclinal nature, however, in its eastern part, on the Southern Uralian (Zilair) and Middle Uralian stretches, there appear ultrabasic and gabbro massifs, serpentinite melange, basic volcanic rocks, and Ordovician–Lower Devonian siliceous-shale formations; these formations are particularly widespread in the south of the Zilair Synclinorium which incorporates the **Sakmara zone** in its eastern limb. This ophiolite association is generally similar in composition and age to that found farther east, in the Tagil–Magnitogorsk zone. Its tectonic position in the Western Uralian zone has been a subject of particularly sharp polemics. The advocates of the ophiolite autochthonous occurrence maintain that these rocks emerge in the cores of mushroom-shaped anticlines. From a different point of view, the ophiolites and accompanying rocks are outliers of a major tectonic nappe which had originated in the Tagil–Magnitogorsk zone and was "thrown over" across the Uraltau Anticlinorium. Recent data (S. V. RUZHENTSEV, see "Tectonics and Magmatism . . .", 1974) indicate that this displacement must have taken place prior to the Late Devonian, but the nappe was subsequently deformed together with the Zilair series, which happened after the Early Permian but before Triassic time.

In the Polar Urals, southeast of the **Yelets zone** of development of the Devonian–Carboniferous carbonate formation typical of the Western Uralian zone, there stretches a band of specific siliceous-shaly deep-water facies with subordinate Ordovician–Lower Permian volcanics deformed into narrow isoclinal folds complicated with overthrusts; this is the **Lemva zone.** It was established as far back as 1945 by VOINOVSKY-KRIGER (1966) that the Lemva zone forms a major nappe which comes into contact with sediments of different facies of the more external Yelets zone.

As indicated by geophysical data, the ancient basement of the Russian Platform continues to the east under the Western Uralian zone and Bashkirian Uplift, where it reaches the depth of 12 km. OGARINOV & SENCHENKO (1974) believe that under the eastern part of the Bashkirian Anticlinorium and Zilair Synclinorium this basement has already experienced certain reworking.

6.1.4 The Central Uralian Uplift zone (Anticlinorium or Meganticlinorium)

This zone is a natural boundary between the miogeosynclinal region of the Urals western slope and the eugeosynclinal region of the eastern slope. In the Southern Urals, it is composed mainly of rocks corresponding to Yurmata (Middle Riphean) and Karatau (Upper Riphean) series of the Bashkirian Anticlinorium, metamorphosed in the green-schist, in places also in the amphibolite facies, with abundant volcanic rocks within their sequence. Vendian and probably also Lower Cambrian deposits play an appreciable role in the structure of the zone farther north. Ordovician–Devonian rocks making up more downwarped areas of the zone overlie with a sharp unconformity these formations incorporated in the nappe structure.

The Central Uralian zone consists of a discontinuous chain of echelon-like narrow linear anticlinoria. In the Southern Urals, this is the Uraltau Range Anticlinorium overthrust along a flat surface on the Bashkirian Anticlinorium in the north and on the eastern limb of the Zilair Synclinorium in the south. South of the latitudinal reaches of the Ural River, the Uraltau Anticlinorium is no longer traceable, whereas in the north it reaches the Ufalei area. In the Middle Urals, the zone is represented by the **Middle Uralian** (Kvarkusha–Kamennogorsk, according to SOBOLEV, 1963) **Anticlinorium,** in the Northern Urals by the **Lyapin-Isov Anticlinorium,** and in the Polar Urals by the **Polar Uralian Anticlinorium.** The middle and especially northern segments of the Central Uralian uplift zone are characterized by an abundance of Baikalian granitoid intrusions.

In the east, the eugeosynclinal complex of the Tagil–Magnitogorsk zone is overthrust at a low angle on the Central Uralian zone. The Central Uralian zone rocks exhibit high-pressure metamorphism (of the glaucophane facies) in the near-contact zone.

6.1.5 The Tagil–Magnitogorsk zone (Synclinorium, Megasynclinorium)

This zone, up to 110–120 km in width, is the main region in the Urals where a typical eugeosynclinal ophiolite association occurs, which initially corresponded to the simatic basement of the Paleozoic Uralian geosyncline. The ophiolitic complex in its least disturbed aspect, closely resembling a normal sequence of oceanic crust, can be found in the south of the Polar Urals, within the Voikar–Syn'ia Massif (see Burtman et al., 1974). In many areas it is represented by individual fragments or is converted into serpentinite melange. The ophiolitic association must be of pre-Devonian, more specifically pre-Ordovician age; it is overlain by differentiated volcanic rocks and Ordovician–Silurian –Lower Devonian silicites. The Middle Devonian is, as a rule, discordant. The Devonian and Lower Carboniferous sequence is saturated with volcanics. In the southern, Magnitogorsk part of the zone, the Middle Carboniferous rocks are the youngest Paleozoic deposits, which sets the age of the last folding as post-Middle Carboniferous. No sediments younger than Lower Carboniferous have been observed in the Tagil part of the zone.

The granitoid intrusions, quite common in the Tagil–Magnitogorsk zone originated partly in the Late Silurian–Early Devonian and partly in the Late Paleozoic (after the Namurian or even Middle Carboniferous). The lower portion of the Paleozoic sequence experienced regional metamorphism.

The Tagil part of the zone differs appreciably in many respects from the southern, Magnitogorsk one which is connected with the former via a narrow "Karabash Corridor" and replaces it in an echelon-like manner, with a displacement toward the east. The structure of the Tagil segment has the shape of a relatively simple and gentle synclinorium which in fact may be a synformal nappe. The Magnitogorsk segment is much more complex in structure, and therefore a whole series of individual synclinoria and anti-clinoria are usually identified within it; the entire zone is termed megasynclinorium. Besides, in contrast with the Tagil segment, where two principal gabbro-ultrabasic belts have been identified (platinum-bearing in the west, along the boundary with the Central Uralian zone and the Serov–Nev'yansk Belt along its eastern boundary), several such belts have been established in the Magnitogorsk segment. Each of them may well lie at the base of an individual nappe; and the extreme western and eastern ones may be linked at depth to form a single base of the Tagil and Magnitogorsk nappes. However, the present occurrence of the bodies of serpentinized ultrabasic rocks must be controlled largely by Late Hercynian subvertical faults.

The zone in question is generally followed along the entire Urals, constituting its main **"greenstone belt."**

6.1.6 The Eastern Uralian Uplift zone (Anticlinorium or Meganticlinorium)

This zone extends from the River Tura (north of Sverdlovsk) to the southern termination of the Mugodzhary. According to the geophysical data available, the zone can be followed northwards to Severnaya Sos'va River or even farther, where it expands to 160 km as against 25–40 km in the exposed part. In the northern part, it is expressed by a series of granite-gneiss domes and arches which form a single or double chain. Their nuclei contain gneisses, migmatites, and other rocks of the amphibolite facies of metamorphism, intruded by intrusive granite massifs and giving place at the limbs to less metamorphosed Paleozoic (Ordovician to Lower Carboniferous) rocks. The structure of the domes is complicated by more recent faults. The abundance of granitoids, mainly Upper Paleozoic, suggests that this zone may be regarded as the main granitic belt of the Urals, just like the Tagil–Magnitogorsk zone is its main greenstone belt. In the south of the zone, in the Mugodzhary Range, there is a vast area where a complex of gneisses, migmatites, and quartzites is exposed, whose Precambrian age has been confirmed radiometrically by MILOVSKY (MILOVSKY & BARANOV, 1971). Similar data were obtained for the rocks making up the core of the Il'menogorsk Uplift; Early Precambrian age of the gneiss complex of other dome structures in the zone may also be assumed by analogy, if these structures are referred to the type of so-called mantled domes, although some of the rocks, metamorphized in the amphibolite facies, formed from younger Late Precambrian and Paleozoic deposits.

Consequently, the Eastern Uralian Anticlinorium is an ancient continental block in the central part of the Uralian Paleozoic geosyncline, which is supported by the DSS evidence. Judging by the composition, thickness and mode of occurrence of the Middle and Upper Devonian, and especially Lower Carboniferous deposits, the zone since then has been a positive element of the Uralian paleogeography and paleotectonics. The Eastern Uralian Uplift zone is not just edged by serpentinized ultrabasic rock belts, but these formations are also quite common within the zone, surrounding the gneiss domes (such as the Sysert Dome) or filling troughs between them. From the point of view of the Urals nappe structure, this may be accounted for only by the allochthonous occurrence of the ophiolitic complex and its subsequent deformation together with the autochthon during the period of dome formation, i.e., in the Late Paleozoic.

6.1.7 The Eastern Uralian Synclinorium zone

This zone extends parallel to the Eastern Uralian Uplift zone from Alapaevsk in the Middle Urals to the Eastern Mugodzhary. Geophysical data and individual boreholes have made it possible to follow the continuation of this zone far northwards, to the polar Transurals. Sedimentary-volcanic Middle and Upper Devonian and Lower Carboniferous formations play a major role in the constitution of this zone. In some places, Silurian formations of similar composition are exposed from under them; besides, terrigenous-

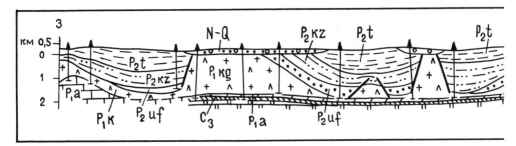

Fig. 38.   Profile across the Uralian Foredeep at latitude of Meleuz Town (after KAMALETDINOV, 1974). Lower Permian formations (Assel', Sakmara, and Artinskian stages = P_1a): 1 = reef type, 2 = depression type, 3 = molasse type.

carbonate deposits of the Upper Carboniferous are occasionally observed within synclines. An abundance of the terrigenous coal-bearing Lower Carboniferous deposits overlying unconformably the Devonian to Lower Tournaisian rocks is a typical feature of the Eastern Uralian zone sequence. Unconformities have also been established at the base of the Middle Devonian, in the middle of the Tournaisian, and at the base of the Middle Carboniferous. The sedimentary-volcanic fill of the Eastern Uralian Synclinorium is deformed into folds of variable morphology and size, complicated by faults and enclosing intrusive bodies of various composition: from gabbro to granites. Owing to the unconformities, the Carboniferous strata form structures of the type of superimposed troughs. Western displacement is generally predominant in the synclinorium, with gentle dips prevailing on the western limb and steep dips on the eastern limb. A belt of serpentinized ultrabasic rocks extends along the boundary of the Eastern Uralian Synclinorium with the uplift zone of the same name.

6.1.8 The Transuralian Uplift zone

This zone, 40 to 100 km or more in width, is hidden under the Meso–Cenozoic sedimentary cover practically along its entire extent. Geophysical and drilling data show that its structure is generally similar to that of the Eastern Uralian Anticlinorium and that it is also a region of extensive manifestation of Late Paleozoic granitoid magmatism, which is confined to dome structures, especially so in the north and east. These structures proper and the linear folds complicating their limbs and intermediate synclines are composed of Precambrian and Lower Paleozoic (?) gneisses and schists, as well as of nonuniformly metamorphized sedimentary-volcanic Silurian and Devonian rocks; just like farther west, the Middle Devonian is unconformable. The synclines are filled with Lower Carboniferous rocks, the deepest of them also with Middle Carboniferous rocks which overlie conformably older formations. Within this zone, like farther west, serpentinized ultrabasic rocks are widespread; the Near Tobol ultrabasic belt is confined to the eastern boundary of the zone; it is the easternmost in the Urals.

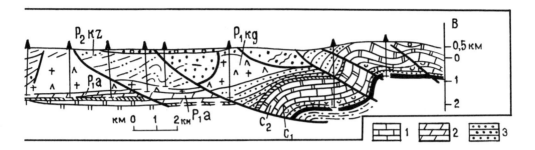

6.1.9 The Tyumen'–Kustanai zone (Synclinorium)

This zone, up to 100–125 km wide, is composed mainly of Lower Carboniferous volcanic and carbonate-terrigenous rocks which experienced fold and fault deformations of moderate intensity. The Lower Carboniferous volcanic rocks make up a quite characteristic formation transitional from late geosynclinal to early orogenic; it has been termed the Valer'yanov series. The Valer'yanov volcanics constitute a volcanic belt which can be followed far south in the direction of the Tien Shan. Devonian and even Silurian rocks form occasional outcrops from under the Lower Carboniferous deposits, and the superimposed troughs are filled with Middle and Upper Carboniferous sediments. Only a limited number of small massifs of basic and granitoid intrusions have been revealed within the zone.

The Tyumen'–Kustanai zone is the easternmost of the Urals and borders on the region of Precambrian–Caledonian consolidation in the west of Central Kazakhstan along the major Tyumen'–Livanov or Central Turgai fault.

Following the above brief description of the Uralian main tectonic zones, it will now be appropriate to discuss their characteristics within individual segments of the Urals. We will start from the south which is the most fully represented and best studied.

6.2 The Southern Urals

The zone of foredeeps is represented here by the cis-Southern Uralian (Belaya) Trough which closes north of Ufa with a specific transverse structure of the Karatau Range and which opens up wider and deepens southwards, until it plunges in this direction under the Meso–Cenozoic cover of the North Caspian Syneclise eastern board. Three structural stages differing in the formations and tectonic style have been identified in the cross-section of the trough (Fig. 38).

The lower stage is composed of Upper Carboniferous and Lower Permian deposits, up to the Artinskian stage. In the axial part of the trough, it is represented by thin (50 to 200 m) "depression" facies: siliceous-carbonate-shale deep-water sediments deposited under the conditions of uncompensated subsidence. At the eastern edge of the trough, these facies are replaced by thick (up to 2,000 m) sediments of the lower gray molasses, at the western edge by reef limestones (up to 1,500 m) which form a barrier reef extending for 750 km in the meridional direction and consisting of a chain of reef massifs. The latter are expressed in the north, in Bashkiria, at the surface as isolated mountains, the so-called shikhans; in the south, in the Orenburg and Aktyubinsk regions, they are buried under younger sediments and have been revealed by drilling. The Upper Carboniferous–Lower Permian barrier reef was formed over a flexure bend of the underlying platform cover deposits; corresponding to the flexure bend in the basement is a fault. The high-porosity carbonate rocks of the reefs contain oil (in the north) and gas (in the south) pools, sealed by Kungurian evaporites.

The salt-bearing series of the Kungurian stage forms the second structural stage of the trough; rock-salt is present only in its axial part, being replaced by gypsum, anhydrite, and dolomites toward the periphery. Active manifestations of diapirism, varied in form are associated with the Kungurian salts: from flat salt pillows to true piercement cores; the intensity of these manifestations increases toward the south, and so does the quantity of anticlinal zones in which diapiric structures group. It is of interest that these structures generally occur within synclines of Artinskian and underlying deposits, where the Kungurian salt reaches maximum thickness (more than 1,500 m in places) whereas over gentle Artinskian anticlines the Kungurian sequence is represented mainly by anhydrites and gypsum, its thickness decreases to 200–300 m, and its attitude reflects the shape of these anticlines partly dissected by faults under the Kungurian deposits.

The third structural stage of the Southern Uralian Foredeep is composed of red Lower Permian–Upper Triassic molasses. The molasses fill irregularly shaped synclines between diapiric injections of the Kungurian salt. Cretaceous and Tertiary platform deposits have been preserved in places over the diapiric intrusions in small depressions which resulted from dissolution or overflowing of the salt.

The eastern slope of the trough, facing the Urals, is composed mainly of Upper Carboniferous and Lower Permian strata which are deformed into linear folds complicated by overthrusts. The latter are particularly distinct in pre-Kungurian deposits and partly attenuate in the saliferous Kungurian. Their surfaces dip eastwards, and the amplitude increases in the same direction up to several kilometers. Confined to frontal parts of the overthrusts are anticlinal folds, not infrequently oil-bearing (KAMALET-DINOV, 1974). The Southern Uralian Foredeep is bounded in the east by a gently dipping overthrust (nappe) whose amplitude, as established by KAMALETDINOV, is at least 15–20 km. This overthrust is particularly distinct south of the Bashkirian Anticlinorium, where its hanging limb is composed of Zilairian deposits.

In the north, the Southern Uralian Foredeep is separated from the Middle Uralian one by the transverse salient of the Karatau Range composed of Upper Riphean, Devonian, Carboniferous, and Lower Permian deposits. The Karatau salient not only "cuts across" the structure of the foredeep, but it also protrudes into the platform. It is limited by faults on three sides; its southwestern end has the character of a wrench fault continuing into the platform (the Asha wrench fault). FREDERIKS (FREDERIKS & EMELYANTSEV, 1932) suggested as far back as the 1930's that the Karatau salient is a nappe slice from the Urals region. Then the opinion of the Karatau horst nature had prevailed for a long time, and only recently M. A. KAMALETDINOV proved the validity of FREDERIKS' idea, pointing out, among other facts, that the assumption of the Karatau horst structure is at variance with the geophysically established deep (10 to 12 km) position of the basement under the Karatau Range. We will add here that the Karatau salient is quite similar to the Cumberland nappe of the Southern Appalachians, which was proved by drilling.

The Bashkirian Anticlinorium (in the north) and Zilair Synclinorium (in the south) lie immediately east of the Karatau Salient and Southern Uralian Foredeep. The **Bashkirian Anticlinorium** is composed of a thick Middle and Upper Proterozoic series overlain transgressively by Ordovician (in the south and east) or directly by Devonian (in the north and west) rocks, which are followed by younger Paleozoic deposits, to the Lower Permian ones. In the north, an ancient, Early Precambrian crystalline basement outcrops from under the Burzyan series of the Lower Ripheau. This is the so-called Taratash Swell. The Precambrian series of the Bashkirian Uplift plunges southwards under the Paleozoic rocks of the Zilair Synclinorium. This plunge may be accompanied, as suggested by KAMALETDINOV (1974), by an ancient cross fault.

The Bashkirian Uplift consists of a series of large folds forming an arc, gently convex toward the northwest, and separated by faults. The largest of them is the Zil'merdak Overthrust separating the western limb of the uplift from its axial part and followed southwards to within the Zilair Synclinorium. KAMALETDINOV (1974) believes that other faults are overthrusts too, which confine five anticlinally and synclinally bent slices thrust over one another from east to west; the mentioned Karatau Salient is located on the continuation of one of the central slices. The Late Precambrian deposits increase discontinuously in thickness and grade of metamorphism from west to east, the main jump, however, taking place along the tectonic boundary between the Bashkirian Anticlinorium and the Uraltau Anticlinorium.

The Zilair Synclinorium starts in the north, at the outflow of the Belaya River, between two anticlinoria, and farther south it covers the Bashkirian Uplift completely. A major role in its structure belongs to a thick (up to 3 km) flyschoid graywacke-argillite Upper Devonian–Lower Carboniferous Zilair series underlain by relatively thin (1 to 2.5 km) carbonate-terrigenous Ordovician–Middle Devonian deposits and overlain in the south of the western limb by a Middle–Upper Carboniferous flysch and Lower Permian molasses. The western limb of the synclinorium, thrust over the foredeep, consists of three overthrust-slices; within the eastern limb, there occur two major nappes[1] (Kraka (Fig. 39) and Sakmara) which are composed of material totally alien to the given zone: rocks of the Ordovician–Lower Devonian siliceous-ophiolitic association and originating, according to some investigators, from the Magnitogorsk zone, i. e., thrown over the Uraltau Anticlinorium. The largest of these nappes, the **Sakmara** nappe, extends in the southeastern part of the Zilair Synclinorium for a distance of over 150 km, the width ranging from 20 to 40 km. The inner structure of the Sakmara nappe is very complicated; RUZHENTSEV (1971) identifies up to ten tectonic slices within it, which consist of Ordivician, Silurian, and Lower–Middle Devonian deposits of different facies, which have been shuffled and reshuffled. These slices are usually deformed into folds (anti- and synforms) varying in size and morphology, from recumbent and isoclinal to brachymorphous and dome-shaped, and produce on the whole a complex synform with a centroclinal closure in the north. The lower slice is usually composed of ultrabasic rocks (serpentinites), gabbro, amphibolites and is largely transformed into a serpentinite melange which often pierces or raises through the overlying slices. RUZHENTSEV (1971) dates the origination of the Sakmara nappe as Early Devonian–beginning of the Eifelian, the principal phase of its

---

[1] Some investigators even at present deny the nappe nature of these areas of the Zilair Synclinorium.

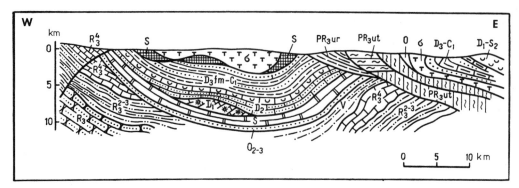

Fig. 39.   Schematic profile across ultrabasic Kraka massifs (after KAMALETDINOV, 1974).

formation as the Eifelian–beginning of the Givetian, and the completion as the Middle Carboniferous, when the Sakmara allochthon was thrust over the Zilair series. The "birthplace" of the Sakmara nappe, according to RUZHENTSEV (1971), is the Sakmara-Voznesensk subzone of the Magnitogorsk zone, situated immediately to the east of the Uraltau Anticlinorium[2]; in this case a minimal displacement amplitude amounts to 40–50 km.

-----

[2] The lithostratigraphic sequence of this subzone, however, differs somewhat from that of the Sakmara nappe.

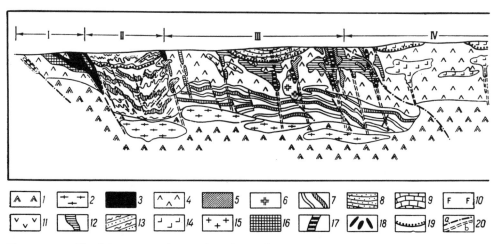

Fig. 40.   Profile of the western slope and axial part of Magnitogorsk Megasynclinorium, based on geologic-geophysical data (after VAKHROMEEV, KLEMIN & SENCHENKO, 1974, simplified). 1 = "basaltic" layer; 2 = fragments of pre-Paleozoic basement; 3 = serpentinized dunite-harzburgite; 4–7 = rocks of early geosynclinal stage; 4 = rhyolite-dacite-andesite-basalt (sodium) association, 5 = late subvolcanic bodies with composition of sodium rhyolite-dacites, 6 = plagiogranites, 7 = terrigenous, siliceous-shale deposits; 8–16 = formations of late geosynclinal stage (S–D): 8 = flyschoid and flysch deposits, 9 = limestone, 10 = andesite formation, 11 = trachyrhyolite-

Another group of investigators recognizes the existence of overthrusts only at the western periphery of the Sakmara zone (and the Kraka area). These geologists are of the opinion that siliceous rocks and magmatites are a natural element in the sequence of the eastern part of the Zilair Synclinorium, which, according to I. S. OGARINOV and G. S. SENCHENKO (see "Tectonics and Magmatism . . .", 1974) is already transitional from mio- to eugeosyncline. There may be a rationale in these views which are also applied to similar areas of the Middle Urals, since there really exists a band with deeper-water facies in the Devonian within the western slope region, which is separated by a barrier reef in the Middle Urals and which is known as the Lemva zone in the Polar Urals. Within this band, which had long remained an uncompensated trough, more specifically corresponding to a continental slope and rise, according to PUCHKOV (1975, 1979), basic volcanic rocks and cherts could well accumulate; however, ultrabasic and gabbro bodies must be alien to it, and their allochthonous positions still seem most probable.

The Central Uralian zone is represented in the Southern Urals by a narrow and strictly linear **Uraltau Anticlinorium.** According to KAMALETDINOV (1974), this anticlinorium consists of two slices thrust over each other from east to west and gently overthrust together on the Bashkirian Uplift in the north and on the eastern limb of the Zilair Synclinorium with its Kraka and Sakmara nappes in the south. The western slice consists of Upper Riphean terrigenous rocks of the Suvanyak complex, metamorphosed in the greenschist facies; the eastern slices are composed of Middle Riphean terrigenous- igneous formations of the Maksyutov complex, metamorphosed in the amphibolite facies. A band of high-pressure metamorphism (glaucophane-lawsonite facies) is observed along the tectonic contact with the Magnitogorsk zone adjacent on the east. The apparent amplitude of the Uraltau zone thrusting over the Bashkirian Uplift and Zilair

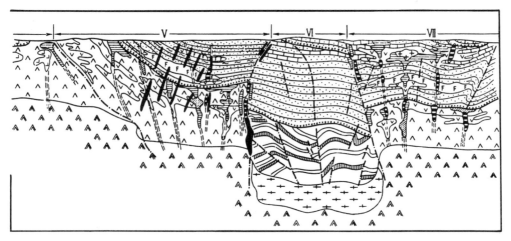

basalt formation, 12 = late subvolcanic bodies with composition of potassium-sodium rhyolite- trachytes, 13 = siliceous shale horizons, 14 = andesite-basalt (potassium-sodium) association, 15 = granitoids, 16 = gabbroids; 17–18 = rocks of orogenic period: 17 = minor intrusions of granite- porphyries and porphyrites, 18 = gabbroids; 19 = angular unconformity surfaces; 20 = fault displacements (a) and zones of jointing (b). Structural zones: I = Sakmara–Voznesensk, II = Peri–Sakmara, III = Tanalyk, IV = Irendyk, V = Uzunkyr, VI = Kizil, VII = Uchaly.

zone is estimated by KAMALETDINOV (1974) at 12–15 km, and the actual one is believed to be much greater.

In the north, at the boundary with the Middle Urals, and in the south, north of the latitudinal cross-section of the Ural River, the Riphean core of the Uraltau Anticlinorium plunges under younger formations (under the transgressive Ordovician in the south). The Kempirsai Anticlinorium had once been considered a southern continuation of the Uraltau zone, south of the Ural River: however, the metamorphic rocks in its core turned out to be partially Ordovician and partially (volcanic rocks of the Lushnikovka series) may arbitrarily be referred to the Cambrian.

The eugeosynclinal **Magnitogorsk zone** is characterized by an extremely variable and complex structure; a profile shown in Fig. 40 gives a certain idea of it, even though it has been constructed from the purely "verticalistic" standpoint. The zone narrows sharply and practically thins out in the north, at the boundary with the Middle Urals, then, in the central part, at the latitude of Magnitogorsk, it expands to 105–110 km, and south of the Ural River narrows again to 30–35 km, disappearing under the Meso–Cenozoic platform cover. In accordance with the conventional treatment, the Magnitogorsk "Megasynclinorium" incorporates two individual anticlinoria and three synclinoria. Predominant in the central, Kizil' Synclinorium are Lower and Middle Carboniferous deposits, in the peripheral synclinoria Middle and Upper Devonian ones, whereas the anticlinoria contain Silurian and Ordovician rocks. The fact that formations of the same age in different areas of the zone differ appreciably in their composition and facies nature, is an indirect proof of their primary disjunction by deep faults and secondary by overthrusts. By now, however, low-angle overthrusts have been established in individual areas of the zone only, and it is a problem for the future to interpret its internal structure in the light of new concepts of the Urals tectonics.

The **Eastern Uralian** (Uralo–Tobol) **Anticlinorium zone** consisting of a series of echelon-like chains of granite-gneiss domes is represented in the Southern Urals by the Chelyabinsk–Suunduk (in the north) and Adamovka–Mugodzhary (in the south) anticlinoria, whereas the similarly built **Transuralian Anticlinorium zone** consists of the Troitsk–Kengusai and Oktyabr'sk–Denisov anticlinoria with the intervening Aleksandrovsk Synclinorium. There is a general tendency toward widening and upheaval of the two anticlinoria zones southwards, with an extensive outcropping of Early Precambrian gneiss-amphibolite series in the Mugodzhary Range. At the same time, the **Eastern Uralian** (Ayat–Irgiz) **Synclinorium** between them gradually wedges out and becomes, in its southern end, a superimposed graben filled with Lower Carboniferous rocks.

The zone in the extreme east of the Southern Urals, the essentially Carboniferous **Kustanai** Synclinorium zone, is completely covered by the platform cover of the Turgai Downwarp, so that its structure is known in general features only.

6.3 The Middle Urals

The Precambrian of the Bashkirian Anticlinorium and of the Uraltau Anticlinorium in the southern part of the Middle Urals plunges under the Paleozoic deposits of the Western Uralian type, and the Magnitogorsk Synclinorium narrows to complete pinching out. As a result, the Western Uralian zone, encircling the depressions of the two anticlinoria, forms an arc slightly convex toward the east, echoed by two zones farther east and by the foredeep (the last in a subdued form). This arc has been referred to in the literature as the Ufa Amphitheater; it is situated opposite the salient of the Eastern European Craton, which is expressed morphologically by the Ufa Plateau, and at the level of the basement by the Perm–Bashkir Arch. The fact that the bend of the Uralian structures is associated with a basement swell of the adjacent platform has first been noted by KARPINSKY (1919) and was fully corroborated by the materials of drilling and geophysical investigations. The basement within the Perm–Bashkir Arch is composed of the most ancient Archaean rocks, and their prevailing strike is latitudinal, i. e., across the Urals. The narrowing of the Uralian system opposite the Ufa Plateau is accompanied by a considerable decrease in the height of its relief, and as a result the Urals here loses nearly completely its character of a highland.

North of the Ufa narrowing, the Uralian system widens again, and Riphean formations, making up the Middle Uralian Anticlinorium, reappear at the surface; the Magnitogorsk Synclinorium gives way to the Tagil one.

The foredeep zone in the Middle Urals is represented by two depressions: **Sylva** (Yuryuzan–Sylva) in the south and **Solikamsk** in the north, separated by the flat Kos'va–Chusovaya Saddle and forming together the **Middle Uralian Foredeep** closed in the north by the Kolva Saddle. The basement in the Solikamsk Depression subsides from 2.5 to 6 km.

In comparison with the Southern Uralian Foredeep the one opposite the Middle Urals is generally shallower and simpler in structure. The western slope of the foredeep is more distinct in the Sylva Depression than in the Solikamsk one; however, in both these depressions it is accompanied by reefs of two generations: Asselian–Sakmarian and Artinskian. The reefs here reach 400 m in height, which is much less than in the Belaya Depression. Enveloping folds have been observed in the Kungurian deposits; one of them is a brachyanticline of the Chusovskye Gorodki where an oil flow, the first one in the history of the Volga–Urals oil-bearing province, was obtained. Structures associated with salt tectonics–arches, usually flat, consisting of brachyanticlines and domes, occur in the axial part of the foredeep. These folds are more extensive and distinct in the Solikamsk Depression which is a major basin of potassium salts. A subsidence of the base of the Kungurian corresponds to the arches in the subsalt deposits. The salt tectonics is poorly expressed in the Sylva Depression and is totally absent in the south, due to a pinching out of the salts; yet, the subsalt deposits form gently sloping arches.

The **Western Uralian zone** in the Middle Urals is 30 to 115 km wide and is composed of a terrigenous-carbonate Ordovician–Lower Permian formation, with a general east to west dip of the fold mirror. The zone has a complex fold-overthrust and folded-nappe structure (Fig. 41). Indications of the existence of nappes in the valley of the Chusovaya River, which is a tributary of the Kama, had long been noted by FREDERIKS (FREDERIKS & EMELYANTSEV, 1932) and later by NALIVKIN (1949). Several large nappes at the western slope of the Middle Urals have been described by KAMALETDINOV (1974). The more outward of them (the **Nizhneserginsk nappe**) is formed of Silurian and Devonian limestones thrust over the terrigenous Middle and Upper Carboniferous of the foredeep eastern limb. The more inner ones (the **Nyazepetrovsk nappe**, in places overlying the Nizhneserginsk, and the overlapping **Bardym nappe**) are already composed of sedimentary-volcanic Ordovician deposits, siliceous shales, and sandstones of the Silurian; they

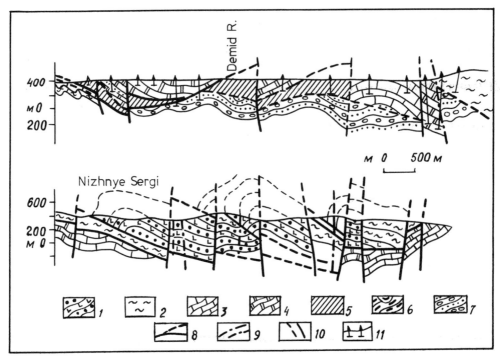

Fig. 41. Geologic profile through the Nyazepetrovsk allochthon (after PLYUSNIN, 1971).
1 = Middle and Upper Ordovician (quartz sandstone, tuffaceous schists, spilites); 2 = Lower
Silurian (phtanites, carbonaceous-siliceous schists, siltstones, limestones); 3–4 = Middle Devonian
(limestones): 3 = Eifelian, 4 = Givetian; 5 = Middle Viséan–Namurian (limestones); 6 = Middle
Carboniferous (sandstones, siltstones, limestones); 7 = Carboniferous–Permian (sandstones,
siltstones); 8 = nappes; 9 = normal and transcurrent faults, sublatitudinal (9); sublongitudinal (10);
11 = boreholes.

must be homologues of the Southern Uralian Kraka and Sakmara nappes. The extreme
eastern **Revda nappe** thrust on the Bardym one and consisting completely of
metamorphosed magmatic rocks (amphibolites and basaltic porphyrites) of Ordovician
(?) age resembles the Kraka and Sakmara nappes even stronger. KAMALETDINOV (1974)
estimates the amplitude of the nappes of the Ufa Amphitheater to equal at least 40–50 km,
based on the presence in the valleys of rivers Chusovaya and Ufa of tectonic semi-
windows of the subplatform Devonian 30 to 40 km east of the Nizhneserginsk nappe
front. The fact that the latter overlies the Upper Carboniferous which experienced fold
deformation indicates quite recent formation of this nappe.

The **Central Uralian zone** in the Middle Urals includes in the south the northern end
of the Uraltau Anticlinorium represented by the Ufalei metamorphic complex thrust
gently over the Bardym nappe of the Western Uralian zone. The Bilimbaevo Saddle
composed of Ordovician–Devonian rocks separates the outcrops of this complex from
the **Middle Uralian** (Vishera–Chusovaya and Kvarkusha–Kamennogorsk) **Anticlinorium** which forms the tectonic axis of the entire Middle Urals and which extends for
about 400 km, the width being up to 60 km. In its core, the anticlinorium is composed of

Upper Riphean and Vendian rocks, and its limbs are made of Ordovician and younger Paleozoic deposits. The Ordovician overlies the Precambrian with a sharp azimuthal as well as angular unconformity, since the Precambrian is dislocated here in the northwestern direction. The anticlinorium in question is gently thrust in the west over the Western Uralian zone (the Chusovaya and Krasnovishera–Nyrob overthrusts). At the northern end of the anticlinorium, the line of this overthrust turns northwest to form the Polyudov Ridge tending to come into contact with the Timan structures and closing the Middle Uralian Foredeep (the Solikamsk Depression), just like a similar Karatau salient overlaps the northern end of the Southern Uralian Foredeep (the Belaya Depression). The amplitude of the Polyudov Ridge overthrust, as estimated by KAMALETDINOV (1974), is at least 10–15 km. The inner structure of the Middle Uralian Anticlinorium is quite complex and consists of numerous folds and overthrusts of western vergence.

The northern part of the eastern limb of the Middle Uralian Anticlinorium is separated by a relatively narrow and elongated Umov–Koiva Syncline filled with Ordovician – Silurian terrigenous-carbonate deposits overlying unconformably the Precambrian rocks, from the southern end of the Northern Uralian or Lyapin–Isov Anticlinorium situated in an echelon-like manner with respect to the Middle Uralian Anticlinorium, with a northeastern shift. The main part of the latter anticlinorium is situated in the Northern Urals.

In the southeast, the Middle Uralian Anticlinorium borders directly on the **Tagil Synclinorium** whose western limb starts with a belt of serpentinized peridotites or dunites, pyroxenites, and gabbro, the latter prevailing. This is the famous Platinonosny (platinum-bearing) Belt of the Urals; in addition to the above rocks which occur with a steep monoclinal eastern dip and alternate in a regular sequence (see above on the oceanic crustal sequence of the Urals), the belt also includes orthoamphibolites and diabases. The platinum-bearing formations make up several major individual massifs of zonal structure. In the west, the belt is thrust over the miogeosynclinal Ordovician of the eastern limb of the Middle Uralian Anticlinorium, pseudoconformably and with the same eastern dip (Fig. 42). Adjacent to it on the east are thick volcanic rocks of the Silurian spilite-diabase, diabase-albitophyre, and porphyrite (andesite-basalt) formations which are replaced by Devonian reef limestones up the sequence. The relationship of the Silurian volcanics with the dunite-pyroxenite-gabbro formation remains unclear: it may either be tectonic or stratigraphically conformable (BURTMAN et al., 1974).

Both the rocks of the Platinonosny Belt and the Silurian volcanics are pierced by intrusions of quartz diorites, plagiogranites, and plagiogranodiorites, as well as apparently somewhat younger syenites. The granitoids are of pre-Eifelian age, since their fragments have been encountered in Eifelian conglomerates.

Upper Silurian and Devonian deposits lie in the axial band of the Tagil Synclinorium in the Middle Urals. As mentioned earlier, these rocks are characterized by a more quiescent structure than the underlying formations. The folds complicating this band have a general western vergence.

The eastern limb of the synclinorium is much more complex, in many areas it is overturned westwards, complicated by folds, of medium or small size, often similar to isoclinal. In the south, it is occasionally sheared by overthrusts, and in the direction from north-northwest to south-southeast is crossed by the Degtyarka–Ufalei regional sinistral wrench fault which was first established by KUZNETSOV (1937). Related to the same limb is a major belt of serpentinized ultrabasic rocks (Serov–Mauk after SOBOLEV, 1963). The rocks making up this belt, in conformity with the general structure, dip steeply east and similarly to the Platinonosny Belt of the western limb of the Tagil Synclinorium, constitute, in all probability, the base of its sequence. The serpentinites had been brought onto the surface as far back as the end of the Silurian–beginning of the Devonian, judging by the presence of their pebbles in rocks of the corresponding age. However, in the south, they form protrusions in younger deposits as well.

The Serov–Mauk serpentinite belt marks the boundary between the Tagil Synclinorium and the **Eastern Uralian Anticlinorium zone.** The latter is represented in the Middle Urals by granite-gneiss domes and arches (Verkhotura, Salda, Verkh-Iset, Sysert, and others), which replace one another from north to south in an echelon-like manner with a general eastern shift (KEILMAN, 1971). Late Hercynian granites and migmatites, as well as gneisses, schists, and amphibolites of Precambrian (probably Early Precambrian) age occur in their cores. A band of similar structures situated farther east is essentially hidden under the Meso–Cenozoic cover.

The **Eastern Uralian Synclinorium zone** is exposed in the Middle Urals even more fragmentarily and mainly in its western limb. This zone is represented here by the Alapaevsk–Techensk Synclinorium composed of sedimentary-volcanic Silurian, Devonian, Lower and Middle Carboniferous formations. Unconformities were detected at the base of the Middle Devonian, Lower (Upper Tournaisian), and Middle Carboniferous. The folded-fault structure is quite complicated, the folds are of numerous orders, with a predominance of western vergence in the western limb and of eastern vergence in the eastern limb. Ultrabasic rock outcrops are observed in the zone bordering on the anticlinorium zone adjacent on the west; farther east there are intrusions of gabbro and granitoids of Early Carboniferous and Late Paleozoic age; large plutons are quite rare in this region.

Several narrow graben-like troughs filled with Triassic–Liassic deposits have been identified in the Middle Urals within the Tagil zone. These strata experienced a quite strong crumpling, and the Paleozoic rocks of the framing were thrust over the graben fill along the faults surrounding the grabens. Structures of this type in the Eastern Uralian zone are encountered in the Southern Urals only.

6.4 The Northern Urals

This stretch of the Uralian fold system extends almost strictly meridionally to 65°N and is exposed in the east only to the Tagil zone. The structure of the Northern Urals is nearly identical to that of the Middle Urals. The differences are the greatest in the foredeep which is expressed here by the Upper Pechora Depression.

The principal difference of the **Northern Uralian (Upper Pechora) Foredeep** from its analogues farther south is the absence in its sequence of the thick Kungurian salt-bearing formation with which manifestations of salt tectonics are connected. A relatively thin gypsum-bearing formation (a formation of gypsum-bearing clays) occurs in place of the salt-bearing formation in the Upper Pechora Foredeep, in its outward limb only and possibly also in the axial zone. This formation is absent in the inner limb, and the gray marine Upper Carboniferous molasse is replaced directly by varicolored continental molasse starting in the Kungurian and encompassing the Upper Permian and the considerable part of the Triassic. The accumulation of molasses in the inner limb started during the Artinskian age and the base of the continental molasses rises to the boundary between the Lower and Upper Permian.

The Lower Permian formation of barrier reefs is also absent in the Upper Pechora Foredeep even though the structural conditions for its origination (the flexure bench at the transition from the platform to the outward limb of the trough) exist here, too. The absence of the salt and reefs in the Upper Pechora Foredeep must have a common cause: the change-over from arid conditions in the Earlier Permian time to mildly humid ones in the northern direction, as was first noted by N. M. STRAKHOV. An additional factor could be a stronger inflow of clastic material, due to a sharper upheaval of the Northern Urals in comparison to the Middle Urals–this recurred during the neotectonic stage; a more humid climate must have also contributed to this. The total thickness of the molasse fill in the Upper Pechora Foredeep reaches 4.0 to 4.5 km.

The inner limb of the foredeep and the transition to the Western Uralian zone, established according to the disappearance of molasses in the synclines, as well as the Western Uralian zone proper are characterized by linear folds and west-oriented overthrusts which pass into nappes. The boundary of the Western Uralian zone with the Central Uralian is also expressed by an overthrust which could not, however, be mapped along its entire extent.

The Northern Uralian Anticlinorium, known also as the Lyapin–Isov (SOBOLEV, 1963) or Vogul–Uraltau one, corresponds to the Central Uralian zone in the segment in question. As pointed out above, in the south this anticlinorium replaces in an echelon-like manner the Middle Uralian one on the east, being separated from it by a narrow and elongated syncline. Notwithstanding the fact that the Northern Uralian and Middle Uralian anticlinoria are so closely spaced, are confined to a single band, have similar inner structures, and are composed of rocks of the same pre-Ordovician, mainly Riphean –Vendian[3], age, their sequences differ so sharply, that this suggests the existence of a deep fault beneath the Ulsov Syncline separating them (PERFIL'EV, 1968). In fact, while the Riphean and Vendian sequences of the Middle Uralian Anticlinorium correlate quite well with the sequence of the Bashkirian Anticlinorium, the Riphean and Vendian of the Northern Uralian Anticlinorium are lithologically so different from the Bashkirian stratotype, that their comparison is virtually impossible.

The principal features of this sequence are, primarily, the absence of the "Mignar" type carbonate series and, secondly, the occurrence of basic and (usually higher up stratigraphically) acid volcanic rocks in its uppermost portion. All the pre-Ordovician rocks have been metamorphosed in the greenschist and partly (at the anticlinorium core) in the amphibolite facies. Yet another feature of the Northern Uralian Anticlinorium is a rather wide occurrence of Baikalian granitoid rocks, which is not the case of the Central Uralian zone farther south. It was established subsequently that the Northern Uralian Anticlinorium is largely thrust over the Ulsov Syncline at its southern end.

In the east, the Northern Uralian Anticlinorium borders on the northern part of the Tagil Synclinorium along an overthrust zone formed along the Main Uralian Deep Fault

---

[3] The upper part of the sequence of the Northern Uralian Anticlinorium may well be of Cambrian age (PUCHKOV, 1975).

– the boundary between the eu- and miogeosyncline. This zone is marked by outcrops of serpentinized ultrabasic rocks of the so-called Salatim Belt; the outcrops are smaller than in the south, within the Platinonosny Belt; however, farther east they are also replaced by massifs of gabbro and amphibolites and then plagiogranites and tonalites of pre-Middle Devonian age. The sedimentary-volcanic fill of the synclinorium in the lower structural stage (Silurian) consists of a spilite-diabase-albitophyre formation, in the upper stage (Middle Devonian–Lower Carboniferous), separated from the lower one by an unconformity and starting with conglomerates, of terrigenous and volcanic rocks (of a more diverse composition). The variation in the composition of these deposits along the strike points to a primary closure of the synclinorium in the northern direction. Its axial part and the eastern limb are hidden under the Meso–Cenozoic platform cover.

The Northern and Polar Urals are connected within the **Kozhim Transverse Uplift** which is a specific tectonic knot in which Northern Uralian meridional structures give way in the east to the structures of the Polar Urals, oriented in the northeastern direction. The zone of foredeeps became narrower in the Kozhim Uplift stripe: the Upper Pechora Foredeep narrowed already farther south, owing to the wedging in of the southeastern end of the Pechora Ridge from the direction of the platform, closes completely at the eastern limb of the Kozhim Uplift, being limited on the east by the Chernyshev Ridge dislocation zone. The same zone serves as a northwestern boundary of the Vorkuta Foredeep which originates at the northwestern limb and northern subsidence of the Kozhim Uplift. The Northern Uralian Anticlinorium first widens in the southern part of the uplift, as there appear the oldest Riphean strata and numerous granite bodies, and then plunges being overlain unconformably by Ordovician and younger deposits. Northeast and east of the Kozhim Uplift there appear Paleozoic deposits in deep-water shale facies, which have hardly been encountered farther south.

6.5 The Polar Urals

This extreme northern segment of the Uralian system proper extends in the general northeastern direction from the Kozhim Uplift to the Baidaratsky Inlet of the Kara Sea. Its structure is quite unusual, particularly so at the northeastern end, and consists in fact of two dissimilar "subsegments": a wider southwestern one (near-Polar Urals) and a shorter (in the exposed part) northeastern "subsegment" (Polar or Transpolar Urals proper), separated by the Sob' transverse uplift situated between Vorkuta and Salekhard and studied in detail by PERFIL'EV (1968).

The broad foredeep of the near-Polar Urals (**Vorkuta** or **Kos'yu–Rogov**) is filled with an extremely thick (up to 8–10 km) Permian–Triassic molasse series. This series is characterized by the occurrence of a paralic coal-bearing formation (coal-bearing molasse) between the lower marine and upper continental molasses. The coal-bearing formation occupies the same stratigraphic level (Kungurian) as the evaporite formation of more southern Uralian foredeeps. The outer zone of the foredeep overlaps the edge of the Timano–Pechora Platform; the inner zone and the southern centrocline overlie unconformably the folded pre-Permian deposits of the Lemva zone (see below) and of the Kozhim Uplift. The foredeep is separated from the adjacent platform by a narrow dislocation zone of the Chernyshev Ridge, distinctly visible in the relief and composed of

Devonian and Carboniferous rocks deformed into rather steep folds which are complicated by overthrusts directed toward the platform.

The inner structure of the Vorkuta Foredeep is relatively simple: its outer limb is
practically monoclinal and the inner one is crumpled into broad, gentle folds. In the
north, the Vorkuta Foredeep is separated from the adjacent Karataikha Foredeep situated
opposite the Pai Khoi Ridge by a dislocation zone of the west-northwest – east-southeast
Pai Khoi trend, which links the Chernov Ridge with the northern boundary of the Sob'
transverse uplift. This dislocation zone has an obvious suprafault nature and, crossing the
entire Polar Urals, it enters the Transuralian region.

The molasses of the Vorkuta Foredeep overlap in the south, on the slope of the Kozhim
Uplift, formations of two structural-facial Ordovician–Carboniferous zones: **Yelets** in
the west and **Lemva** in the east; the latter zone is thrust over the former one (VOINOVSKY-
KRIGER, 1966). The Yelets zone plunges completely under the foredeep, whereas the
Lemva zone is only partly covered by the foredeep and forms its eastern framing
throughout the foredeep's extent. It was pointed out above that the Ordovician–Carboniferous deposits in the Yelets zone are represented by a shelf limestone-dolomite
formation with quartzose sandstones at the base; this formation is quite thick (up to 5–6
km). In the Lemva zone, the same deposits make up the deep-water limestone-siliceous-
shaly formation, just 1,000 to 1,200 m in thickness. Both south and north of the Vorkuta
Trough, between the two structural-facial zones there is an up to 2.5 km thick band of the
sandy-shale formation (Tisvaiz and Verkhnesob'); this formation seems to correspond to
sediments of the upper part of the continental slope.

Structurally speaking, the Yelets zone of the Polar Urals western slope, which is a
direct continuation of the Western Uralian zone, is characterized by mild deformations
expressed as a system of coffer-box folds inclined slightly westwards and complicated by
steep overthrusts dipping eastwards. The Lemva zone is totally different. It has an
isoclinal-imbricate "Schuppen"-structure with a persistent northwestern vergence; on
the whole, the zone is a very complex synclinorium up to 250 km long and up to 40 km
wide. As far back as 1945 VOINOVSKY-KRIGER (1966) suggested that the Lemva complex
overlies the Yelets one in a nappe manner. This opinion was controverted later, in the
1950's and 1960's by some investigators who pointed to the existence of facies transitions
between the two complexes. A recent general revision of the Uralian tectonics concepts
has put VOINOVSKY-KRIGER's interpretation back at its foremost position as the most
likely one. This was asserted by KAMALETDINOV (1974) and PUCHKOV (1975) who
pointed out the existence of tectonic windows within the Lemva nappe, of the Lower
Permian autochthonous deposits traversed by a borehole beneath the allochthonous
formations, as well as of probable outliers of the Lemva nappe within the Yelets zone.
These data make it possible to estimate the nappe amplitude at dozens of kilometers (at
least 15 km) and date it as Late Permian–Early Triassic.

As one assumes the nappe nature of the Lemva complex, one should also give some
thought to the origin of this nappe, to the "birthplace" of the Lemva facies. An analysis of
the map shows that the Lemva facies extend in the southeast beyond the northern end of
the Northern Uralian Anticlinorium in the Kozhim Uplift, as though continuing toward
the northern end of the Tagil Synclinorium. It is noteworthy that volcanic rocks (diabases
and keratophyres) similar to those of the Tagil zone appear in the Ordovician at the
southeastern termination of the Lemva Synclinorium.

The facies nature of the Lemva deposits corresponds, most probably, to the environment of the continental rise; these deposits must have formed at the very edge of a continental block, at or near the boundary with the region of oceanic crust starting with the Tagil zone. The presence of polymictic conglomerate lenses from Central Uralian pre-Ordovician rocks in the southeastern limb of the Lemva Synclinorium and at the base of its sequence indicates that the Lemva sediments had been primarily deposited on the continental crust and that the zone of their accumulation must have initially been limited by the uplift of this crust.

Corresponding to this uplift in the recent structure must be the axial anticlinorium zone of the near-Polar Urals – the **Polar Uralian Anticlinorium** – which is thrust along a gentle surface over the Lemva complex on the southeast. This anticlinorium does not constitute a direct continuation of the Northern Uralian one, but rather is displaced eastward from it. The Polar Uralian Anticlinorium is composed of the upper part of the pre-Ordovician metamorphic complex and, like its southern homologues, exhibits a complex west-vergent inner structure.

The **Voikar Synclinorium,** which is a continuation of the Tagil Synclinorium displaced eastwards is in turn thrust over the eastern limb of the Polar Uralian Anticlinorium. The Voikar–Syn'ya gabbro-peridotite complex – ultrabasic rocks, gabbro, amphibolites, and diabases – lies at the base of the overthrust. This complex provides an oceanic crust sequence which is best of all preserved in the Urals. Farther east, there occur Lower Devonian diorites and tonalites, as well as younger quartz diorites and granodiorites piercing the gabbro-peridotite complex (see Fig. 42). The sedimentary-volcanic fill of the Voikar Synclinorium, emerging farther east, is represented in its deepest part (Silurian–Lower Devonian) by a spilite-diabase-albitophyre formation in association with chert and tuffites. Higher up, at the level of the Middle and Upper Devonian, also separated by a hiatus, in the northwestern limb of the synclinorium, there lie unconformably volcanic rocks of a somewhat different diabase-porphyritic (andesite-basaltic) composition, which are replaced partially by tuff-rudaceous formations of the same volcanic-siliceous material in the opposite limb. The amplitude of hiatuses and unconformities in the sequence increases at both synclinorium closures, which indicates a primary nature of these boundaries.

The **Polar Urals** proper, separated from the near-Polar Urals by the mentioned Sob' transverse uplift, had been poorly studied until recently. Its structure was then discussed by OKHOTNIKOV (1968, 1973) and OKHOTNIKOV & STREL'NIKOV (1974). The continuation of the axial anticlinorium of pre-Ordovician series in the Polar Urals has been termed the **Kharbei Anticlinorium** (PERFIL'EV, 1968). This anticlinorium is bounded on the

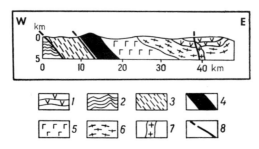

Fig. 42. Profile across the southern part of Polar Urals at latitude of the Kok–Pel'e River (after BURTMAN et al., 1974). 1 = Devonian porphyrite and tuffaceous formation; 2 = Ordovician shale formation; 3 = Riphean (?) metamorphic schists; 4 = ultrabasic rocks; 5 = gabbro-amphibolites and amphibolites; 6 = tonalites and plagiogranites; 7 = granodiorites; 8 = faults.

west by an overthrust which reaches the vicinity of the Baidaratsky Inlet in the north, where it forms several branches, two of which extend in an almost latitudinal direction. These two branches confine in the north the pericline of the pre-Ordovician Gerdiz Uplift which ends the Kharbei Anticlinorium. The core of the Gerdiz Uplift contains amphibolites and gneisses which it is most logical to date as Early Precambrian rather than Riphean, according to recent data of L. L. PODSOSOVA. As the uplift plunges, west of the Baidaratsky Inlet, there occurs an Ordovician terrigenous-volcanic series whose folds change their strike from northeastern to northern and north-northwestern toward the Pai Khoi. On the east, thrust over the axial anticlinorium are eugeosynclinal formations of the **Shchuch'ya Synclinorium** which is a continuation of the Voikar Synclinorium, including large massifs of ultrabasic rocks (Rai-Iz in the south and Syum-Keu in the north). Small bodies of ultrabasic rocks have also been encountered along the western fault (adjacent to which is the Rai-Iz Massif) and may well be outliers of the nappe thrown over the Kharbei Anticlinorium, i. e. klippes.

The lower part of the sequence of the Shchuch'ya Synclinorium is composed of basic volcanic rocks and Silurian–Lower and Middle Devonian reef limestones which are overlain unconformably by coarse terrigenous Upper Devonian–Carboniferous deposits. Toward the Baidaratsky Inlet, the synclinorium experiences a nearly complete narrowing, due, on the one hand, to a buried swell of a hypothetically Early Precambrian basement (the Loborov Swell), which has a northwestern "antiuralian" strike, and, on the other hand, due to the bundle of folds and faults of sublatitudinal strike branching eastwards from the pericline of the Gerdiz axial uplift.

Thus, the Uralian fold system near the Baidaratsky Inlet experiences natural degeneration and attenuation and does not continue to the Kara Sea area. Its axial pre-Ordovician uplift plunges, turns toward the east, branches out and becomes concealed under Ordovician and younger formations; the western miogeosynclinal zone changes its northeastern strike to a northwestern one and continues into the Pai Khoi; the eastern eugeosynclinal zone becomes virtually closed, even though there are interlayers of volcanic rocks in the terrigenous Ordovician of the Polar Urals northern termination.

As pointed out justly by PERFIL'EV (1968), the attenuation of the Uralian system is connected directly with the growing role of faults and structures which exhibit northwestern trends across the Urals; this is expressed in the formation of the Kozhim, Sob', and Laborov transverse uplifts; these faults become particularly numerous at the subsidence of the Kharbei Anticlinorium; extending into the Pai Khoi, they become longitudinal and determining in the orogenic evolution of the Pai Khoi system. We will discuss this system below.

6.6 Pai Khoi and Novaya Zemlya, structure of the Barents Sea

The Pai Khoi–Novaya Zemlya fold system is a northern continuation of the Urals in the recent structure. It extends as a huge 1,500 km long arc convex toward the west across the Pai Khoi Range, Island Vaigach, Novaya Zemlya Islands, and probably reaches the islands of the Arctic Institute in the Kara Sea and (or) Severnaya Zemlya Archipelago (Fig. 43). This fold system differs, however, from the main part of the Urals in several features. These are the major of them.

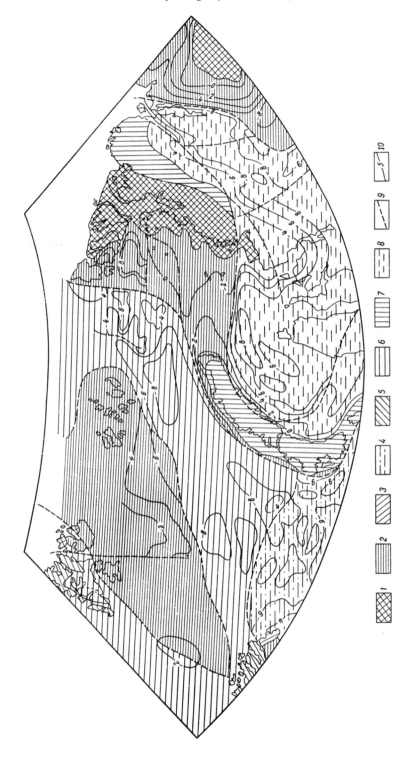

Fig. 43. Tectonic diagram of the Barents–Kara region. 1 = ancient (pre-Baikalian) massifs, outcrops; 2 = the same, under the cover; 3 = Baikalian fold complex, outcrops; 4 = the same, under the cover; 5 = Caledonian fold complex, outcrops; 6 = the same, under the cover; 7 = Late Hercynian fold complex, outcrops; 8 = the same, under the cover; 9 = major faults; 10 = isobaths of equal elevation of the foldbasement.

1) The Pai Khoi–Novaya Zemlya system, at least its northern part, does not comprise the continuation of the main most typical zone of the Uralian system proper: the eugeosynclinal zone of the eastern slope. The latter, as was shown above, plunges in the north under the waters of the Kara Sea Baidaratsky Inlet; geophysical evidence indicates that it may continue farther to the latitude of the Yugorsky Shar or even Kara Gate straits; identification of its very shortened analogues along the Novaya Zemlya is quite problematic. Consequently, the Pai Khoi–Novaya Zemlya system is a continuation of only the western, miogeosynclinal zone of the Uralian system; that is in perfect conformity with the facial nature of the deposits; hence, it is of an intracratonic ensialic type, and this enables us to speak of the Uralian system degeneration in the northern direction.

The previous assumption of a possible connection between the Polar Urals and Taimyr fold zones has not been confirmed by the geophysical investigations or drilling in the north of the Western Siberian Plain. It became obvious that the northern end of the Uralian eugeosynclinal zone is bounded on the east by a major Nurma uplift composed of Precambrian gneisses.

2) Permian deposits are quite widespread within the Pai Khoi–Novaya Zemlya system; they cover this system over a considerable area (Southern Island of the Novaya Zemlya). This is why V. S. ZHURAVLEV and A. S. PERFIL'EV (ZHURAVLEV et al., 1965) believed that the Pai Khoi had originated in the region of the Urals pericratonic subsidence. The beginning of the major deformation in the Pai Khoi–Novaya Zemlya system is now assumed to correspond to the end of the Triassic–beginning of the Jurassic, i. e., to the Early Cimmerian epoch of tectogenesis; this is also the age of the most recent granitoid intrusions. Therefore, the Pai Khoi–Novaya Zemlya orogen is Early Cimmerian (like the not so distant Taimyr farther east!) rather than Hercynian or even Late Hercynian. As one goes north, the features, which make it different from the Urals, accumulate; it is in this direction that the intensity of the Caledonian diastrophism increases and the thickness and dislocation intensity of the Late Paleozoic deposits decrease.

Three structural complexes are involved in the structure of the Pai Khoi–Novaya Zemlya system: Baikalian, Caledonian, and Hercynian–Early Cimmerian. The Baikalian complex is exposed in the cores of anticlinoria of the southern part of the arc and is composed of strongly deformed (in the northwestern, in the north in the west-north-western and near-latitudinal directions) and weakly metamorphosed carbonate, terrigenous, and volcanic (basic to acid) formations, at least 6 km in thickness. This complex looks very much like the complex of axial anticlinoria of the Northern and Central Urals, which is of the same age and which is considered to be an eugeosynclinal Baikalian complex.

The Caledonian complex encompasses deposits from Upper Cambrian to Middle Devonian, which make up two formations: the lower, coarse terrigenous and Cambrian in the south; in the north of the Novaya Zemlya also Ordovician; and the upper, carbonate or terrigenous-carbonate, containing basic volcanic rocks in the Upper Silurian. The upper formation is 3 to 3.5 km in thickness. In the north of the Novaya Zemlya, the Upper Silurian becomes terrigenous, conglomerates are found in it, whereas the Lower and Middle Devonian in the extreme northeast fall out of the sequence.

The Caledonian complex overlies the Baikalian one with a general structural unconformity: both the angular and the azimuthal unconformity reaches 60°, the latter is particularly conspicuous in the southern part of the Novaya Zemlya. The degree of deformation of the Caledonian complex proper decreases northwards; in the northeastern part of the Novaya Zemlya it has a terrigenous-carbonate composition and a gentle monoclinal attitude.

The Hercynian complex, in turn, is separated from the Caledonian one by an unconformity which is especially distinct on the Novaya Zemlya, where it is azimuthal (15–30°) as well as angular. A granitoid intrusion took place on the Novaya Zemlya at the level of this unconformity. This happened, in particular, in the Matochkin Shar Channel region. The Hercynian complex includes deposits ranging from Upper Devonian to Permian and, similar to the Caledonian one, consists of two formations. The lower, miogeosynclinal formation is of Upper Devonian and Carboniferous age, carbonate (limestones and dolomites) composition, and is 1.5 to 2.5 km thick. In the Matochkin Shar region, the carbonate formation contains flows and sills of basic volcanic rocks in the west and is superseded by a terrigenous formation comprising conglomerates toward the east. The upper, early orogenic formation is Lower Permian and constitutes a 2 to 2.3 km thick paralic coal-bearing molasse series in the Pai Khoi and a 1 to 1.3 km marine molasse on the Novaya Zemlya. The accumulation of the complex terminated with a folding and formation of minor granitic and granodioritic plutons whose age is estimated at 240–250 m.y. Later, at the end of the Triassic–beginning of the Jurassic (190 m.y. B.P.) granite and alaskite intrusions were formed which correspond to the late orogenic stage of development (or reactivation).

The folded structure of the Pai Khoi–Novaya Zemlya arc consists of three echelon-like anticlinoria separated and surrounded by synclinoria. The **Pai Khoi Anticlinorium** is situated in the extreme southeast, extends along the northeastern slope of the Pai Khoi to the Yugorsky Shar, and probably attenuates in the Kara Sea. The anticlinorium is composed of all the three structural complexes described above and is asymmetrical, with a steeper and stronger dislocated southwestern limb; the folding here reaches isoclinal style, and southwest-directed overthrusts are not infrequent. The **Baidaratsky (Kara) Trough,** filled with Lower Permian molasse, is adjacent to the northeastern, more quiescent, limb. In the northwest, the Pai Khoi Anticlinorium borders on the **Vaigach** (Kara or Southern Novaya Zemlya) **Anticlinorium,** which is part of an echelon-like arrangement with the former, along a major fault. The Vaigach Anticlinorium extends from the northwestern part of the Pai Khoi toward Island Vaigach and the Southern Novaya Zemlya, the trend changing gradually from northwestern to meridional. This anticlinorium is composed of the same series as the Pai Khoi one; however, the shale formation of the Middle Paleozoic is here replaced by a carbonate formation, and the structure is somewhat less complicated. In the southwest, on the continent, along the two anticlinoria, there extends the **Korotaikha Trough** filled by Permian and Triassic deposits; this is yet another link in the system of cis-Uralian Foredeeps, with a characteristic asymmetry. The Korotaikha Trough is bounded on the west by a marginal uplift of the Sorokin (or Varandei) Range whose structure is similar to that of the Chernyshev Range uplift.

The northern slope of the Vaigach Anticlinorium is separated by the **Karmakul' Saddle (Synclinorium)** composed of folded Lower Permian molasses from the **Novaya Zemlya (Northern Novaya Zemlya) Anticlinorium,** which starts in the northern part of Island Yuzhny and occupies the entire Severny Island of the Novaya Zemlya. An appreciable part of this anticlinorium is hidden from observations by an ice sheet. The folds in the west of the Karmakul' Saddle are overturned toward the Barents Sea, in the east toward the Kara Sea. The northern Novaya Zemlya Anticlinorium is generally symmetrical in structure and consists of straight linear folds; in the south, however, in the region of the Matochkin Shar, the western limb becomes steeper and the folds are overturned westwards. The anticlinorium core comprises Cambrian–Ordovician rocks pierced through by granitic intrusions; farther north there are Silurian deposits; the limbs are composed of Devonian and Carboniferous rocks. A major fault passes along the northwestern limb of the anticlinorium (the Major Novaya Zemlya Fault, after ROMANOVICH, 1964), along which the Silurian of the anticlinorium core is thrust over the Devonian of its southwestern limb. The outcrops of Permian deposits along the western coast of the Novaya Zemlya mark the continuation of the Uralian Foredeep zone along the entire coast and shelf.

The Novaya Zemlya Anticlinorium, over its extent toward the north, changes its trend gradually from north-northeastern through northeastern to east-northeastern, i. e., near latitudinal, and this extent persists, according to gravimetry data, to the Arctic Institute Islands in the Kara Sea. The upper part of the Paleozoic cross-section becomes gradually less deformed, and the predominant rocks become increasingly more terrigenous and coarse clastic. In the extreme northeast, the Silurian deposits (Llandoverian–Wenlock) lie quite smoothly, filling in a major superimposed trough cut away from the Novaya Zemlya Anticlinorium by a latitudinal fault. The Wenlock beds are here of a molasse nature. Consequently, this part of the Novaya Zemlya is pre-Hercynian rather than Hercynian, according to the age of the main deformations. This makes it similar, as noted by POGREBITSKY (1971), to the northwestern part of the Severnaya Zemlya Archipelago, 600 km northeast of the Zhelaniya Cape.

The northwestern part of the Severnaya Zemlya Archipelago, according to the data of EGIAZAROV (1957), is composed of Cambrian to Devonian deposits. The Cambrian rocks constitute a rather thick (about 2 km) and coarse terrigenous formation; the Ordovician has a carbonate-terrigenous composition and contains gypsum, its thickness is about the same; the Silurian is mainly carbonate; and the Devonian consists of red beds, is lagoonal-continental, with a thickness of about 2 km. This complex, whose thickness totals 6–7 km, lies, generally speaking, conformably; however, its deformation is quite nonuniform. Linear folds are encountered in fault zones; confined to them are trap dykes and small granitoid bodies. Outside the fault zones the folding is brachymorphous in character, especially so in Devonian deposits filling major troughs.

The above pattern indicates that there is no typical Caledonian geosyncline on the Severnaya Zemlya, contrary to what was hitherto believed. Rather, one deals here with a platform margin – a zone of pericratonic subsidence, or else with an intracratonic zone which originated at or near the termination of a Caledonian geosyncline. In either case, this region must be close to the zone of final attenuation of the Uralian and Novaya Zemlya Paleozoic geosynclinal system.

Geologists of NIIGA (Scientific Research Institute of Arctic Geology), starting with B. S. ROMANOVICH (1964) provided grounds for the assumption that there exists in the Barents Sea a Caledonian fold system of a northeastern and in the east of an east-northeastern trend, which branches off the Scandinavian Caledonides in the region of the Tana Fjord across the Barents Sea and continues farther toward the Severnaya Zemlya.

This system, termed **Scandinavian–Severnaya Zemlya** is delineated on the Tectonic Map of the Basement in the USSR (1974). Its identification within the water area was based on the data of gravimetry surveys, as well as on the bottom relief and on the results of dredging operations on submerged banks and now appears doubtful.

The facies distribution and composition, as well as the size of clastic material in the Novaya Zemlya Silurian–Devonian deposits are indicative, according to BONDAREV (1979), of the supply from the Barents Sea region, i. e., from the area of the hypothetical Caledonides. In spite of the non-existence of the Scandinavian–Severnaya Zemlya system, the fact that a broad band of Baikalides occupies the entire area from the Kanin Peninsula and Kola Peninsula to the southern part of the Novaya Zemlya, makes one fully reject the concept of SHATSKY & BOGDANOV (1961) and STILLE (see STILLE, 1955) that there exists in the central and eastern parts of the Barents Sea and in the continental tundra a single large block with an Early Precambrian basement – **Barentsia** –, which is a fragment of the Eastern European Craton. It now seems more likely that there exist in the region just two or three small ancient blocks of the type of microcontinents. The largest of them seems to occupy the northeastern part of the Barents Sea and an adjacent part of the Kara Sea; this is the Barents Sea Massif proper. Its crystalline basement is exposed on the Northeastern Spitsbergen and on Island Belyi, where it is, however, reworked by Caledonian reactivation (to which the radiometric dating of rocks corresponds). It is reached now by a borehole under Upper Paleozoic–Mesozoic cover on the Franz-Josef-Land Archipelago. On the remaining territory of the massif, the basement occurs at a depth of 3 to 5 km. The cover of the massif emerges on the surface in the southern part of the Northeastern Land and on the King Carl Land (Spitsbergen Archipelago), and also in the Franz-Josef-Land Archipelago. It consists of a Carboniferous carbonate formation, Permian red beds and evaporites, and shallow marine and paralic, partly coal-bearing Triassic–Jurassic–Lower Cretaceous formations. Associated with the latter on the Franz-Josef-Land are Lower Cretaceous traps. The cover lies subhorizontally, however, on the Franz-Josef-Land it is strongly broken up by faults of mainly northeastern and latitudinal orientation. Two median massifs have been assumed to exist in the southern part of the Barents Sea Basin: the **Malozemel'sky (Kolguev)** and **Bol'shezemel'sky (Khoreiver)** massifs.

6.7 Main stages of evolution

6.7.1 Pre-Middle Riphean history

It is now quite obvious that the geosynclinal belt, at whose western margin the Uralian system later formed, could not originate earlier than the Middle Riphean, which is the time of accumulation of the Bashkirian Yurmata series and its equivalents. Before that the western Uralian zones up to the Central Uralian zone had been a direct continuation of the Eastern European Craton and its epi-Karelian granite-gneiss basement. The Taratash Swell of this basement in the northern part of the Bashkirian Anticlinorium is a direct confirmation of the above statement. As concerns the eastern zones, the presence of similar formations in the Mugodzhary Mountains and farther north in the Eastern Uralian and possibly Transuralian anticlinoria, as well as in the Polar Urals indicates that relics of the same basement which occupies a vast territory in the Central Kazakhstan Kokchetav Massif are preserved here, too. At the same time, there is substantial evidence for the assumption that both the Tagil–Magnitogorsk and more eastern synclinorium zones had originated in the Paleozoic on an oceanic crust. These data may be correlated only if one assumes that the Early Precambrian continental crust had primarily occupied the entire territory of the present Uralian fold system and was then secondarily destroyed within its Paleozoic eugeosynclinal troughs. This assumption is supported also by the sublatitudinal, deviating to northwestern or northeastern, trend of the basement structures in the Russian Platform, which is practically perpendicular to the orientation of the Uralian Geosyncline.

The terrigenous-carbonate deposits of the Bashkirian Anticlinorium Burzyan series are the youngest pre-Middle Riphean formations in the Urals. These deposits are regarded as a stratotype of the Lower Riphean. They must be deposits of an ancient platform cover rather than miogeosynclinal formations, if one judges by their formational nature. The platform conditions of development in Bashkiria during the Burzyanian time are attested to by the alkaline-basaltic composition of volcanic rocks which preceded the accumulation of the Burzyan series, and also by the typically platform character of the rapakivi granite Berdyaush Pluton, intruding it. Thus, platform conditions had existed on the territory of the future Urals during the Middle Proterozoic, as well as in the beginning of the Late Proterozoic. At the same time, it is very probable, in accordance with an assumption of GARAN' (1960), the first investigator of the Uralian Riphean, that the Burzyan series fills a structure of the aulacogen type, i. e., of a rift oriented in the direction of the future Uralian Geosyncline. The same opinion was recently expressed by S. N. IVANOV (1981).

6.7.2 The Grenville and Baikalian stages (Middle and Late Riphean, Vendian, and Early Cambrian)

An abrupt change in the evolution of the Urals took place with the onset of the Middle Riphean, when relatively vigorous subsidence started along the entire western slope and in the Central Uralian zone; this subsidence continued in the northwest within the western part of the young Timan–Pechora Platform. The onset of the subsidence was accompanied in the east of the Bashkirian Anticlinorium and in the Central Uralian zone by an outburst of volcanic activity which led to the formation of spilites, diabases, and albitophyres of the Mashak suite and its analogues which overlie the Burzyan series rocks in Bashkiria with a sharp unconformity. The overlying sediments of the Middle Riphean Yurmata series form a quite independent sedimentation cycle starting with a terrigenous (quartzite-shale) and terminating with a carbonate formation. They are separated by a hiatus and a certain unconformity from the deposits of the overlying Upper Riphean Karatau series. The unconformity points to weak fold and fault deformations; concurrently with them basic (gabbro and gabbro-diabase dykes and plutons) and then also acid (the Guben, Ryabinovka and other massifs)[4] magma intruded. There are indications of

---

[4] According to some data these granites are older (1,350 m.y.) corresponding thus to the Gothian epoch of diastrophism.

the Grenville uplifts and deformations, possibly also granite formation in the Polar Urals too (PUCHKOV, 1975). Radiometric dating in the Mugodzhary Range also bears witness to the manifestations of the Grenville diastrophism.

This implies that the Grenville diastrophism was more vigorous in the Urals than in the region of the Russian Platform, however, it did not result in the formation of a mountainous relief here either, which is indicated by a total absence of the molasses. The Zil'merdak quartzites lying at the base of the Yurmata series, just like the similar Zigal'ga quartzites at the base of the Yurmata series, are the products of supply from the Eastern European Platform rather than from the Uralian inner uplifts. This suggests that the Grenville stage of the development of the Uralian system was not complete, lacking the final orogeny.

The situation is different in the case of the **Baikalian stage.** The Upper Riphean deposits, unlike the Middle Riphean ones, form two sedimentation cycles consisting of a terrigenous (below) and carbonate (above) formations. Repeated appearance in the sequence of terrigenous sediments (the Inzer suite) must be connected with a certain revival of uplifts at the level of the first phase of the Baikalian tectogenesis. Basic and intermediate volcanism was expressed in the Central Uralian zone within the Middle, Northern, and Polar Urals. The region of sediment accumulation expanded in the Late Riphean westwards, in the direction of the Eastern European Platform.

Baikalian sedimentation culminated in the deposition of the Late Vendian Asha molasse series lying on different Upper Riphean horizons (GARAN', 1960; BEKKER, 1968). The Asha series together with the underlying Riphean formations was later deformed into folds and broken up by faults (overthrusts, by KAMALETDINOV's (1974) determination). This happened in the pre-Ordovician, probably even as early as pre-Late Cambrian time, i. e., during the Salairian epoch of tectogenesis. Baikalian and Salairian diastrophism intensified eastwards and northwards, reaching maximum in the Northern and Polar Urals, as well as in the Timan where the Riphean–Vendian deposits experienced considerable metamorphism, up to the amphibolite facies, and are pierced by granitoid intrusions. The Baikalian and Salairian uplifts were responsible for the extremely limited occurrence of Cambrian sediments in the Urals: the Lower Cambrian was detected only in two areas of the Southern Urals: in the Sakmara zone and in the valleys of rivers Uya and Sanarka at the eastern slope (Transuralian Anticlinorium), and also in the Polar Urals; no Middle Cambrian deposits could be found in the Urals; the Upper Cambrian is known in eastern zones of the Southern Urals and in the Polar Urals; in either of these cases it is closely associated with the Ordovician which, in the absence of the Upper Cambrian, overlies Precambrian formations with a sharp unconformity.

All this evidence indicates that, unlike the Grenville cycle, the Baikalian cycle has reached completion in the Urals as well as in the Timan. The Baikalian-Salairian movements resulted in the formation on the territory of the future Timan–Pechora Platform and the Urals, as well as in the southern half of the Pai Khoi–Novaya Zemlya system, of a folded highland system which was called "pre-Uralides" by KHERASKOV (KHERASKOV & PERFILIEV, 1963). The trend of this system within the Middle and a greater part of the Southern Urals coincided in general with the trend of the Hercynian Urals ("Uralides" as termed by KHERASKOV), however, in the Northern and Polar Urals and especially in the south of the Novaya Zemlya it disgressed northwestwards and in the extreme south, probably, southwestwards, as was suggested by KHERASKOV & PERFIL'EV (1968), and

ZHURAVLEV (1972). Thus, the "pre-Uralides" enveloped the eastern protrusion of the eastern European Craton with a culmination in the region of the Ufa Plateau.

Hence, there is little doubt that there once existed the Baikalian–Salairian "pre-Uralides" fold system on the territory of the present Urals. Much less clear is the problem of the tectonic nature and structural pattern of the precedent region of the Middle and Late Riphean sedimentation. Is it true that the Middle and Upper Riphean deposits of the Western and Central Uralian zones are geosynclinal? Some investigators (PRONIN, 1971) believe them to be of platform nature, which opinion is not without foundation: relatively small thickness, obvious indications of accumulation within shallow-water, shelf environments (cross lamination of quartzites, abundance of stromatoliths in carbonate rocks, etc.). Yet, if one follows variations in the composition and thickness of the Riphean from the central regions of the Russian Platform to the Cis-Uralian region and Urals, one can clearly observe a transition from the aulacogenic environment to those of pericratonic subsidence and farther on to a zone, hypothetically miogeosynclinal, of linear downwarping (the Bashkirian Anticlinorium) at the outer periphery of the shelf and in the Central Uralian zone, probably, to the continental slope. This implies deep-water environments must have existed in the Eastern Uralian zones, which is corroborated by the absence of any substantial supply of clastic material from the east at the western slope (with the exception of the Burzyan series) and in the axial zone of the Urals. The situation is, however, not quite clear in what concerns Riphean deposits at the eastern slope of the Urals. The Chulaksai suite (Middle Riphean) identified by MAMAEV & CHERMENINOVA (1973), which is composed of carbonaceous siliceous shales, quartzites, and amphibolites, may represent sediments of this basin. It is overlain unconformably by the Zymnik tuffite-shale suite of probably Vendian age (BEKKER, 1974).

It is only in the Northern and Polar Urals that there are substantial grounds to identify a eugeosynclinal as well as a miogeosynclinal zone in the Late Precambrian. The eugeosynclinal zone has been established in the Central Uralian zone according to the abundance of volcanic rocks in the cross-section of both Middle and Upper Riphean and Vendian; in the latter the volcanic rocks are of acid composition. However, the true extent of manifestations of the Late Precambrian volcanism in these regions is not yet precisely known, so that identification of an eugeosyncline of the corresponding age here is quite arbitrary. PRONIN (1971) believes that the Timan–Uralian Riphean geosyncline had been generally oriented in the northwestern-southeastern direction, and that a continuation of its eugeosynclinal zone should be sought in Central Kazakhstan, which, however, is not confirmed by the evidence for the latter region, where the Baikalides are present on a limited scale (see next chapter).

The paleotectonic environment in the Urals throughout the Cambrian period is not quite clear either. It is possible, in the opinion of several researchers, particularly with respect to the Middle, Northern, and especially Polar Urals, that the Vendian conditions had still prevailed at the beginning of the Cambrian, and then, during the Lena age and during the Middle–beginning of the Late Cambrian upheavals predominated. At the same time, in two Southern Uralian zones (in the Sakmara zone in the west and in the Transuralian zone in the east) a spilite-diabase formation with lenses (biostromes) of limestones containing a Lena age archeocyathean fauna has been found. It is true that these rocks are strongly tectonized and are present in the Sakmara zone only in the melange, yet they enable one to assume that the Paleozoic geosyncline could originate in the south of the Urals as early as the Cambrian or even Vendian or else there could exist here Cambrian troughs of short-term development which had terminated in the Salairian epoch of tectogenesis (the latter, however, is less likely). If one considers the Urals as a whole, then the beginning of the major stage in its geosynclinal evolution should be referred to the end of the Cambrian–beginning of the Ordovician.

Concluding the discussion of the pre-Ordovician history of the Urals, we must stress once more the vagueness of the knowledge of a Mid-Riphean–Cambrian tectonic regime and its particularly conditional interpretation as geosynclinal. Recently, IVANOV (1981) put forward fairly weighty arguments for continental-rift development of the Urals (and Timan) at that stage. Angular unconformities indicative of folding, as well as regional metamorphism and granitization, especially in the north, fit in this conception worst of all. However, it is possible that no continuous development of one rift system took place here, but recurrent opening and closing, due to compression, of early, middle and late Riphean rifts or intracratonic geosynclines (R_2 and R_3). In any case, the absence of ophiolites and typical sedimentary geosynclinal formations testify against true geosynclinal conditions in the Urals in pre-Ordovician times.

6.7.3 The Caledonian stage (Late Cambrian–Ordovician–Early Devonian)

The Ordovician formations, starting from the Tremadocian stage, occur transgressively and with sharp unconformities on the underlying rocks practically throughout the Urals. Only the extreme north–Polar Urals region–presents a certain exception. There, in the deposits lying conformably with the Ordovician rocks, Upper Cambrian fauna was found. The coarse clastic composition of the Lower Ordovician indicates that the deposition of these rocks was preceded by upheaval, and the wide occurrence of acid volcanics in the directly underlying Upper Precambrian–Lower Cambrian formations, also throughout the Urals suggests that these upheavals were accompanied by vigorous volcanic activity.

The situation of the Paleozoic eugeosyncline of the Urals eastern slope between the Eastern European Craton, which used to extend to the present Central Uralian zone, and the Central Kazakhstan microcontinent is a sure indication of the fact that the Uralian Geosyncline had formed in Early Paleozoic over a continental crust, in all probability, in the process of rifting (NALIVKIN, 1972). This process spread here from inner areas of the Uralo–Siberian belt where it had started as early as the Late Precambrian. IVANOV (IVANOV et al., 1975), PERFIL'EV (1979), and other investigators identify Upper Cambrian–Ordovician arkose-graywacke sediments with sills and dykes (not infrequently parallel) of diabases, often subalkaline, and also with acid tuffs, as witnesses of this initial rifting stage of evolution of the Uralian Paleozoic geosyncline.

Rifting later caused the formation of a vast and deep basin on an oceanic crust with strong manifestations of tholeiite-basalt volcanism which led to the origination of a spilite-diabase formation. The Eastern Uralian Sea was situated at the periphery of the Uralo–Siberian Ocean. Miogeosynclinal sediments of the Western Uralian zone deposited along its western continental margin; in zones farther east, alongside basic volcanic rocks, cherts are very common. The most vigorous volcanism was confined to the Tagil–Magnitogorsk zone; its expression was much weaker in Uralian zones farther east.

The Central Uralian Uplift, which had separated the mio- and eugeosynclinal regions, had been an island in the Early Ordovician only and was later covered by the sea. It had retained this submerged position throughout the Middle Paleozoic (SMIRNOV, 1971) or at least during its greater part, with the exception of the upheaval and deformation epochs.

The first compression phase started in the south before the Silurian, in all probability, at the end of the Ordovician (the Taconian phase of Caledonian tectogenesis). The pre-Silurian unconformity in the Southern Urals, east of the Zilair zone is so distinct that some investigators (ABDULIN, 1971) draw a line of demarcation between two stages of Paleozoic evolution of the Urals–Caledonian and Hercynian–along this boundary and even refer the Ordovician to the "pre-Uralides". This may have been associated with the proximity of Central Kazakhstan with its Caledonian tectogenesis, particularly distinct at this boundary. The angular unconformity between the Ordovician and Silurian west and north of the Southern Uralian eastern zones is expressed only locally; somewhat wider spread is an erosion unconformity with a hiatus in the Upper Ordovician, and in many cases both systems occur conformably with a gradual lithological transition between them (PRONIN, 1971). The fact that the areas of conformable and unconformable occurrence alternate without any apparent regularity within the same longitudinal structural-facial zone has been interpreted by PRONIN in the sense that the structural-facies zonality in the Ordovician had been oriented essentially according to the "Timan" pattern, i.e., in the northwestern direction, and only starting from the Silurian longitudinal, Uralian trends took over (other investigators believe that the rearrangement took place even later). This is probably why the unconformity between the Ordovician and Silurian is expressed best of all in the north of the Uralian western zones, near the boundary with the Timan–Pechora Platform (the Polyudov Range and the Chernov Ridge). Two facies zones formed during the Silurian in the north of the Western Uralian zone, starting from the northern part of the Middle Urals: the outer, where the shelf terrigenous-carbonate sediments continued accumulating, and the inner, which must have corresponded to the continental slope and rise and in which much less thick siliceous-clayey sediments deposited under the conditions of presumably uncompensated subsidence. In the Polar Urals, the first zone is known as Yelets, the second as Lemva zone. V. N. PUCHKOV (1975) believes that these zones separated out as far back as the Ordovician.

Volcanic activity continued within the eastern Uralian zones in the Silurian and even reached its maximum intensity there. The Tagil and Magnitogorsk zones were the arena for its particularly vigorous expression, as was also the case during the Ordovician. Farther east, the thickness of the volcanic rocks decreases and units of siliceous-clayey graptolite shales become more important. The composition of the volcanic rocks, especially in the west, had changed noticeably by the end of the Silurian: the uniform spilite-diabase series were replaced by differentiated continuous andesite-dacite-quartz albitophyre or contrasting (bimodal) diabase-quartz albitophyre series with a predominant andesite-basaltic rather than basaltic nature of the magma. Linear effusions were superseded by central-type eruptions with a considerable role of explosions, in all probability, within island arcs formed by that time. There exist different petrological and tectonic interpretations of this variation in the composition of magmatic products. Some petrologists believe that this change results from the evolution of subcrustal magma chambers, while others, FROLOVA (FROLOVA & BURIKOVA, 1977) in particular, maintain that andesites and more acid lavas may appear only as a result of formation of crustal magma chambers in a sialic crust. The latter opinion implies that a continental crust must lie at the base of the entire Uralian Eugeosyncline, and this may be accounted for by the beginning of its oceanic crust obduction on the western continental framing, possibly, with a detachment from the lower, mantle part of the lithosphere and its concurrent underthrusting beneath the same framing along a Zavaritsky–Benioff zone.

At the end of the Silurian and especially in the Early Devonian tectonic movements greatly intensified in the Urals under the conditions of sharply growing compression. This epoch of tectonic activity corresponds to the global Late Caledonian epoch. It was the strongest in the Eastern Uralian Anticlinorium and Transuralian zones in the Southern and Middle Urals which have played the role of a stable positive element in the Urals structure ever since (LUCHININ, 1972; SMIRNOV, 1971). It was only the Eastern Uralian Trough separating them which retained a tendency towards subsidence. The Middle and Upper Devonian within the Eastern Uralian and Transuralian uplifts overlie unconformably the metamorphosed Precambrian and Cambrian–Lower Devonian basement which comprises plutons of the gabbro-plagiogranite complex of post-Silurian–pre-Upper Eifelian age. The Middle Devonian deposits include both red molassoids and basalt-andesite-rhyolite volcanic rocks of subsequent type, which are accompanied by subvolcanic comagmatic rocks. One cannot but see expressions of the powerful Late Caledonian tectogenesis of the near-by Central Kazakhstan in all these phenomena.

The Late Caledonian movements also had an appreciable effect on an almost entire territory of the Urals. The Tagil eugeosynclinal trough had closed on a greater part of its extent by the Middle Devonian. This did not involve its northern end (north of Severnaya Sos'va River), as noted by SHTREIS (1951). Major deposits of bauxites are associated with the pre-Middle Devonian hiatus in the Northern Uralian bauxite-bearing region. The Upper Ludlow–Lower Devonian deposits show a rather mild degree of deformation in this region, and the volcanic rocks of this age belong to the trachybasaltic formation. The Sakmara and other nappes in the eastern part of the Zilair zone must have formed during the same Late Caledonian epoch of movements and deformations, in accordance with recent data (RUZHENTSEV, 1971). This assumption is based on the fact that these nappes incorporate Silurian and even (?) Lower Devonian rocks, but the deposits of the Zilair series (D_3–C_1t) overlap these nappes transgressively and unconformably. Over the entire extent of the Middle, Northern, and Polar Urals, on both sides of the Central Uralian uplift zone and farther south in the Bashkirian Anticlinorium, the Middle Devonian is lying transgressively and with a more or less sharp unconformity over the deposits ranging from Lower Devonian and Silurian to Riphean and with conglomerates, not infrequently quite coarse, to several dozens of meters in thickness, or with sandstones (the so-called Takata sandstones) at the base. Clastic material was supplied both from the Central Uralian zone which must have been an island for some period of time, and from the Volga–Uralian Anteclise of the Russian Platform. Dolerite and diabase dykes and sills were formed in the Northern Urals during the same phase.

Thus, the Caledonian stage of the Urals evolution culminated in upheavals almost throughout the territory, with probably the only exception of axial parts of some troughs (Magnitogorsk and Eastern Uralian), in rather strong deformations, in the Southern Urals or probably even in the Urals as a whole (KAMALETDINOV et al., 1978), which led to the formation of nappes and also to the appearance, although on a relatively smaller scale, of molasse-like formations and orogenic (subsequent) magmatic rocks, including granitoids. All this evidence provides sufficient grounds for the identification of the Caledonian stage in the tectonic history of the Urals; however, it does not make it possible to consider it fully completed, at least for the entire fold system. The Late Caledonian upheavals were short-lived and failed to produce a true highland; neither did they lead to total disappearance of geosynclinal subsidence which resumed soon afterwards and was quite vigorous.

6.7.4 The Hercynian geosynclinal stage (Middle Devonian–Early Carboniferous)

This stage started with resumed subsidence practically throughout the Urals; this subsidence had gained force until the beginning of the Carboniferous. Shallow-water carbonates still prevailed in the Western Uralian zone during the Devonian and Tournaisian. This was not the case only in the Lemva zone of the Polar Urals and its southern continuation, where deep-water siliceous-clayey sediments continued to accumulate. The Bashkirian and Central Uralian uplifts were again covered by the transgressing sea. Volcanism, first of basic and then, starting from the Late Devonian, of increasingly more differentiated composition, gained force again in the Magnitogorsk and Eastern Uralian troughs as well as in the northern continuation of the Tagil Trough. This picture, which then recurred in the Early Carboniferous, enabled investigators of Uralian volcanism to talk of a cyclic multifold recurrence in its history of alternating undifferentiated, diabasic volcanism and differentiated (diabasic-quartz albitophyre) one. IVANOV et al. (1975), however, introduced an essential amendment into this concept. They showed that the composition and character of volcanic manifestations changed at different times in different zones. For example, in the west, in the Tagil zone, this change took place as early as the Silurian (in the west of the Tagil zone even in S_1) and in the east (the Magnitogorsk zone) in the middle of the Devonian[5].

Volcanic activity in the Magnitogorsk zone attenuated markedly at the beginning of the Late Devonian and ceased almost completely in the Famennian time. The thick series of volcanic rocks which had accumulated here since the Ordovician or even Cambrian, was superseded by the Zilair graywacke-clayey flyschoid series which was particularly extensive and thick (up to 3 km) farther west, in the Zilair Trough, and which initially overlapped the Uraltau Uplift.

That this clastic series formed in the vicinity of the platform in the Western Uralian zone is an obvious indication of intensified upheavals in the inner Uralian zones, primarily, in the Eastern Uralian Anticlinorium zone (SMIRNOV, 1971). At the same time, volcanic rocks, ranging in composition from basic to acid, continued accumulating alongside graywackes in the Eastern Uralian Trough.

On the whole, during this earliest stage of the Urals Hercynian development, the major subsidence and respectively the greatest thicknesses were confined to the northern part of the Tagil zone, to the Zilair, Magnitogorsk, and Eastern Uralian (synclinorial) zones, whereas the Eastern Uralian Anticlinorium zone experienced the greatest upheaval.

---

[5] Initial relationships between these two zones were distorted by their subsequent tangential movements: the Magnitogorsk zone advanced westwards farther than the Tagil zone.

A new phase of upheaval and deformation revival affected the eastern zones as early as the beginning of the Tournaisian time. There are indications of vigorous upheavals during this period of time also in the Transuralian uplift zone. On its western periphery, in the Eastern Uralian Trough, there deposited a paralic coal-bearing formation with minor acid volcanics in the Late Tournaisian–Early Viséan. This formation filled the structures of the type of superimposed troughs (the Poltava–Bredy Syncline is a typical example), which overlap unconformably older formations up to the Ordovician. The ascending movements of the Early Carboniferous affected the Eastern Uralian Uplift zone no less strongly. In all the three zones, which behaved as an integral unit since then, these movements were accompanied by the formation of a gabbro-granitic intrusion complex. Similar intrusive activity took place in the Tagil Trough.

Early Carboniferous upheavals affected partly the Magnitogorsk zone, however, they spread to the zones farther west only in the Early Viséan time. In the Early and Middle Viséan this territory as well as the Eastern Uralian Trough saw the accumulation of a coal-bearing formation. The clastic material of the latter was supplied from the west, from the Russian Platform which rose in the Viséan time above sea level and was dissected by the Kama–Kinel fluvial system. Smirnov (1971) believes that the Eastern Uralian Trough acquired its individuality by this particular time.

The phase of tectonic activity in the Early Carboniferous corresponds generally to the Bretonian phase of Hercynian tectogenesis in Western Europe. This activity started as early as the Late Devonian, when the Zilair Synclinorium initially formed, and, therefore, it corresponds to a certain extent also to the Appalachian Acadian tectogenesis. On the other hand, the completion of this epoch of tectomagmatic activity in the Urals coincided with the inter-Viséan phase of movements, established in the Altai und Tien Shan, and which also found expression in the Moroccan Meseta.

In the Late Viséan and Namurian, the Uralian Geosyncline experienced a brief period of weakened tectonic activity and broad sea transgression, with deposition of essentially carbonate sediments. Volcanic activity recommenced as early as the Late Tournaisian in the eastern part of the Magnitogorsk Trough, though it affected mainly zones farther east: the Eastern Uralian Trough and Transuralia. The Valer'yanov volcanic belt along the eastern boundary of the Uralian Geosyncline formed during the Viséan–Namurian time. The products of volcanism had everywhere a very differentiated composition at that time: from basalts and andesites to rhyolites and trachytes.

6.7.5 The Hercynian orogenic stage (Middle Carboniferous–Early Triassic)

Following a brief Viséan–Namurian "rest", the Uralian Geosyncline entered its final, orogenic stage of evolution. This stage started with another phase of deformations at the end of the Namurian–beginning of the Middle Carboniferous, the phase corresponding to the Sudetian tectogenesis in Western Europe. This phase was particularly pronounced in the Eastern Uralian Uplift zone, where it started the period of formation of gneiss domes and granitic batholiths. However, in the axial parts inherited from the geosynclinal stage of the Magnitogorsk and Eastern Uralian troughs as well as in some graben-like depressions within the Eastern Uralian Uplift, the Namurian limestones are superseded conformably by thin Bashkirian ones.

The upheavals intensified in the Late Carboniferous; the eastern zones of the Urals and also the Central Uralian zone and the eastern part of the Zilair zone rose above ocean level; it was only the narrow graben-troughs which were filled with coarse continental molasses partially red in the south.

At the same time the outermost zone of the geosyncline, the Western Uralian one together with the western part of the Zilair zone, continued to subside steadily under the conditions of the persisting marine environments. In the Middle Carboniferous, however, the accumulation of carbonate sediments gave way to the accumulation of a terrigenous flysch formation, which proceeded during the Late Carboniferous and Earlier Permian and spread to the future eastern limb of the Uralian Foredeep. The western limb was part of the platform at that time, however, at the end of the Carboniferous it started experiencing increasingly sharper flexure subsidence with the formation of a barrier reef above the flexure. A deep-water trough filling with a thin layer of carbonate-siliceous-clayey sediments had formed between this reef and the zone of flysch accumulation by the Permian time. This is how the Uralian Foredeep came into being.

The upheaval of the Uralian inner zones intensified in the Early Permian, having produced a mountainous relief and having arrested further sedimentation, and advanced to the Western Uralian zone. The coarse clastic material reached the inner limb of the foredeep in the Artinskian time, and the flysch formation gave way to the marine gray molasses. In the Kungurian time, this trough squeezed between the paleo-Urals highland and the emerging Russian Platform and also bounded in the south by the newly formed fold structures of the Northern Caspian and Southern Emba, became an arena of accumulation of evaporites which are represented by rock and potassium salts in the axial part. The accumulation of evaporites must have started either in deep-water environment (A. L. YANSHIN and M. P. FIVEG) or under the conditions of an intracontinental depression with the bottom several hundred meters below ocean level (V. I. KOPNIN), as is presently assumed for the case of the Late Miocene (Messinian) Mediterranean Sea. Within the extreme northern link of the Uralian Foredeep in the Vorkuta (Kos'yu –Rogov) Depression, humid climate prevented salt accumulation and promoted replacement of the salt-bearing molasse by coal-bearing molasse making up the Pechora (Vorkuta) coal-bearing basin.

The Uralian Foredeep, filled with evaporites almost to ocean level during the Late Permian and Early Triassic epoch, became a region of accumulation of red lagoonal-continental molasses. This indicates another intensification of upheaval in the Urals following its marked weakening in the Kungurian time. In the Middle Triassic, a greater part of the trough underwent complete drainage, and only in the extreme north did the sediments of the same formation accumulate further.

The thrust-fold Uralian structure was finally shaped during the Permian and Triassic periods; this applies, in particular, to the outer zones. The principal features of this structure in the inner zones must have originated much earlier. As pointed out above, the Upper Ludlow–Lower Devonian strata are characterized by flat-dipping beds in the Tagil zone (its southern part). In the Eastern Uralian Anticlinorium zone this applies to deposits starting from the Middle Devonian, in the Eastern Uralian and Kustanai synclinoria and in the Transuralian Anticlinorium starting from the Upper Viséan–Namurian, and in the Magnitogorsk Synclinorium starting from the Middle Carboniferous. Block movements along subvertical faults[6] in the Eastern Uralian zones and in Transuralia, superimposed on the folding, started no later than the Late Carboniferous, judging by the age of the rocks filling the adjacent near-fault troughs. The main (or concluding) phase of formation of the Eastern Uralian and Transuralian granitic batholiths took place in the Early Permian, judging by the sharp maximum in the radiometric dating (17% out of 36% for all Hercynian granitoids, PRONIN, 1971).

In contrast to the eastern zones, the structure of the western zones began taking shape only in the Permian, although Late Caledonian movements were clearly expressed at the periphery of the Central Uralian zone. The phase at the boundary of the Artinskian and Kungurian ages must have been the first phase of Late Hercynian tectogenesis in the Urals, if one takes into account the greater degree of deformation of the Artinskian and older deposits in comparison with the Kungurian ones in the eastern limb of the Uralian Foredeep. Coinciding with this phase, which corresponds to the Saalian phase in Western Europe, is the mentioned peak of granitization in the east of the Urals; it is very distinct also in the Hercynides of the southern framing of the Eastern European Platform (the Donets–Caspian zone). It is during this particular phase that nappes originated in the region of the Ufa Amphitheater, in all probability, of gravitational nature (ZHIVKOVICH, 1980).

The next phase must have taken place at the end of the Permian, however, it is poorly documented. One can only assume, after SOBOLEV (1963), that by that time the formation of the fold-overthrust structure of the Western Uralian zone had been completed. As for the greater part of the Uralian Foredeep, its final deformations took place in the Middle Triassic. The age of these deformations in the foredeep in front of the Polar Urals and in the dislocation zone of the Chernyshev Range has been estimated as the end of the Triassic; the Triassic deposits here include basaltic flows. Of even greater significance were the movements at the end of the Triassic–beginning of the Jurassic within the Pai Khoi–Novaya Zemlya system, where granitic intrusions are also referred to this epoch.

In the Urals, just like in the Hercynian fold systems of Europe and Middle Asia, the orogenic period witnessed major transcurrent displacements which were described in great detail by PLYUSNIN (1971). The transcurrent faults here are mainly longitudinal and oblique, which had prevented their detection; the displacements are sinistral.

In Transuralia, in the Eastern Uralian Synclinorium and particularly on the Transuralian Uplift, the Paleozoic basement in the Late Permian–Early Triassic time was broken up by subvertical faults along which basalts and rhyolites of the Tura series outpoured. These rocks were then preserved in the near-fault depressions. Volcanic rocks alternate with continental sediments of molassoid type. These formations may be regarded as transitional from orogenic to taphrogenic. Taphrogenic deposits proper are Rhaetian – Liassic rocks filling another graben system. The formation of faults bounding the grabens and the deformation of their sedimentary-volcanic fill reflect the same tectogenic epochs – Middle Triassic and Early Jurassic (Early Cimmerian), which are responsible for the folding deformations on the Urals western periphery. It is significant that the faults along the graben edges were transformed during the Early Cimmerian phase into overthrusts with an amplitude of up to 4–5 km.

---

[6] These particular faults were recorded on the DSS profiles, although inclined faults have also been detected recently.

6.7.6 The Platform stage (Jurassic–Eocene)

The Triassic was generally transitional from the orogenic to platform stage in the evolution of the Urals. The platform stage proper started in the Jurassic after the fold system underwent complete peneplanation. Starting from the Middle Jurassic, the eastern part of the system became involved in the increasing subsidence of the Western Siberian Megasyneclise, the southern part in the subsidence of the Turanian Platform, the anticlinal zones in this process showed a tendency towards lagging behind. This differentiation was particularly distinct in the Southern Urals, where depressions filled with Rhaetian–Jurassic and younger sediments (the Orsk–Primugodzhary Depression, etc.) alternate with uplifts on which these sediments did not deposit. Repeated movements along older faults proceeded on the depression slopes. In the Senonian, the greater part of the Urals was covered by the sea; in the extreme north and south, the seas of the Russian Platform linked across the Urals with the seas of Western Siberia and Turgai. A considerable transgression took place also in the Eocene, when this linkage may have occurred in the Middle Urals, at the latitude of Sverdlovsk.

6.7.7 The Neotectonic stage (Oligocene–Quaternary)

The recent (neotectonic) stage in the Urals evolution–the stage of tectonic reactivation – started in the Oligocene. By the end of the Oligocene, the entire territory of the Urals rose above ocean level. Upheavals intensified in the middle of the Miocene–beginning of the Pliocene, in the middle and end of the Pliocene, in the second half of the Quaternary period: their total amplitude reached 500–700 m. The uplifts were of an arch-block, somewhat differentiated character, which is indicated, in particular, by the existence of tectonic scarps on both sides of the Central Uralian Uplift, on the Bashkirian Uplift, at the boundary between the Urals and its foredeep, Middle Uralian Foredeep (Solikamsk–Sylva Depression), and the Ufa Plateau of the Russian Platform, the Urals and the Western Siberian Platform. The latter scarp is especially pronounced in the Northern Urals, where it reaches a height of up to 150 m and can be followed over a distance of 500 km; marine sediments of the Paleocene–Eocene along it are strongly deformed.

Recent upheavals affected mainly the zone of the Central Uralian and Bashkirian uplifts and only partly the Western Uralian, Tagil–Magnitogorsk and Eastern Uralian Anticlinorium (Mugodzhary) zones. The eastern zones generally retained the tendency toward relative subsidence, connected with the evolution of the Western Siberian Megasyneclise. This must be associated with their greater consolidation in the process of Paleozoic granitization and regional metamorphism. A similar explanation suggests itself for the transverse subsidence of the Urals at the latitude of Perm' and Sverdlovsk, though here one should take into account a proximity to the surface of the ancient, Early Precambrian basement.

In the recent stage, the Uralian Foredeep showed a clear tendency toward renewal of active subsidence, especially so on its middle and southern parts. This found expression in the accumulation of continental coal-bearing Oligocene–Miocene deposits in the Belaya Depression, and then in the advance of the Akchagyl (Late Pliocene) Sea into the Kama and Belaya depressions; Quaternary deposits also have an increased thickness in these depressions.

Generally speaking, recent reactivation of the Urals upheaval did not cause the greater part of the region to lose its platform character; in a few areas only (near-Polar Urals) it acquired an alpine relief. As a matter of fact, this is not a true epiplatform orogen, not a regenerated but only rejuvenated highland, a part of the Epihercynian platform, nonuniformly reactivated.

6.8 Some conclusions and problems

In view of the enormous extent of the Uralian geosynclinal system, one does not wonder why this system shows certain changes along the strike, on the contrary, its striking persistence, especially contrasts between individual longitudinal zones are spectacular. It is, therefore, only natural that the Urals has become a classic example of a linear geosynclinal system separated into mio- and eugeosynclinal regions across the strike.

The fact of the origin of "greenstone" Uralian zones on the oceanic crust does not leave much doubt at present, particularly so since even in the present Uralian cross-section the DSS data point to the occurrence of late geosynclinal volcanic rocks in these zones directly on the crustal "basaltic" layer. The principal problem, which still remains unsolved, concerns the interpretation of the salients of the granite-gneiss basement, which are observed on the eastern slope of the Urals, and of their relationships with the ophiolite belts. Three solutions to this problem are possible: 1) The Uralian Paleozoic geosyncline could form within the region with an older continental crust, and, consequently, the basement outcrops may well be regarded as fragments of this crust, i. e., as microcontinents. 2) They may be remobilized protrusions of the granite-gneiss basement—eastern continuation of the basement of the Eastern European Craton, which lies under the ophiolite nappes produced in the process of obduction. 3) These protrusions may be regarded as foci of formation of continental crust in the primary oceanic space.

The first assumption is contradicted, primarily, by the great quantity of ophiolitic belts within the Urals and Transuralia; it is difficult to regard each of them as a relic of an individual oceanic gap and each band with outcrops of the granite-gneiss basement as an independent microcontinent. Secondly, a quite specific ophiolite, mainly serpentinite fringing of granite-gneiss domes and arches can be observed within the Eastern Uralian and Transuralian anticlinoria. This circumstance suggests rather the existence of major tectonic overlapping of the ancient continental basement by ophiolite nappes. The DSS data, however, do not reveal continuous occurrence of this basement under the Tagil and Magnitogorsk "greenstone" zones, unless one assumes that this basement enters into the composition of the geophysical "basaltic" layer, as a result of deep metamorphism, which cannot be ruled out either. The least probable seems to be the third assumption, which is contradicted by the discordant cutting of the inner deep structure of the Uralian miogeosynclinal zone along the Main Uralian Fault, which indicates the initial continuation of this basement eastwards as well as its renewed emergence in Central Kazakhstan. Also conspicuous is the repeated recurrence of upheavals and granitization within the same zones, which is indicative of the surprisingly stable high heat flow. The concept of a "hot spot" (or a band of such spots) in the mantle under the corresponding areas accounts for this concentration of the heat flow best of all.

Various interpretations of the origin of the calc-alkaline Paleozoic volcanic rocks and also granitoids result in particular solutions to the given problem. These Paleozoic magmatites may either be products of melting of the granite-gneiss layer of the ancient continental crust, more specifically, mixing of these products with the basaltic magma rising from the mantle (FROLOVA & BURIKOVA, 1977) or result from melting out from the oceanic crust undergoing subduction in the Benioff zone. The existence of such zones in the Urals during the Paleozoic can hardly be doubted, however, reconstruction of their location is made difficult, among other things, by the insufficient exposure of the easternmost zones. The Benioff zone in the earliest stage of existence of the Paleozoic geosyncline must have extended along the Main Fault of its western margin; glaucophane metamorphism in the zone of this fault and, probably, Vendian–Early Cambrian acid volcanism must be connected with this Benioff zone; this must have been an Andean-type margin. Subsequently, obduction could take place in this band at upper levels, whereas subduction could continue at the lower levels. In this case acid volcanic rocks and granitoids may have a source in the contamination of the continental crust remobilized under the effect of the high heat flow issuing from the Benioff zone. If one assumes that the more recent andesite-rhyolite volcanism was associated with subduction alone, then it is necessary to admit that starting from the Silurian there has functioned a Benioff zone inclined eastwards rather than westwards (HAMILTON, 1970) and confined, in all probability, to the eastern slope of the Tagil and later also Magnitogorsk basins with oceanic crust. If this is the case, then how is one to account for the origin of the ophiolite belts in the Uralian zones farther east? Are they relics of independent, most probably earlier "oceanic gaps" or outliers of nappes originating from the Tagil and Magnitogorsk zones? In the latter case, however, obduction as well as subduction must have been involved.

The work by KAMALETDINOV, KAZANTSEVA & KAZANTSEV (1978) has introduced the concept of the nappe occurrence and obducted position of the Uralian ophiolites most consistently. The entire recent structure of the Urals is, in the opinion of the above authors, allochthonous; this gigantic allochthon lies on the subsided margin of the Russian Platform. Ultrabasic belts form frontal zones of individual major nappe sheets which were consecutively overthrust westwards in the Early Ordovician–Late Silurian (the westernmost belt, the Main Uralian Overthrust), in the Middle–Late Devonian, and Early Carboniferous. Glaucophane metamorphism in the frontal parts of the ultrabasic belts coincided with the formation of these nappes. The fact that the oceanic crust overlies tectonically the continental crust, causing its deep downwarping, leads to melting of the continental crust, starting from its basaltic layer and to volcanism ranging from basic to acid, to an increased alkalinity, accompanied by migration of the foci of this volcanism from west to east. The continental crust also produced fluids which led to metasomatism and sialification of the obducted oceanic crust and to the formation of the Uralian Paleozoic granites.

This concept is logical enough, though it does not solve all the problems associated with the complicated evolution of the Uralian Geosyncline and ignores subduction.

7. Paleozoic geosynclinal fold system of Central Kazakhstan and Northern Tien Shan

7.1 Boundaries and main structural subdivisions, state of geological knowledge

The fold geosynclinal region of Central Kazakhstan and Northern Tien Shan (Fig. 44) occupies the western central part of the Uralo–Okhotsk (Uralo–Mongolian) geosynclinal belt, in the rear of the peripheral Late Hercynian systems of the Urals and Southern Tien Shan, between these systems and the Late Hercynian Ob'–Zaisan system which occupies an axial position in the belt. Geographically, this region is situated between the Western Siberian Plain in the north, Turgai Trough, Lake Aral, and Central Kyzyl Kum in the west and southwest and Southern Tien Shan in the south, the valley of river Irtysh in the east and northeast; it includes the Kazakhstan hilly area and the highland system of Northern Tien Shan. The geological (tectonic) boundaries pass along major faults which are partly expressed on the surface but generally hidden under the young sedimentary cover so that they could be detected only by geophysical methods. In the west, the **Central Turgai Fault** and its continuation in the Eastern Aral region and in the southern part of the Western Siberian Platform is such a feature. In the northeast, the boundary **Kalba–Chingiz Fault** extends on the Irtysh left bank, along the northeastern limb of the Chingiz Anticlinorium, between the latter and the Chara zone of the Ob'–Zaisan system. Farther northwest, it was traced by geophysical methods in the south of the Western Siberian Platform. Here, it links with the continuation of the Central Turgai Fault, bounding, together with the latter a buried northern promontory of the Kazakhstan–Tien Shan region which became part of the Western Siberian Platform in the Middle – Late Jurassic. The southern boundary of the region is the least distinct. It should probably be drawn along the Southern Ferghana Fault and the latter's continuation into the Kyzyl Kum and eastern framing of the Ferghana Depression. The region within the above boundaries is shaped like an isosceles triangle with rounded angles, extended slightly eastwards. The base of this triangle is approximately 2,500 km long; the height of the triangle is about 1,800 km. The greater part of the region is situated within the Soviet Union, mainly in the Kazakh and Kirghiz SSR, only its southeastern corner enters the territory of the People's Republic of China.

The Kazakhstan–Tien Shan region has a very complex and heterogeneous structure. The following features make it more or less unified: 1) the relatively wide occurrence of Precambrian, in particular, Early Precambrian formations; 2) Caledonian tectogenesis which finds expression practically throughout the region (with the exception of the Dzhungaro–Balkhash system?) and which is determining here; 3) common features of the structural pattern, more specifically, the fold systems form a large arc which is flatly convex first eastwards, then southwards, is inflected sharper in the center and contracted in the southeast by the Chinghiz–Tarbagatai chord.

The structure of the folded basement of Central Kazakhstan and Northern Tien Shan was shaped by several epochs of tectogenesis: the end of the Archaean, the end of the Early or Middle Proterozoic (Karelian epoch), the end of the Middle Riphean (Grenville or Issedonian, after ZAITSEV, 1972), Baikalian (in the southwest and south), Caledonian, and Hercynian. Individual fold systems and median massifs are identified within the region, depending on the relative role of deformations and granite magmatism of different age. The **Ishim–Naryn fold system** of a miogeosynclinal nature is the westernmost one; in the north, there lies the Ishim–Baikonur (Talas) zone of Caledonian (Early Caledonian) age; and in the south, the Karatau–Naryn zone which is a two-stage Caledonian–Hercynian structure, was delineated. This latter zone, which is sometimes called the **Median Tien Shan,** is not considered part of the region in question by some investigators. In view, however, of the major significance of Caledonian tectogenesis for this zone, we agree with BOGDANOV (1965) who referred it to the same region as the Tien Shan zones further north. West of the Ishim–Naryn system, the **Turgai–Syr Darya Median Massif** was identified largely hypothetically, since it is completely hidden under a young cover. Further east, there lies the **Kokchetav–Muyunkum Massif.** Deeply metamorphosed and granitized Precambrian formations play a major role in the structure of the above two zones. The Kokchetav–Muyunkum Massif is bounded by an eugeosynclinal Late Caledonian system on the east. The **Dzhungaro–Balkhash Hercynian fold system** is situated inside the steep arc formed by the above system. The **Chinghiz–Tarbagatai Caledonian–Hercynian fold system** of a persistent northwestern trend occupies an extreme northeastern position in the region.

Following the Hercynian tectogenesis, the entire region of Central Kazakhstan and Northern Tien Shan entered in the Mesozoic the stage of platform development. However, at the end of the Paleogene the platform conditions on the territory of Central Kazakhstan and Dzhungarian Alatau had given way to the orogenic conditions, owing to which this region became part of the huge Central Asian orogen, whereas Central Kazakhstan experienced only a relatively weak (in places moderate in intensity) reactivation, which preserved it as a shield of the young Eurasian Platform. The common history and structure of the Central Kazakhstan and Northern Tien Shan fold basement seem more significant than the differences in their recent development to an extent that they justify (other investigators support this point of view, too) the identification of these two zones as one tectonic region.

The region (within the boundaries of the USSR) has been studied quite extensively by geological and geophysical methods, although some fundamental problems are still open to discussion and await their solution. This concerns, primarily, the age of the oldest metamorphic rocks and ophiolite complexes, the role of the latter in the regional structure, the extent of development of the overthrust-nappe structures, the structure of the areas hidden under the young cover, etc.

The works by KASSIN (1960) and NIKOLAEV (1933) had been of fundamental importance in the geological and, in particular, tectonic investigation of Central Kazakhstan and Northern Tien Shan. N. G. KASSIN's pupils and followers, primarily BORUKAEV (1955) and SHLYGIN (1977), later headed the Kazakhstan school of geologists. The Moscow academic team under the guidance of N. S. SHATSKY, has initiated a new trend in tectonic studies of Central Kazakhstan. A joint team from MGRI (Moscow Institute of Geological Exploration) and MGU (Moscow State University), headed by BOGDANOV

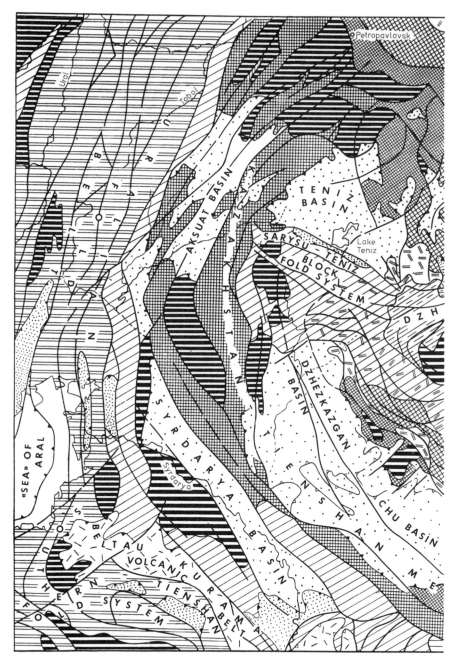

Fig. 44. Tectonic regions and zones of the pre-Mesozoic Basement of the Central Kazakhstan fold region (after ZAITSEV, 1972b). 1 = the largest salients of the Precambrian granitic-metamorphic basement; 2–5 = regions of Caledonian folding: 2 = Early Caledonian (Late Ordovician) fold systems – eugeosynclinal (a), "hemieugeosynclinal" (b), 3 = Late Caledonian (Middle Silurian) fold systems – eugeosynclinal, 4 = Late Caledonian (Devonian) marginal volcanic belt, 5 = epi-Caledonian deformed cover of the Kazakhstan–Tien Shan Median Massif; 6–11 = regions of Hercynian folding: 6 = Middle Hercynian (end of the Middle Carboniferous) fold systems – miogeosynclinal (a) and eugeosynclinal (b), 7 = Late Hercynian fold systems (folding at the end of

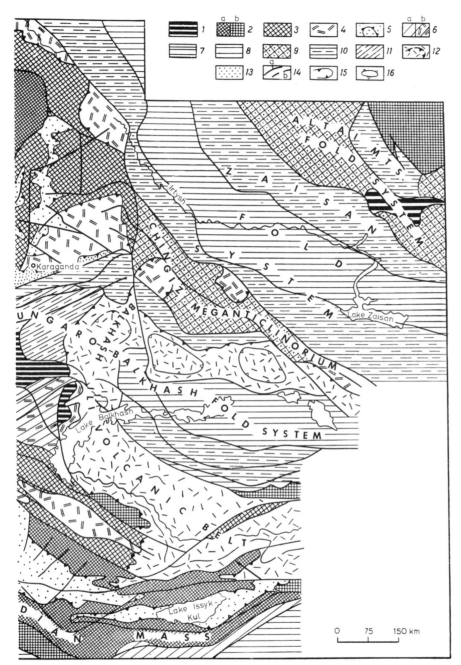

the Carboniferous) – eugeosynclinal and "hemieugeosynclinal", 8 = regions of earlier Saurian (Middle Viséan) folding inside Late Hercynides, 9 = geanticlinal (Late Caledonian fold complexes reworked by Hercynian deformations), 10 = eugeosynclinal and "hemieugeosynclinal", 11 = miogeosynclinal; 12 = Late Hercynian marginal volcanic belts; 13 = Hercynian orogenic (molasse) basins; 14 = major faults: a) in all folded complexes, b) in the Caledonian basement; 15 = outlines of superimposed inner basins, minor foredeeps, and volcanic belts; 16 = outlines of uplifts inside basins of the sedimentary cover.

(1963, 1965), has been working fruitfully in this region since then. The main tectonic studies within Northern Tien Shan have been conducted for many years by KNAUF (1981) and KOROLEV (1981). The geologists of VSEGEI (All-Union Scientific Research Institute of Geology) have also contributed much to the investigations in this region. We will start our review of the structure of the Kazakhstan–Tien Shan region from its oldest "nucleus": the Kokchetav–Muyunkum Massif.

7.2 The Kokchetav–Muyunkum Median Massif

It was mentioned above that a median massif with a basement composed of Early Precambrian deeply metamorphosed rocks, extending as a band 300 to 500 km in width in the submeridional direction from Northern Kazakhstan to Northern Kirghizia, i. e., for a distance of about 1,500 km, is the oldest steadily consolidated element in the structure of the Kazakhstan–Tien Shan region. It may be conjectured that initially, prior to the onset of the Middle or Late Riphean, this massif had been an integral whole with the buried Turgai–Syr Darya Massif further west, i. e. it was twice as wide as at present.

The Precambrian basement of the massif is now exposed in three areas of Central Kazakhstan: in the Kokchetav, Ulutau, and Eastern Betpakdala massifs, as well as in the cores of fold structures of the Northern Tien Shan northern ranges. The Kokchetav Block, which is frequently referred to as massif, is the largest of these blocks preserving its monolithic character (SHLYGIN & SHLYGIN, 1964).

The **Kokchetav Block,** situated in the northwest of Central Kazakhstan, has a shape close to isometrical, extended somewhat in the latitudinal direction (up to 340 km wide and up to 200 km long). The northern part is covered by the Meso–Cenozoic cover of the Western Siberian Platform. The oldest basement formations of the block, identified as the Zerenda series and represented by gneisses and schists of the amphibolite facies of metamorphism, must be of Archaean age, though their radiometric datings do not exceed 1,600 m.y. The Zerenda series forms major uplifts of the type of granite-gneiss domes which are surrounded by synclinorial zones filled with metavolcanic rocks (acid and intermediate at the bottom and more basic at the top) of the Efimov series. The Efimov series is at least 1,400–1,300 m.y. old; it belongs to the Lower Proterozoic (ZAITSEV & FILATOVA, 1972) rather than to the Lower Riphean (ROZEN, 1971).

These hypothetically Early Precambrian formations are overlain with a sharp unconformity by Riphean deposits: acid volcanic rocks, sericite-chlorite-quartzose and carbonaceous shales and dolomites of the Middle Riphean as well as Upper Riphean quartzite-sandstones (the Kokchetav series). A presumably Vendian Nikol'sk–Burluk sedimentary-volcanic series occurs at the top of the massif's Precambrian sequence mainly on the periphery of the massif and in small patches only in its central part. There are no Cambrian deposits in the massif, and its sequence is completed by relatively thin Ordovician terrigenous deposits filling flat-lying superimposed troughs. More than half of the area of the massif is occupied by major plutons of Late Ordovician granitoids (Zerenda, Borovskoye, and others).

The massif, along its entire perimeter, is bounded by Caledonian synclinoria filled mainly with thick Ordovician deposits and separated from the massif by major syn-sedimentary faults: ultrabasic bodies sometimes occur along them. These rocks are occasionally encountered inside the massif, where they are in contact with deposits ranging to Vendian; their tectonic position remains insufficiently well studied.

The Kokchetav Block is divided by Late Caledonian, partly older and partly younger faults into several blocks which are characterized by specific structural features.

The above data indicate that the cratonization of the Kokchetav Block had started as far back as the pre-Riphean time, that it had been essentially consolidated by the Late Riphean and completely consolidated, following a certain reactivation in the Ordovician, by the end of the Ordovician. The prolonged reactivation of the faults bounding the block is emphasized by the occurrence of Devonian (D_{2+3}) orogenic volcanic rocks in its western and southern framing and of near-fault synclines, filled with Middle–Upper Paleozoic rocks, along the same sutures (BABICHEV et al., 1968). The Kokchetav Block is separated from the parts of the median massif further south by the latitudinal Kalmykkul' Caledonian Synclinorium.

The **Ulutau Block,** in contrast to the Kokchetav one, extends in the meridional direction, in accordance with the strike of the Lower Proterozoic series prevailing within it. These series make up blocks of the extended **Karsakpai Synclinorium** in the basement, which is a typical example of the Karelian jaspilite-volcanic suture fold systems. Being just 10–20 km wide, this synclinorium has been followed over 350 km on the surface and for another 100 km further south, under the cover of the Dzhezkazgan Basin. Its sedimentary-volcanic fill consists (FILATOVA, 1976), in the lower part, of a dacite-keratophyre (leptite) sedimentary-volcanic formation (the Aralbai series), about 7 km in thickness and in the middle part of a jaspilite-spilite greenschist formation (the Karsakpai series), 4.5 km in thickness. The sequence is completed with a rather thick series of acid metavolcanic rocks–porphyroids (the Zhiida series). A volcanoclastic formation of a similar composition (the Maityube series), which lies unconformably higher must be of Middle Proterozoic age. The inner structure of the synclinorium is characterized by a complex and tight folding with eastern vergence; the folds are often isoclinal in shape. In the west, the synclinorium is thrust along the Karsakpai Fault over the adjacent Maityube zone ("anticlinorium") which is composed of a Middle Proterozoic porphyroid formation. This zone in turn is superseded along the fault by the Caledonian **Baikonur Synclinorium;** the boundary of the latter is hidden under the young Turgai Trough.

In the east, the Karsakpai Synclinorium bounds on the **Ulutau Anticlinorium,** in which there occurs a crystalline schist series of Archaean age, which is probably the oldest in Ulutau. It forms relatively large folds–gneiss ridges, usually complicated by tight folding and crenulation. Karelian granite-gneisses form the axial part of the anticlinorium, and bodies of serpentinized ultrabasic rocks occur along the faults bounding it on the west and east. On the other side of the Eastern Ulutau Fault, there extends a synclinorium filled with the Lower Proterozoic Aralbai series (see above) which is intruded by Ordovician granites and which is characterized by a very complicated structure, becoming simpler somewhat in the easterly direction.

The Precambrian basement of the Ulutau Uplift, together with the Baikonur Caledonian fold complex surrounding it on the west, is exposed in the cores of major brachymorphous and dome-like uplifts of Hercynian age and of essentially platform type. The Hercynian complex filling the depressions between these uplifts is deformed much more vigorously than at the limbs and forms linear folds complicated by overthrusts of eastern direction. On the whole, the folding of the Hercynian complex here is connected with movements of the basement blocks and is either of a reflected ("plis de revêtement") or near-fault and suprafault character. In the north, the Ulutau Block is disrupted and truncated by northwest-trending faults, in the south it is hidden under the cover of the Dzhezkazgan Basin, in the east it is linked at the front with a block-folded zone of the Sarysu–Teniz water divide, and in the west it is fringed by the young Turgai Trough.

The Early Precambrian basement within the northern half of the median massif occurs also in the southeastern framing of the Dzhezkazgan Basin in the central Betpakdala. This is the so-called **Chu Block** composed mainly of a greenschist-jaspilite-porphyritoid series, probably similar to the Karsakpai one, deformed into complex disharmonic folds of northwestern trend. On the northeast, the Chu Block borders, apparently, along a deep overthrust with the Caledonian Zhalair–Naiman fold zone which is part of the Yerementau–Chuili system (see below).

Further southeast, the ancient basement of the massif in question occurs on both sides of the Chu–Sarysu Basin at the northwestern end of the Kendyktas Range, in the western part of the Kirghiz Range (the Makbal Horst), and in the eastern end of the Basin in the Ak Tyuz region, between the Transilian Alatau and Terskei Alatau. The Issykkul' Block (median massif) farther southeast must be a fragment of the same region of epi-Karelian consolidation.

Geophysical investigations have shown that the ancient crystalline basement in the southern part of the Chu–Sarysu Basin lies at a shallow (1–2 km) depth. The same applies to the area of linkage between the Dzhezkazgan and Chu–Sarysu basins. In other areas of the massif, however, the Early Precambrian massif underwent considerable reworking, breakdown, and subsidence. This was especially pronounced in the upper reaches of river Ishim, where protruding into the body of the massif between the Kokchetav and Ulutau blocks in the west is a rather broad band of Caledonides, inside of which usually are distinguished the **Kalmykkul' Synclinorium** (in the north) and **Dzharkainagach Anticlinorium** (in the south). Geosynclinal siliceous-terrigenous Upper Precambrian (R_3 + V)-Ordovician formations completed with an Upper Ordovician flysch or volcanic rocks are involved in the structure of this Upper Ishim Caledonides band. Ultrabasites and Early Caledonian (Late Ordovician) granitoids have been encountered in the zone of the tectonic contact with the Kokchetav Block; the granitoids are also intruded into the core of the Dzharkainagach Anticlinorium, whereas younger, Late Caledonian (D_{1-2}) granites occur in the eastern closure of the Kalmykkul' Synclinorium. The fold and fault deformations are much stronger in the anticlinorium (narrow linear folds, extended longitudinal faults) than in the synclinorium.

In the east, the Caledonian fold zone of the Upper Ishim region plunges under the superimposed Teniz Basin. A most specific **Sarysu–Teniz block-fold zone** extends between this basin and the Dzhezkazgan one further south and east of Northern Ulutau in a near-latitudinal direction. It consists of alternating relatively narrow but long horst-anticlines and graben-synclines (Fig. 45). The former have in their cores the Precambrian and Caledonian basement, the latter consist generally of a thick Upper Devonian carbonate formation folded disharmonically to form tight folds. This formation is separated from the Caledonian fold complex by volcanic molasses of the Lower and Middle Devonian. The total thickness of the Hercynian complex reaches 6–8 km. One observes in the given zone the sharpest difference between the structural pattern of pre-Devonian formations, on the one hand, and the younger Paleozoic deposits, on the other. The

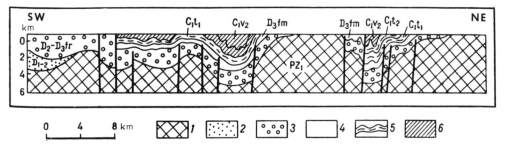

Fig. 45.   Cross-section through the Aidagarma and Kagyl' graben-synclines and Kuyandy horst-anticline of the Sarysu–Teniz zone (after TIKHOMIROV, 1975). 1 = Lower Paleozoic; 2 = Lower–Middle Devonian; 3 = Middle Devonian–Frasnian; 4 = Famennian; 5 = Tournaisian; 6 = Viséan.

Caledonian structural pattern has mainly a meridional orientation, the Hercynian structural pattern the northwestern orientation. In the east, the Sarysu–Teniz block-fold zone abuts against an arc of the Devonian volcanic belt (see below).

The Sarysu–Teniz zone separates two major Late Hercynian basins superposed on the Kokchetav–Muyunkum Massif: the Teniz Basin in the north and the Dzhezkazgan Basin in the south.

The **Teniz Basin** of irregularly isometrical outlines, 300 to 350 km across, elongated somewhat in the latitudinal direction, overlies unconformably the Precambrian – Caledonian structures of the massif and is filled with Upper Paleozoic secondary-orogenic formations: the red Middle and Upper Carboniferous, the gray and variegated Permian ones. The former is underlain by a quasi-platform carbonate formation of the Upper Devonian and Lower Carboniferous, which takes also part in the structure of the Sarysu–Teniz zone, but is there a little thinner (in the north). Total thickness of the Middle and Upper Paleozoic deposits reaches 7 km in the south and west of the basin; the Upper Paleozoic accounts for 3 to 3.5 km. The structure of the basin is complicated by flat-lying synsedimentary brachyfolds which are obviously formed over faults in the southwest and which inherited folds of the basement on the rest of the territory.

The **Dzhezkazgan Basin,** in contrast to the Teniz one, is clearly elongated in the northwestern-southeastern direction, reaching 800 km in length, its width being up to 300 km. It is filled with a similar complex of Middle and Upper Paleozoic deposits up to 4–5 km in thickness. The only difference consists in the presence of a salt-bearing formation at the boundary between the Devonian and Carboniferous, locally also in the top portion of the Lower Permian. Manifestations of the salt diapirism in the southwestern part of the basin, the so-called Sarysu domes, are associated with the first of these formations. The rest of the area is characterized by usual brachyfolds of meridional strike in the west (west of the buried continuation of the Eastern Ulutau Fault) and latitudinal in the east. Copper ore deposits, including the well known Dzhezkazgan deposit, which is situated in the nodes of intersection of buried faults, are associated with the Middle and Upper Carboniferous red-bed formation. The central and southern parts of the Dzhez-kazgan Basin are covered by a thin (hundreds of meters) layer of Upper Cretaceous and Cenozoic, mainly continental deposits. The Dzhezkazgan Basin together with the adjacent Chu one (on the southeast) forms a single Chu–Sarysu Basin, if one judges by the above deposits. In the Paleozoic, they are separated by a buried transverse uplift which lies on the continuation of the basement's Chu Swell.

The **Chu Basin,** which is an extreme southeastern link in the system of superimposed Late Paleozoic depressions of the Kokchetav–Muyunkum Median Massif, stretches in the northwestern direction, acquiring a nearly latitudinal trend in the south. This part is overlain by the **Frunze Foredeep** which is a marginal structure of the alpine Tien Shan.

The Precambrian and partly Caledonian basement lies at a depth of 3–4 km in the Chu Basin and is dissected rather vigorously by faults into uplifted and lowered blocks over which there occur branchyanticlinal folds combined into arches; salt domes (south of the Chu Block) have also been encountered.

If one forgets for a moment the thin Meso–Cenozoic typically platform cover, which covers a small territory, then the following structural complexes in the structure of the Kokchetav–Muyunkum Median Massif can be identified: 1) an integral Early Precambrian complex consisting of Archaean blocks and Early Proterozoic linear protogeosynclinal fold zones cementing the blocks; 2) an orogenic (?) Middle Proterozoic volcanic complex; 3) an Early (?)–Middle Riphean geosynclinal complex; 4) a quasiplatform complex of the upper part of the Middle and lower part of the Upper Riphean (the Kokchetav series and analogues thereof); 5) an Upper Riphean–Vendian–Lower Paleozoic (Early Caledonian) geosynclinal complex which does not occur continuously; 6) a Lower and Middle Devonian (Late Caledonian) orogenic volcanic complex which occurs even more sporadically; 7) an Upper Devonian–Lower Carboniferous (Early Hercynian) quasiplatform[1] complex; 8) a Late Hercynian Upper Paleozoic teleorogenic complex. The massif must be a fragment of an epi-Karelian continental platform, which was partially involved in a geosynclinal reworking during the first half of the Riphean, then reconsolidated; it experienced partial reworking again in the Caledonian stage, and repeated consolidation at the end of this stage, was affected nonuniformly by Late Hercynian deformations (which were the strongest in the Sarysu–Teniz zone), and became ultimately stabilized, together with the entire Uralo–Siberian Belt by the beginning of the Mesozoic.

The **Northern Kirghizian zone.** This northernmost zone of Northern Tien Shan, which may be called Northern Kirghizian and which includes the Kirghizian, Transilian Alatau, Kungei- and Terskei Alatau ranges is in fact a southeastern continuation of the Kokchetav–Muyunkum Massif, strongly reworked in the Caledonian and least so in the Hercynian cycle. The Early Precambrian basement here makes up two relatively large blocks essentially buried under younger deposits. These are the **Transilian** (after KNAUF, 1981) or **Muyunkum** (after NIKOLAEV, 1933 and KOROLEV, 1981) and **Issykkul'** (after KNAUF) or **Kungei–Terskei** (after KISELEV & KOROLEV, 1972) blocks. This basement was also found in the **Makbal Horst** in the western part of the Kirghizian Range. In the Early or rather Middle Riphean there started in the Kirghizian–Terskeian band downwarping accompanied by accumulation of sandy-clayey-calcareous-dolomitic sediments, similar, according to V. G. KOROLEV et al., to the Southern Uralian Riphean deposits and, like the latter, having a miogeosynclinal (?) character and a considerable thickness. These formations, consisting of three series separated by unconformities, are in turn overlain with a

---

[1] ZAITSEV (1972a) considers this complex as miogeosynclinal in the Sarysu–Teniz zone.

sharp unconformity by the Upper Riphean Terskei series. This latter unconformity must be Grenvillian; ZAITSEV (1972b), who called it **Issedonian**, showed that it is characteristic of the entire Kokchetav–Muyunkum Massif. The Upper Riphean Terskei series, unlike the underlying ones, consists mainly of volcanic rocks of basaltic composition, including spilites associated with silicites, i. e., it is already eugeosynclinal in appearance. It fills depressions on both sides of the Central Terskei Uplift which had formed during the Grenvillian epoch, and had been deformed into folds prior to the Vendian superseded conformably by the Cambrian. The terrigenous Vendian, also containing conglomerates, is regarded as the Baikalian molasse which occurs in depressions on both sides of the Terskei Uplift (this is the Chatkal–Naryn zone) which had expanded owing to the accretion of the Baikalian complex.

In the Early Paleozoic, between the Transilian and Issykkul' massifs the **Kemin Trough** was formed, bounded by faults, strongly compressed and deformed at the end of the Ordovician. This trough is filled with a Cambrian (?) terrigenous formation and volcanosedimentary rocks of the Middle and Late Ordovician, with the thickness totalling up to 5 km. A thin cover of terrigenous-carbonate sediments of the same age (C–O_2) overlay the Riphean fold complex on the massifs with a sharp unconformity. Flysch or andesitic volcanic rocks accumulated in the Middle and Late Ordovician along the periphery of the Kemin Trough as well as in the central part of the Muyunkum Massif; the fill of these Late Caledonian troughs and especially the Cambrian–Ordovician cover experienced only relatively weak deformations. In the late Ordovician and probably Early Silurian there formed red molasses filling troughs on the limbs of which they overlie unconformably the Cambrian–Lower Ordovician strata ("Tectonics of pre-Paleozoic . . .", 1970). During the same epoch there finally formed huge granitoid batholiths which presently occur at summits of the Kungei–Alatau and Terskei–Alatau ranges, along the periphery of the Issykkul' Block, as well as in the western part of the Kirghizian Range.

These particular granites and the Upper Ordovician "red molasses" are responsible for the Caledonian appearance of the Northern Kirghizian zone, despite the fact that Lower Paleozoic geosynclinal formations do not play any significant role in its structure.

At the end of the Caledonian and in the Hercynian stage, Northern Tien Shan experienced general upheaval, however, there appeared depressions along its periphery and partly in its inner parts, which were filled with Lower and Middle Devonian subaerial volcanic rocks (not everywhere), Middle and Upper Devonian red-beds, Upper Viséan– Namurian terrigenous-carbonate, Middle and Upper Carboniferous terrigenous and in places Permian subaerial trachyandesite-basaltic formations. Total thickness of this sequence of formations is 2 to 3 km, reaching 6–8 km in individual areas. Its deformation shows an obvious connection with tilt and block movements of an older basement and do not go beyond saxonotype. The Sonkul' Depression which formed at the southern boundary of the zone, along the "Nikolaev line" is the largest of these features. The Sonkul' Depression is in fact a small foredeep of the Hercynides in front of the region of epi-Caledonian consolidation. It has a sharply asymmetrical structure with a flat-lying northern and a very complex southern limb upthrown northwards and bounded in the south by an overthrust of Famennian–Lower Carboniferous carbonate series of the Karatau–Naryn zone of Median Tien Shan. This zone narrows gradually in the eastern direction as one passes to the territory of China, where it is cut by the northern boundary of the Tien Shan Hercynian fold system.

7.3 The Ishim–Talas zone
of Early Caledonides

This zone is very long, but relatively narrow (up to 50–60 km). It surrounds the Kokchetav–Muyunkum Massif on the west and south. In the north, the zone is represented by the **Mar'evka Synclinorium,** by central parts of the Kalmykkul' Synclinorium and Dzharkainagach Anticlinorium described earlier (these are situated in an echelon-like manner with respect to the Mar'evka Synclinorium) and by the **Baikonur Synclinorium.** The western part of the zone is here hidden under the young cover of the Turgai Trough, and its prevailing trend is close to meridional (except for the Upper Ishim Branch). The southern trend of the zone changes to southeastern and then sublatitudinal; it includes here structures of Lesser Karatau and the Talas Range bounded on the southwest by the major **Karatau–Talas Fault** ("the most significant structural line in Tien Shan," as NIKOLAEV, 1933, put it). The northeastern limb of the Karatau fold structure is hidden under the Chu Depression; farther east, the **Ichkeletau–Susamyr Fault** forms the northern boundary of the zone. The zone wedges out in the extreme east of the Soviet Tien Shan.

The period of active geosynclinal development of the zone had started in the Early or Middle Riphean, first, most probably, in the south and continued until the Late Ordovician. There are indications in the southern part of the zone that the subsidence was interrupted, at least twice–in the beginning of the Vendian and before the Late Vendian–by upheavals and moderate-intensity fold deformations. This is attested to by unconformities between individual Late Precambrian series and by the appearance of coarse clastic sediments which may be regarded as the Baikalian molasse. Less distinct are the hiatuses and unconformities inside the Riphean: a reflection of the Gothian and Grenvillian epochs of diastrophism. Notwithstanding the unconformities in the Riphean and Vendian, the Late Precambrian formations are deformed according to the same plan as the Lower Paleozoic ones.

The Caledonian stage proper started in the Late Vendian. This epoch, like the Cambrian, is characterized by the accumulation of thin layered and relatively deep-water carbonaceous-siliceous-phosphate and clayey rocks. A similar environment prevailed in the first half of the Ordovician, with a carbonate rock accumulation in the Lesser Karatau. In the Late Ordovician, intensified upheavals produced increasingly more coarse terrigenous sediments. The troughs closed, and their fill experienced strong fold-fault deformations; granitic plutons intruded along the fault zones.

Generally speaking, the Upper Precambrian and Lower Paleozoic sequence of the zone being studied is composed of carbonate-terrigenous, partly flyschoid or even flysch (Middle and Upper Riphean, after KISELEV & KOROLEV, 1972) formations. Volcanic material is present at individual levels only (Vendian, Upper Ordovician), mainly in the form of pyroclastic rocks and in clearly minor quantities. This imparts a character closer to miogeosynclinal rather than eugeosynclinal to the zone, which was earlier noted by BOGDANOV (1965).

Recent structure of the Ishim–Talas zone is characterized by linear folds of several orders, complicated by overthrusts and forming groups of anticlinoria and synclinoria. A southeastern vergence prevails in the exposed part of the Mar'evka Synclinorium, whereas the Baikonur Synclinorium exhibits a western vergence. The Talas zone shows an antivergent structure, the Riphean deposits here being dislocated much stronger, up to isoclinal folding; these deposits experienced initial greenschist metamorphism.

At the boundary between the Ishim–Talas and the next to the south Karatau–Naryn zone belonging already to the Median Tien Shan, i. e. approximately along the "Nikolaev line" in the western part of the Kirghiz Range, on the southern side of the Susamyr Range and in the eastern part of the Terskei Alatau, an ophiolite zone was ascertained recently, formed before the Vendian, apparently at the end of the Riphean (MAKARYCHEV & GES', 1981). The ophiolites occurring mainly in the melange underly here a Vendian – Lower Cambrian volcanic andesite and andesite-basalt suite, overlain by thick olistostrome formation of Middle Cambrian–lowermost Arenig. The latter is overlain by Ordovician (up to the lowest O_3) flyschoid terrigeneous formation with limestone intercalations. The whole eugeosynclinal complex is pierced by Late Ordovician granodiorites and Silurian potassic granites and overlain with sharp unconformity by Devonian–Lower Carboniferous red molasse. All this indicates that this zone is connected with the Northern Tien Shan.

7.4 The Karatau–Naryn Caledonian–Hercynian fold zone

The Karatau–Naryn[2] zone encircles the Ishim–Talas zone on the southeast (Greater Karatau Range) and south (Median Tien Shan Ranges) and is separated from the latter zone by the Karatau–Terskei Fault (the Nikolaev line). The southwestern boundary is, in the west, hidden under the cover of the Syr Darya Depression and is barely detectable by geophysical methods, and then, within Tien Shan, it can be drawn with a sufficient degree of arbitrariness. In the northwest, the zone continues into the Turgai Trough, and here it probably pinches out. Yet, some investigators believe the Ishim – Baikonur zone to be a continuation of the Karatau–Naryn one rather than Talas, and the latter is assumed to thin out in the northwestern direction (ZAITSEV & FILATOVA, 1972).

The following are the most typical features of this zone: 1) late beginning of geosynclinal evolution–since the Vendian and, respectively, total absence of the Riphean geosynclinal complex; 2) bicyclic structure of the Vendian–Paleozoic geosynclinal orogenic sequence: the Caledonian complex is overlain by an approximately equivalent Hercynian complex. The Karatau–Naryn zone outside the Riphean geosyncline of the southwestern and southern framing of the Kokchetav–Muyunkum Massif, was situated at the southwestern and southern periphery of the Caledonian geosyncline of Northern Tien Shan and at the northeastern and northern periphery of the Southern Tien Shan Hercynian geosyncline. Caledonian subsidence in this zone had ceased in the Silurian rather than Ordovician. It was so weak in the Greater Karatau that one can talk of the Vendian–Lower Paleozoic deposits of a type close to platform.

The Early Precambrian basement overlain by a thick (up to 3.5–4 km) subaerial-volcanic porphyroid formation (the Greater Naryn series) of Upper Riphean age lies along the southern periphery of the zone, in the east, in the **Sarydzhaz Massif.** This formation, with which granitoid plutons are associated, is overlain unconformably by

---

[2] Or Karatau–Chatkal, if its easternmost stretch is disregarded.

Vendian tillites (or tilloids) which start a Caledonian geosynclinal complex. The latter includes a Vendian–Cambrian carbonate-carbonaceous-siliceous-shale formation similar to the formation of the same age adjacent in the north, a flyschoid terrigenous Ordovician formation and terminates with a molasse or molassoid Upper Ordovician – Silurian formation. This formational sequence is overlain, with a sharp unconformity and after a considerable hiatus, by a Devonian (D_2^2–D_3^1) red continental molasse[3] which in turn is overlain conformably by a thick (up to 4–5 km) Famennian–Lower Carboniferous carbonate formation. The latter is either directly or via an intermediate flyschoid terrigenous-carbonate formation of the top of the Lower–Middle Carboniferous overlain unconformably by Late Hercynian molasses of the Upper Carboniferous and Permian, partly volcanic (andesitic). The Caledonian complex is everywhere deformed much stronger than the Hercynian one, however, the degree of its deformation is smaller than that observed in more northerly zones, pointing to the attenuation of Caledonian deformations in the southerly direction. The Caledonian complex forms anticlinorium cores, whereas the less deformed Hercynian complex (D_3^2–C_1) makes up their limbs and fills synclinoria. BOGDANOV (1965) referred the Karatau–Chatkal zone also to the region of Caledonian consolidation, assuming its Hercynian complex to be fully orogenic, which, however, is contradicted by the presence of a thick Upper Devonian and Lower Carboniferous carbonate formation. The major Hercynian deformations took place prior to the Middle Carboniferous and were accompanied by granite intrusions. The Upper Paleozoic molasse fills individual troughs; it in turn experienced folding in the vicinity of faults.

The zone is subdivided into three distinct structural segments. The northwestern segment coincides with the Greater Karatau Range and extends along the Major Karatau Fault. Further northwest, beyond the end of the Greater Karatau Range, the Hercynian miogeosynclinal complex must wedge out, so that the Caledonian complex alone remains in the basement of the Turgai Trough. This conclusion, however, is based on geophysical evidence only.

The structure of the **Greater Karatau** is quite specific and is characterized by a regional decollement of the Famennian–Lower Carboniferous carbonate complex along the surface of the Lower Famennian plastic shale member underlain by rigid Frasnian molasses. An additional surface of decollement emerged at the top of the Famennian between the latter and the carbonate Lower Carboniferous. This decollement caused formation of Famennian–Lower Carboniferous nappes which were first observed here by V. V. GALITSKY in 1936. V. V. GALITSKY's concepts came under vigorous attack by BRONGULEEV (1957) at the end of the 1950's and beginning of the 1960's, yet were confirmed by further studies. The nappe structure can be followed along the Karatau Fault over 450 km, and the amplitude of horizontal displacements reaches 20 km; it increases towards the Karatau Fault, which points to a connection of overthrusting with displacement movements along this fault. Consecutive overthrusting of up to three nappe slices is observed in places. The slices themselves are deformed in brachysynclinal folds between which diapiric intrusions of Lower Famennian shales are not infrequently observed. According to V. V. GALITSKY, displacements of the Famennian and Lower

---

[3] The dual position of this formation is noteworthy. It completes the Caledonian series in terms of composition and starts the Hercynian series in its structural position.

Carboniferous nappes relative one another and relative to the autochthon proceeded from northwest to southeast parallel to the Major Karatau Fault which GALITSKY believes to be a sinistral shear ("Geology of the USSR", Vol. 40, Geological Description, 1971). At the same time the majority of other investigators regard this fault as well as its Talas–Ferghana continuation as a dextral shear. It is possible that the Caledonian basement of the Greater Karatau experienced displacement of the type of dextral wrench fault, i. e., towards northwest and plunged under the cover which moved in the opposite direction relative to this basement. Yet, in places nappes were encountered in the Caledonian complex itself, and therefore only the oldest Precambrian basement may be assumed to be an autochthon.

The next segment of the zone in question, the **Chatkal** one, encompasses the Chatkal and adjacent ranges. Its constituent fold structures are characterized by a northeastern trend which is anomalous for Tien Shan, but near the Talas–Ferghana Fault, which bounds the segment on the east, change it into the east-northeasterly trend, forming a gentle arc convex towards northwest. It is quite obvious that the formation of this folded arc, complicated by northwestern overthrusting, is connected with a displacement along the Talas–Ferghana Fault, which took place mainly in the Late Paleozoic, mainly in the Middle Carboniferous and is estimated at about 200 km. Permian granitic intrusions, which occcur in the fault zone, did not experience shearing displacements.

The eastern, **Naryn segment** of the Karatau–Naryn zone is situated east of the Talas–Ferghana Wrench Fault and is therefore displaced south relative to the central Chatkal segment. It includes the Dzhetymtau, Moldotau, and Naryntau ranges in which Caledonian anticlinoria occur in combination with Hercynian synclinoria. The predominant virgation in the Hercynian complex is northern.

7.5 The Turgai–Syr Darya Median Massif

This massif has been identified somewhat arbitrarily west and southwest of the Caledonides region of Central Kazakhstan and Northern Tien Shan, according to both geological and geophysical data (KUNIN, 1968). The basement and the Paleozoic cover of the massif over a greater part of the territory are hidden under the Meso–Cenozoic formations and are exposed only in the region of Central Kyzyl Kum desert (Auminzatau, Bel'tau, Tamdytau, and Bukantau Mts.) as well as in the south of the Chatkal Range and in the Kurama Range of Middle Tien Shan.

The rocks of the Auminza series of Central Kyzyl Kum and of its analogues, metamorphosed in the amphibolite facies, could be Early Precambrian. These rocks are amphibolites, gneisses, and crystalline schists. They occur also in the Kassan Uplift in the south of the Chatkal Range where marble is also present in the sequence. The greenstone apoterrigenous-volcanic Karabulak series may be a little younger. The Riphean (Middle and Upper Riphean) and Vendian formations occur with a distinct unconformity over the ancient basement, are metamorphosed in the greenschist facies and, judging by the small thickness (R_2 is less than 200 m, R_3 is up to 1,000 m, and V ranges from 700 to 1,000 m) in the Kyzyl Kum and persistent lithological composition, belong already to the massif cover. These are mainly terrigenous, initially sandy-clayey deposits with a high content of quartz in the sand fraction ("Precambrian of Middle and Southern Tien Shan", 1975).

The Lower Paleozoic complex within the massif is strongly reduced to complete thinning out, which is observed in the south of the Chatkal Range. The Silurian deposits in the same region, on the contrary, possess a considerable thickness, however, andesitic volcanic rocks play an appreciable role in their composition. The Devonian strata lie unconformably, and are represented by acid to intermediate subaerial volcanics at the bottom, and towards the top by red-beds and gypsum replacing the lower part of the Chatkal zone carbonate formation. Geophysical data indicate a shallow occurrence of the pre-Paleozoic basement directly below the Meso–Cenozoic cover in Eastern Kyzyl Kum and in the lower reaches of Syr Darya (KUNIN, 1968).

A specific feature of the southwestern and southeastern framing of the massif consists in the accumulation of a huge series of subaerial volcanics, which started in the Namurian with mainly basic effusions and finished at the end of the Permian and beginning of the Triassic with acid lavas and tuffaceous rocks. Late Hercynian granites, which together with the volcanics constitute a single volcano-plutonic association, are also widespread here. These formations make up the region of the Bel'tau Mountains in Central Kyzyl Kum and the Kurama Range, the extreme southwestern branch of Median Tien Shan. The Kurama Range magmatites must be connected with transverse faults of northeastern "Kurama" trend, which cut off the southeastern termination of the Turgai–Syr Darya Massif. The earthquake of December 6, 1966 was associated with one such fault passing through the city of Tashkent.

BOGDANOV (1965) drew parallels between this **Bel'tau–Kurama volcanic belt** and the Valer'yanov Belt in the western margin of the Transuralian buried Kustanai Synclinorium, maintaining that these belts, similar to the more easterly Devonian marginal volcanic belts of Central Kazakhstan mark a natural boundary of the Caledonian Kazakhstan–Tien Shan Massif. It, however, appears more appropriate, after A. K. BUKHARIN, V. G. GAR'KOVETS, and K. K. PYATKOV ("Tectonics of the Uralo–Mongolian ...", 1974), that the hypothetically single belt existing here extends along the boundary of the Turgai–Syr Darya massif (which was not identified by BOGDANOV, 1965) with the Hercynides of Southern Tien Shan.

Let us now proceed to the eastern part of the Kazakhstan–Tien Shan region.

7.6 The Yerementau–Chuili Caledonian eugeosynclinal fold system[4]

This system which is shaped like an arc concave towards the east, goes around the Kokchetav–Muyunkum Massif in the east and extends from the region of Tselinograd–Pavlograd through Betpakdala to the Transilian Alatau and farther on to China, had the major eugeosynclinal region of Central Kazakhstan and Northern Tien Shan confined to it at the end of the Precambrian and beginning of the Paleozoic. The serpentinized ultrabasic and gabbro rocks quite common here and presently occurring under complex tectonic conditions along faults seem to correspond to the base of the

---

[4] It includes in the west the peripheral Stepnyak–Betpakdala zone which is sometimes identified separately.

geosynclinal complex. Their contacts with younger to Ordovician deposits, which had given basis for ascribing Late Ordovician age to these rocks, are, in all probability, tectonic, the more so since their fragments, after the data of ANTONYUK (1974), are encountered already in the Lower Ordovician and even Cambrian deposits ("Tectonics of the Uralo–Mongolian ...", 1974). A thick (occasionally up to 7.5–8 km) spilite-diabase-jasper association whose age is still open to discussion corresponds to the upper part of the eugeosynclinal sequence; it includes, most reliably, the Vendian and Early Cambrian. Volcanic rocks in the upper part of this series give way to silicites and there appear limestone lenses and interlayers. Starting from the Middle Cambrian the volcanics acquire andesitic-basaltic and then andesitic composition, and graywackes become quite widespread. This must have been the stage of island arcs. Such arcs first emerged above sea level in the beginning of the Ordovician (pre-Arenig), which was also accompanied by folding deformations referred to the Salairian tectonic epoch. The volcanic rocks are replaced by flysch-type deposits along the western (Kokchetav) and eastern peripheries of the zone. Deformations became more vigorous at the end of the Ordovician, in the Taconian epoch, and were accompanied by upheavals which led to the substitution of flysch accumulation by a marine molasse. Concurrently, the major Caledonian intrusive complex formed: the Krykkuduk one which is composed of a tonalite-granodiorite formation. The formation of the main iron, copper and gold ore deposits of Northern Kazakhstan is connected with this complex, in particular, with the scarns accompanying it.

As a result of upheavals in the end of the Ordovician and in the Silurian, subsidence continued only on limited areas in the eastern part of the central (between the upper reaches of the rivers Ishim and Sarysu) and northern parts of the southern segments of the system (north of the Transilian Alatau). The Silurian deposits consist of marine sandy molasses, not infrequently variegated or red-colored.

The Devonian continental red-bed and volcanic formations, corresponding to the Upper Caledonian molasse and overlying all the older rocks with a sharp unconformity, make up superimposed synclinal structures which merge along the eastern, inner part of the arc to form the **marginal volcanic belt of Central Kazakhstan.** The Famennian – Lower Carboniferous carbonate formation is preserved by places in these superimposed troughs concordantly with the Devonian molasses.

The present structure of the system, which developed consecutively as a result of the fold and fault deformations at the end of the Cambrian and beginning of the Ordovician, the end of the Ordovician, and the end of the Silurian and beginning of the Devonian, is characterized by an alternation of anticlinoria and synclinoria. The former are usually composed of Upper Precambrian–Cambrian rocks (the Salairian stage), the latter mainly of Ordovician (the Taconian structural stage), partly of Silurian rocks (in the east in the central segment, in the north in the southern segment, see above). The cores of some anticlinoria contain blocks of deeply metamorphosed rocks referred to the Early Precambrian. The presence of these blocks, provided their ancient age is confirmed, is essential in connection with a recent interpretation of the geosynclinal system being studied as the one which formed at the end of the Precambrian over a crust of a presumably oceanic type. One such block lies in the north within the core of the **Ishkeol'mes Anticlinorium** separated from the Kokchetav Massif by the **Stepnyak Synclinorium.** This block was found to contain outcrops of analogues of three series of the Kokchetav Massif: Zerenda (A?), Efimov (PR_1^2), and Kokchetav (R_{2-3}?). These parallels seem quite justified; it may therefore be assumed that the Stepnyak Synclinorium, which protrudes bay-like into the body of the Kokchetav Block from southeast, formed on a continental crust at the subsided part of this block, whereas the Ishkeol'mes Anticlinorium emerged on its uplifted edge.

Strongly deformed actinolite-chlorite schists, amphibolites, and porphyritoids regarded as analogues of the Kokchetav Block Efimov series are exposed in places in the core of the next easterly **Niyaz Anticlinorium** in the northern segment of the Yerementau system, which is separated from the Ishkeol'mes Anticlinorium by the **Selety Synclinorium.** This comparison requires a more cautious approach, since the given rocks may also be looked on as belonging to a Riphean crust of oceanic type.

The outcrops of granite-gneisses with a radiometrically determined age of 1,410 (Pb-isotopic method) and $1,500 \pm 150$ (α Pb method) million years in the Aktau–Mointy Uplift (ZAITSEV & FILATOVA, 1972) may be regarded as belonging to the Balkhash – Dzhungarian Massif (microcontinent) in the rear of the Yerementau–Chuili geosynclinal system. The same may apply to the Anrakhai paragneiss and granite-gneiss swell in the Chu–Balkhash Anticlinorium which accompanies the Zhalair–Naiman (Chu–Ili) Synclinorium in the northeast. The latter synclinorium is the main ophiolitic trough in the central segment of the system. The eastern continuation of the system, acquiring a latitudinal trend, extends along the northern slope of the Transilian Alatau Range to the Ketmen' Range and farther on to China, constituting the northern zone of the Chinese Tien Shan.

Returning now to the northern part of the Yerementau–Chuili system, it may be noted that the **Maikain Anticlinorium** of northeastern trend is its easternmost exposed anticlinorium. A major ophiolitic belt of Central Kazakhstan is confined to its axial part. The Maikain Anticlinorium is separated from the Yerementau–Niyaz one by the **Shiderty Synclinorium** and from the northwestern plunge of the Chinghiz Meganticlinorium by the **Ashchisu Synclinorium;** in the south they merge into a single **Bayanaul' Synclinorium.** Residual troughs filled with Silurian–Lower Devonian molasse are confined to axial zones of the Selety (see above) and Shiderty synclinoria.

The northern continuation of the Yerementau–Chuili system, buried under the Meso–Cenozoic cover, surrounds the northern, also buried, promontory of the Kokchetav Block and passes further in the northwestern direction where it probably links with the eugeosynclinal system of the eastern slope of the Urals and Transuralia.

To conclude this brief description of the Yerementau–Chuili system, it is necessary to note that its structure cannot be considered sufficiently understood yet. The wide occurrence of the ophiolitic complex, with which major horizontal movements are usually associated and the presence of protrusions of the ancient sialic basement suggests that the apparent anticlinorium-synclinorium structure of the given system came into being only at the very end of its geosynclinal evolution and that in fact these anticlinoria and synclinoria are complex anti- and synforms superimposed on an earlier nappe structure. The existence of nappes is most likely along the western boundary of the system, expecially at its boundary with the Kokchetav–Ulutau and Muyunkum massifs; these nappes moved probably from east (northeast) to west (southwest).

7.7 The Chinghiz–Tarbagatai Late Caledonian and Hercynian fold system

This system constitutes an extreme northeastern element in the Paleozoic folded region of Central Kazakhstan, separated by the **Kalba–Chinghiz Fault** from the Late Hercynian Ob'–Zaisan system. On the whole, it forms a meganticlinorium extending for more than 700 km in the northwestern-southeastern direction. In the northwest, the Chinghiz – Tarbagatai Meganticlinorium is bounded by the submeridional **Central Kazakhstan Deep Fault** (of transcurrent type), discordant with respect to its inner structure, along which it links with the northeastern part of the Yerementau–Chuili Caledonian system–the **Ashchisu Synclinorium.** Northwest, in the direction of this fault, the Chinghiz–Tarbagatai system expands and a wider branch taps from it: the **Akbastau Anticlinorium** between which and the Chinghiz Anticlinorium there lies the **Abraly Synclinorium.** In the southeastern direction, the system narrows and is represented by the **Tarbagatai Anticlinorium** alone, which plunges under the geosynclinal Middle Paleozoic on the Chinese territory.

Occurring in the cores of anticlinoria within horsts cut out by Hercynian faults are complexly deformed rocks of the diabase-jasper Vendian (?) – Cambrian formation, metamorphosed in the greenschist facies, and probably the initially underlying serpentinites. Intrusions of the gabbro-plagiogranite formation are connected with the same Salairian complex. Upper Ordovician and Lower Silurian deposits which represent the terrigenous, not infrequently flyschoid[5] and porphyrite formations, occur in intermediate graben-synclines, on depressed areas of anticlinoria and on their limbs, overlying unconformably Cambrian rocks. These deposits were deformed at the end of the Silurian to form broad and flat folds, the simplest ones in the Abraly Synclinorium where a thick and complete Ordovician sequence is observed. Formation of major granite plutons is also connected with this phase of Caledonian tectogenesis. The Devonian and Carboniferous formations–the D_{1-2} volcanic molasse and the terrigenous-carbonate, partly D_3–C coal-bearing formation form superimposed synclines, graben-synclines and volcano-tectonic troughs. Formation of faults of a transcurrent nature and the block movements of the Caledonian basement and its Hercynian cover are associated with Late Hercynian movements.

Consequently, the Chinghiz–Tarbagatai zone should be referred to Caledonian structures, however, with a predominant role of Late Caledonian (Late Silurian) movements in comparison with Early Caledonian (Late Ordovician) ones. In the southeastern direction, this uplift acquires the character of a geanticline within the Hercynian geosynclinal region. This is due to the narrowing of the Tarbagatai Uplift and convergence of the Dzhungaro–Balkhash and Irtysh–Zaisan geosynclinal systems bounding the uplift on the south and north, respectively. Further east, the Caledonian complex disappears from the surface, and there appear Devonian subaerial volcanic rocks at its place.

7.8 The Dzhungaro–Balkhash Hercynian fold system

This is the youngest system in the region being studied, where it occupies the central and southeastern parts, being surrounded by the Yerementau–Chuili system on the south-

---

[5] Not infrequently too it is regarded as marine molasse.

west, west, and south and by the Chinghiz–Tarbagatai Caledonian systems on the northeast. In the east, on the Chinese territory, its continuation merges with the southeastern continuation of the Ob'–Zaisan system and, together with the latter, extends to Southern Mongolia. Within the Chinese territory, the eastern part of the Dzhungaro–Balkhash system is overlapped by the major **Dzhungarian Basin** filled with thick Meso–Cenozoic continental molassoid deposits. It is usually assumed that a large block of the Precambrian basement occurs beneath the Dzhungarian Basin, however, this has not been proved.

The boundary between the Dzhungaro–Balkhash system and the surrounding Caledonides is not sufficiently distinct everywhere, and some areas of Central Kazakhstan (the Atasu–Mointy Uplift and the Tekturmas Anticlinorium) are assigned to the given system by some investigators, and to the Caledonides, by others. The point is that a hiatus and an unconformity between the Silurian and Devonian are expressed in the outer zone of the Dzhungaro–Balkhash system, and there are numerous outcrops of Lower Paleozoic and Upper Precambrian formations in the cores of anticlinoria. This unconformity attenuates as one goes farther deep into the system, and the volcanics and red molasses of the Lower and Middle Devonian pass into marine gray (or rather green as a result of metagenesis) terrigenous (mainly graywacke) formation which practically merges with a similar Silurian formation. The outer outlines of the system may be drawn approximately along the boundary of the marine Lower Devonian. The boundary between the region of Caledonian consolidation and the region of regeneration or continuation of geosynclinal evolution in the Hercynian cycle is even more pronounced, as there formed an **Early–Middle Devonian marginal volcanic belt,** which was first identified by BOGDANOV (1965) by analogy with the Okhotsk–Chukotka Belt at the boundary between the Mesozoides and Alpides in the Soviet Northeast. The Devonian volcanic belt, shaped like a gigantic horseshoe, fringes the Dzhungaro–Balkhash system on the west, northwest, and northeast, extending for about 1,500 km and having a width of up to 70 km. Lying conformably with respect to the structural-facial zones of the Dzhungaro–Balkhash system, it overlies with a sharp unconformity the Caledonides inner structure. A latitudinal stretch of the belt, passing through the region of Karaganda, has been studied the best (CHETVERIKOVA, 1970).

The Devonian volcano-plutonic association, making up the belt, consists of subaerial volcanic rocks ranging from andesite-basalt to rhyolite in composition, and also of granites. Rhyolites (and ignimbrites) with which granites are associated predominate in the outer zone of the belt; in the central zone, the rhyolites are replaced by andesite-basaltic rocks, and in the inner zone adjacent to the Hercynian geosyncline products of destruction of the volcanic rocks of the central zone were redeposited. It is assumed that the andesite-basalts and rhyolites were produced by magma chambers which occurred at different depths. Total thickness of the volcanic complex reaches 5 to 7 km. In the direction of the Caledonides, the complex is superseded by a red continental molasse with a minor role of volcanic rocks, whereas in the direction of the Hercynides by a volcano-terrigenous and then purely terrigenous and marine formations. The structure of the volcanic belt is characterized by volcano-tectonic features: domes (with granites and ignimbrite vents in the center), as well as by depressions filled with ignimbrites. Late Hercynian movements transformed depressions into uplifted blocks, and salients of the Caledonian basement turned into graben-synclines. In the region of Karaganda, a Hercynian depression known as the **Karaganda marginal synclinorium** overlies the axial zone of the volcanic belt. This depression extends for up to 350 km, its width being up to 60 km. In its situation at the boundary between the epi-Caledonian massif and Hercynian fold system, in the asymmetry of its structure and distribution of formations, as well as in the latter's composition, this synclinorium is quite similar to foredeeps, as was noted earlier by KABANOV (1972), I. V. ORLOV, and others. The Karaganda Synclinorium is filled with a Givetian–Frasnian graywacke formation, a siliceous-shale-siltstone Famennian–Tournaisian formation (the latter is replaced by a carbonate formation farther north, which lies transgressively over a volcanic complex), a terrigenous Viséan formation of the Culm type, a paralic Viséan–Namurian coal-bearing molasse, a limnic Middle Carboniferous and continental Upper Carboniferous molasse. These series, dislocated perfectly conformably, are overlain subhorizontally by a thin cover of continental coal-bearing Jurassic rocks.

The sufficiently quiescent structure of the northern limb and axial part of the synclinorium is in sharp contrast with the linear-imbricate structure of its southern limb which bears traces of strong compression and displacement towards the north (Fig. 46). In the south, this limb is overlain by a tectonic nappe composed of Famennian–Tournaisian deposits with a lithofacies composition different from that of the deposits making up the principal part of the Karaganda coal Basin. The existence of this nappe, which was first detected by RUSAKOV in 1930, has now been proved by drilling, and its minimal amplitude is estimated at 15–20 km, the extent being up to 200 km. Farther south, there extends around the Karaganda Basin a most specific **Spassky zone** in whose southern part the strongly and linearly deformed volcano-sedimentary Devonian rocks are overthrust by blocks, probably klippes of Lower Paleozoic and Upper Cambrian rocks, as well as gabbro and serpentinites similar to those making up the **Tekturmas Anticlinorium** separated from the Spassk zone by the **Nura Synclinorium.** It is most likely that the pre-Silurian rocks of the Spassk zone belong to an overthrust sheet with which rocks of the Tekturmas zone travelled for dozens of kilometers north.

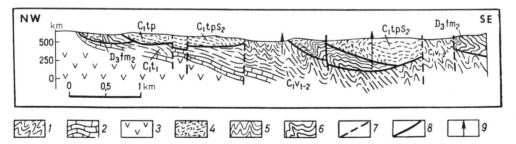

Fig. 46.   Cross-section through the northern part of the Altynsu tectonic nappe, Karaganda region (after KABANOV, 1972). Autochthon: 1 = carbonaceous siltstones and mudstones of the Lower and Middle Viséan; 2 = Famennian and Tournaisian limestones; 3 = andesite-basaltic Middle Devonian porphyrites; allochthon: 4 = siliceous siltstones of the upper member of the Tournaisian *Posidonia* beds; 5 = platy calcareous siltstones of the lower member of the Tournaisian *Posidonia* beds; 6 = Famennian shaly nodular limestones; 7 = faults and upthrusts; 8 = overthrusts; 9 = boreholes.

The Spassk zone is now separated from the Nura Synclinorium–a relatively quiescent structure composed mainly of a Silurian–Lower Devonian graywacke formation as well as of D_{1+2} red-beds and D_{2+3} volcano-sedimentary series–by the Baidauletov Fault, more specifically a Late Hercynian transcurrent fault.

The east-north-east trending **Tekturmas Anticlinorium** situated south of the Nura Synclinorium is characterized by an extremely complicated structure. Its core contains a Vendian (?)–Cambrian siliceous-basaltoid Urtyndzhal' series in association with ultrabasic rocks and gabbro.

The intensity of deformations is very high: linear folding, cleavage, numerous faults. There are indications of a fan-like structure of this core and traces of its thrusting over the Nura Synclinorium; it is quite possible that this overthrust represents the near-root part of the Spassk nappe.

The limbs of the Tekturmas Anticlinorium are composed of Ordovician terrigenous-volcanic and Silurian terrigenous rocks; these deposits overlie unconformably (the Salairian phase) the Urtyndzhal' complex; yet, they had also experienced considerable deformations associated with the Tel'besian phase of Late Caledonian tectogenesis. Subsequent Hercynian movements cut out the ancient core of the Tekturmas Anticlinorium in the form of a horst and caused displacement along its axis (SUVOROV, 1973). The Tekturmas Anticlinorium is accompanied in the south by the **Atasu–Uspensk Synclinorium** of the same trend: a large complex trough which includes a thick (5 to 7 km) Silurian flyschoid terrigenous, a Middle–Upper Devonian sedimentary-volcanic, and an Upper Devonian–Lower Carboniferous carbonate-terrigenous formation. The unconformity between the Silurian and Devonian attenuates in the southern direction, and the structure becomes more complicated up to a large (15 to 20 km in amplitude) overthrust of the Silurian–Devonian slice northward along the **Uspensk zone of intense shearing.** This is the second nappe structure in Kazakhstan, parallel to the Spassk nappe, lying more to the north. The Uspensk nappe rocks belong to the next zone further south: the **Zhaman–Sarysu Uplift** whose core again contains Silurian rocks at the surface. Further south and east, the Middle Paleozoic is hidden under the huge superimposed Upper Paleozoic **Tokrau Basin.** The above group of sublatitudinal structures lying between the Karaganda and the Tokrau basins have another feature in common, in addition to the same trend and marked tangential displacement of northern direction: they are the region where the Tel'besian phase in the Middle Devonian was the main phase of deformations. It may, therefore, be worthwhile identifying them as an independent Late Caledonian folded Southern Karaganda zone.

In the west, the structures of the Southern Karaganda zone come into frontal contact with structures of the Northwestern Balkhash region, which have different orientations, but mainly northwestern-southeastern ones. These structures constitute the **Atasu – Mointy Uplift** separated from the Southern Karaganda zone by the **Akzhal–Aksoran** Silurian–Devonian **Trough** (synclinorium) of the same trend.

The oldest rocks of the Atasu–Mointy Uplift–schists and quartzites–may be of Lower Riphean or even more ancient age. In any case, the Uzunzhal' granites intruding them, with which a porphyry formation is associated, have been dated at 1,410 million years by the lead-isotopic method. The porphyry formation is overlain unconformably by Upper Riphean quartzites and then by Vendian–Cambrian–Ordovician terrigenous-carbonate deposits. Their moderate thickness and the wide occurrence of shallow-water limestones (oölitic, oncholitic, and algal) and dolomites suggest that this region belongs to the western part of the Balkhash–Dzhungarian Median Massif (ZVONTSOV, 1973).

The eastern continuation of the massif is exposed at the surface in the **Tekeli zone** of the Dzhungarian Alatau southern range and in its continuation in the Borokhoro Range. A quartz-sandstone formation[6] lies here at the base of the Paleozoic rocks; then there follow siliceous-carbonate, siliceous-carbonaceous-shale, and carbonate formations. These deposits had experienced vigorous deformations before the Silurian and include a complex of granitoids (plagiogranites, granites, and granodiorite-porphyries) of the same age. The greater thickness of the Lower Paleozoic deposits, the strong Caledonian deformations, and the granitization suggest that this zone is a northern periphery of the epi-Caledonian Tien Shan Massif and that it may be identified as the **Southern Dzhungarian Anticlinorium.** The Caledonian complex is overlain unconformably by andesite-dacite volcanic rocks and coal-bearing Lower Carboniferous deposits making up flat troughs but intruded by Middle Carboniferous normal and post-Permian (?), probably Early Mesozoic alkaline granites. Farther south, being separated by a major fault, there extends the broad Ili Basin which overlies the Chuili–Ketmen zone by its southern limb. This depression is part of the Balkhash–Ili Late Paleozoic volcanic belt which will be discussed later in this book.

The inner part of the Dzhungaro–Balkhash system is characterized by a great complexity of its structural pattern; the anticlinoria and synclinoria identified here have quite variable strikes; wide fields are covered by Upper Paleozoic volcanics which fill independent Late Hercynian depressions. In the region north of Lake Balkhash, the **Northern Balkhash Anticlinorium** is particularly well pronounced. It forms a steep arc convex towards the south, which abuts against the northern shore of the lake's eastern part, and is almost halved by the southern continuation of the Late Hercynian Central Kazakhstan Fault. The core of the Northern Balkhash Anticlinorium again contains a strongly deformed Late Precambrian–Cambrian spilite-jasper complex which here too is accompanied by ultrabasic rocks. Judging by the descriptions available[7], the rocks of this association not infrequently form real tectonic melange. The spilite-jasper complex is overlain unconformably by Upper Ordovician and Silurian deposits which are deformed a little weaker, yet sufficiently strongly to form typically linear folds. The Famennian–Lower Permian deposits form superimposed structures of brachymorphous type, complicated by zones of shearing along the numerous Late Hercynian faults. In contrast, the thick terrigenous Silurian–Lower Permian series on the northern periphery of the anticlinorium lie relatively quietly.

The **Central Dzhungarian Anticlinorium** is a southeastern continuation of the western Tyul'kulam branch of the Northern Balkhash Anticlinorium on the other side of Lake Balkhash and of the young Balkhash Basin. It extends in the latitudinal direction along the axial part of the Dzhungarian Alatau Range to the Dzhungarian Gate in the east.

---

[6] Metamorphic rocks which may correspond to the basement of the massif are exposed over small areas in the northern part of the Ili Basin adjacent on the south.

[7] For example, the Geology of the USSR, Vol. 20, p. 284, reads: "In many cases ... the band of the Itmurundy series rocks are a tectonic macrobreccia composed of plastic basic volcanic rocks, ultrabasic rocks, and rigid competent siliceous beds which experienced a vigorous process of complex crumpling and crushing."

This anticlinorium is composed mainly of a thick and uniform Upper Silurian – Lower and Middle Devonian terrigenous series which experienced initial metamorphism and is deformed into narrow linear folds complicated by overthrusts. They are overlain unconformably by Lower Carboniferous superimposed troughs and intruded by large isometric granitoid plutons of Middle–Upper Carboniferous, partly Permian age. A more vigorously deformed and metamorphosed Caledonian complex–Cambrian, Ordovician, enclosing Late Ordovician granitoids, emerges in some areas from under the Hercynian folded complex.

The narrow **Borotala Synclinorium** separates the Hercynian Central Dzhungarian Anticlinorium from the Caledonian Southern Dzhungarian one. The synclinorium is filled mainly with Devonian–Lower Carboniferous deposits which overlie unconformably the Caledonian basement, and has been subdivided into two structural stages. The lower, flyschoid-terrigenous pre-Upper Devonian stage is strongly deformed (the Tel'besian phase) forming linear, in places overturned folds. The upper-stage lower molassic Upper Devonian–Lower Carboniferous strata underwent deformation in the Middle Carboniferous and are characterized by a brachymorphous structure. In places, the formations of the geosynclinal complex are overlain unconformably by upper volcanic molasses of the Upper Paleozoic, which experienced only very weak deformations. In the east, the Borotala Synclinorium widens abruptly, and Upper Paleozoic intermediate and acid volcanic rocks forming a large superimposed depression become widespread in its central part, in the region of Lake Ebi-Nur. Further east, on the continuation of this depression, there lies a deep Meso–Cenozoic Dzhungarian Basin mentioned previously.

The Central Dzhungarian Anticlinorium is surrounded in the north by the broad **Northern Dzhungarian Synclinorium.** Just like the Borotala Synclinorium, it is composed mainly of the Devonian siliceous-schist rocks with minor quantities of basic volcanics and of Lower–Middle (lower part) Carboniferous rocks. Permian deposits fill near-fault graben-synclines over small areas. Major fold deformations in the southern block of the synclinorium took place in the Late Devonian and Early Carboniferous, those in the northern block in the Middle Carboniferous. All these deformations produced large linear folds complicated by an additional tight folding, in the south sometimes overturned northwards and in the north generally symmetrical. Middle–Upper Carboniferous post-folding granite plutons, usually oval in shape, are confined to major anticlines.

Both the Northern Dzhungarian Synclinorium and the Central Dzhungarian Anticlinorium are truncated crosswise in the east by a major Late Hercynian **Dzhungarian Wrench Fault**–a displacement which had been active until the Recent epoch. On the west, this displacement limits the Dzhungarian Gate Graben and continues within the Chinese territory where it merges with the marginal suture of the Dzhungaro–Balkhash Hercynian system (Bush et al., 1968). The **Alakol' Wrench Fault,** extending in the north-northwestern direction past Lake Alakol' and linking farther north with the Chinghiz Fault of northwestern trend, forms the eastern boundary of the Dzhungarian Gate Graben filled with Jurassic and younger sediments. These two wrench faults are dextral and form a fault system more than 1,500 km in extent (Samygin & Tret'yakov, 1971).

The **Maili–Dzhair Anticlinorium** is a continuation of the Central Dzhungarian Anticlinorium east of the Dzhungarian Fault on the Chinese territory, whereas the

Urkashar Synclinorium composed mainly of Lower Carboniferous deposits is a continuation of the Northern Dzhungarian Synclinorium.

In the west, the Northern Dzhungarian Synclinorium continues into the Northern Balkhash region where it is intersected by the latitudinal (Tyul'kulam) branch of the Caledonian Northern Balkhash Anticlinorium with the ophiolitic belt accompanying it. The ophiolitic complex emerges, in all probability, at the base of the nappe moved northwards. A specific major Late Paleozoic volcano-tectonic depression – the **Northern Balkhash Basin** separates the Northern Dzhungarian Synclinorium from the Chinghiz–Tarbagatai Caledonian Meganticlinorium. According to KOSHKIN (1969) this depression constitutes the northern segment of the **Balkhash–Ilian volcanic belt** ("Tectonics of the Uralo–Mongolian . . .", 1974). Volcano-tectonic features, mainly negative, but partly also volcanic domes predominate in its structure. Some of them had developed during the entire Late Paleozoic.

The Northern Balkhash Depression is bounded on the west by the Central Kazakhstan Fault. The **Tokrau Depression** is its analogue and continuation on the other side of the fault, however, this depression extends longitudinally rather than latitudinally and is a link of a Late Paleozoic volcanic belt. The outlines of this depression are clearly discordant with respect to the adjacent anticlinoria and synclinoria structures of the Lower and Middle Paleozoic. The inner structure is characterized by an abundance of volcano-tectonic, also mainly negative and properly volcanic (up to 20–30 km in cross-section) features; gentle folds of tectonic origin are of minor significance. Young faults of largely northwestern, partly northeastern trend form a dense network. Alongside volcanic rocks, there are numerous Upper Paleozoic granite plutons with which copper porphyry ores of Kounrad, etc. are associated.

The third major link in the Late Paleozoic volcanic belt, which continues the Tokrau Depression southeast of Lake Balkhash, is the **Ilian Depression** separating the Caledonian structure of the southern Dzhungarian Alatau reworked in the Hercynian stage, with which it conjugates along the major **Southern Dzhungarian Fault,** from the Northern Tien Shan Caledonides. In contrast to the Tokrau Trough, the Ilian Depression is inscribed more or less concordantly into the structure of the Caledonides and had experienced partial revival in the Meso–Cenozoic. The depression is filled mainly with two thick complexes of subaerial volcanic rocks: Lower Carboniferous andesite-dacite (1.5–2 km) and Upper Paleozoic–Lower Triassic andesite-rhyolite (2–3 km) separated by an unconformity. The former complex occurs in combination with a coal-bearing lower molasse, the latter with an upper continental molasse. Both these complexes are deformed into brachymorphous folds which are steeper in the Lower Carboniferous and more gentle-dipping in the Upper Paleozoic. There also are numerous faults of normal or strike-slip nature. Very common are orogenic granitoids: Middle Carboniferous granodiorites and granites, Permian–Lower Triassic rocks, more varied in composition, from gabbro and diorites to monzonites, syenites, and granite-porphyries.

Judging by exposures in the cores of some anticlines, the Hercynian orogenic volcanic rocks overlie the Caledonian folded, metamorphized, and granitized basement which is characterized by a sequence, more reduced as compared to that in the Southern Dzhungarian zone, and by shelf facies; hypothetically Precambrian metamorphic rocks are encountered in places. This suggests that the axial zone of the Balkhash–Dzhungarian Median Massif used to reach here from the Atasu–Mointy area of the Northwestern Balkhash region.

7.9 The platform structure of Central Kazakhstan and the orogenic structure of Northern Tien Shan

During the Mesozoic and Cenozoic, Central Kazakhstan was a shield of the Central Eurasian young platform, separating the Western Siberian and Turanian platforms which were linked by the Turgai Trough. Individual negative structures, partly inherited and partly superimposed, evolved against the background of general upheaval. Several such structures emerged at the end of the Triassic–beginning of the Jurassic and developed during the entire Jurassic period, with an accumulation of a continental coal-bearing formation. The **Leont'ev Graben of Karatau,** confined to the zone of the Major Karatau Fault, the **Graben of the Ferghana Range (Yarkend–Ferghana),** and also probably the **Alakol'** and **Kuraili (Ayaguz) depressions** within the zone of the Alakol'–Dzhungarian faults bear a pull-apart graben character. The Jurassic deposits in the Leont'ev Graben are deformed into folds which are tighter and complicated with upthrusts in the Lower and Middle Jurassic layers and more gentle in Upper Jurassic strata. The Jurassic rocks are deformed even stronger in the Ferghana Graben. The **Maikube Coal Basin** in the northern part of the shield and the **Karaganda Basin,** which inherited the Hercynian trough of the same name, possess a flat synclinal structure. A little more complex gently folded structure must be characteristic of the **Dzharkent Basin** inscribed into the Ili Late Hercynian intermontane trough, but bounded by faults, especially in the north. The Dzharkent Trough recommenced its subsidence in the Cenozoic, simultaneously with vigorous mountain formation in the Dzhungarian Alatau and Northern Tien Shan. The Issykkul' and Naryn depressions have similar histories; the former emerged in the center of a Paleozoic median massif. Slight upheavals prevailed on the territory in the Cretaceous–Early Paleogene time. Absolute downwarping took place in the Chu Depression alone.

At the end of the Paleogene and beginning of the Neogene, the southern and eastern parts of the Kazakhstan–Northern Tien Shan fold region were involved into increasingly strong upheavals, which led to their transformation into the mountainous edifices of Northern Tien Shan and Dzhungarian Alatau. This process expanded in a weaker form and probably with a delay to Karatau, Kendyktas, Chu–Ilian Mountains, Tarbagatai, and Chinghiz. Recent structure of the orogenic region is of an arch-block character; in its trend and partly in the position of individual elements it is concordant with the inner structure of the folded basement, the differences being in details only, especially where the structures are the oldest and orogenesis relatively weak. Recent structure is well expressed in the relief, especially in the deformations of the widespread Cretaceous erosion surface. The slopes are inclined by several degrees, less frequently by more than 10°, the amplitude of recent faults amounts to several hundred meters, occasionally to more than 1,000 m. In places, the Paleozoic basement is thrust somewhat over the Cenozoic rocks or older horizons of the Cenozoic cover are thrust over younger ones up to Upper Quaternary (the Alakol'–Dzhungarian fault zone, etc.).

Piedmont troughs formed in places in front of the recent orogenic region. Particularly distinct is the **Southern Chu** or **Frunze Basin** with a three-kilometer series of Cenozoic molasse. Among other uplifted troughs are the **Southern Balkhash** and pre-Dzhungarian depressions with Cenozoic molasse only a little more than 1,000 m thick. The largest intermontane depressions include the Ilian one with the **Alm Ata** and **Dzharkent**

troughs, the **Issykkul' Depression** with its eastern continuation – the **Tekes Depression,** and the **Naryn Depression.** The molasses in these troughs are 3 to 4 km thick; they experienced folding and in places also fault deformations. The fold deformations are partly near-fault and suprafault and partly probably gravitational in origin.

7.10 Main stages of evolution

7.10.1 Pre-Baikalian history

A major problem in the tectonic history of Central Kazakhstan and Northern Tien Shan is the one which concerns the type of the crust on which the Paleozoic (Late Precambrian –Paleozoic) Kazakhstan–Tien Shan geosynclinal region formed and during which period. This problem has now become particularly acute in connection with the new concepts of geosynclinal evolution, and most recent works tackle it from opposing positions. A group of researchers, including geologists of Kazakhstan, Kirghizia, and Uzbekistan as well as of Moscow University and VSEGEI (All-Union Scientific Research Institute of Geology), who refer the oldest metamorphic series of the region to the Early Precambrian, assume that the Kazakhstan–Tien Shan geosynclinal region emerged at the end of the Precambrian–beginning of the Paleozoic over a continental crust which had formed by the end of the Early Precambrian. This does by no means rule out the possibility that formation of geosynclinal troughs could and should have been accompanied by appreciable destruction of the ancient continental crust, its partial oceanization, either in the form of sliding apart, mantle diapirism and subsequent spreading or in the form of basification in the process of subsidence. Fragments of continental crust, preserved from oceanization, remained in the course of further geosynclinal evolution inside the geosynclinal region in the form of median massifs. The structural pattern of Central Kazakhstan and Northern Tien Shan was predetermined by this particular mosaic of the median massif ancient blocks with geosynclinal systems in between. The foundations of this concept were outlined by KASSIN (1960) who should be considered the founder of the Central Kazakhstan geology; another outstanding investigator, NIKOLAEV (1933), expressed similar views for the case of Northern Tien Shan.

An essentially different concept is expressed most comprehensively in the works by N. A. SHTREIS, G. I. MAKARYCHEV (1974, 1978), A. I. SUVOROV, and M. V. MURATOV (1974), from the Geological Institute of the USSR Academy of Sciences. These scientists deny the presence of Early Precambrian basement relics in the Kazakhstan–Tien Shan region and refer all metamorphic rocks there to the Riphean, identifying analogues of oceanic crust at its base. They assume accordingly that the Kazakhstan–Tien Shan geosynclinal region formed in the Riphean time over an oceanic crust, possibly primary, and that the granitic-metamorphic layer formed and expanded in it gradually, starting from the middle of the Riphean. This concept was developed in great detail by MAKARYCHEV (1974), however, its roots may be traced back to the nuclear theory of POPOV (1960), based on the material from Central Asia, and to similar views by VASIL'-KOVSKY (1964).

The supporters of either concept hasten unduly in considering their ideas reliably proved. This follows clearly from comparing the following two quotations:

"N. A. SHTREIS suggested as far back as 1954 that there are no deposits older than Riphean in the Paleozoic folded region of Kazakhstan, even though nothing was known at that time about the basement on which the Riphean geosynclines formed. This assumption may now be considered an established fact" (MAKARYCHEV & PAZILOVA, 1973).

"The available materials contradict the concept of an oceanic or early oceanic stage of evolution in the Archaean or Proterozoic, even though some investigators (N. A. SHTREIS, G. I. MAKARYCHEV, M. D. GES', and others) still share this opinion. Neither is there any confirmation of the concept of the eugeosynclinal type of sequence in the Riphean, lying directly on the oceanic bed (M. V. MURATOV et al.). On the whole, the hypothesis of M. M. TETYAEV, V. A. NIKOLAEV, A. V. PEIVE, and N. M. SINITSYN[8] on the presence of an ancient (Karelian) platform in the Precambrian, further developed by V. N. OGNEV, V. G. KOROLEV, and others, is valid" ("Earth's Crust ...", 1974).

There is yet a third, intermediate point of view which accepts consecutive expansion of the consolidated massif of Central Kazakhstan, starting from individual ancient nuclei, for which, however, Early Precambrian rather than Late Precambrian age is assumed. This point of view was propounded by ARKHANGEL'SKY (1941), SHATSKY (1938), and BOGDANOV (1965). It has now been accepted by MAKARYCHEV (1975) who admits the existence of pre-Riphean continental crust relics in the Transilian Alatau.

Datings of about 2,500 million years have presently been obtained for the Aktyuz and Kemin series of Transilian Alatau and about 2,000 million years for the Bekturgan series of the Ulutau and Makbal series of the Kirghizian Range. The same age may be assumed for the lowermost portion of the Kassan (Tereksai series) and Karatau (Bessaz series) sequences and their analogues. Radiometric datings of about 1,700–1,800 million years are available for granite-gneisses of Ulutau and for the Kirghizian series of the Makbal Horst in the western part of the Kirghizian Range, the intrusion of these granite-gneisses in Ulutau having been preceded by the formation of a thick Karsakpai complex. A specific lithological appearance of the Karsakpai jaspilite formation suggests that the latter complex should be referred to the Early Precambrian rather than Riphean. Completely different are the Middle Riphean deposits of Northern Tien Shan which experienced only initial metamorphism. Vigorous migmatization observed in the Kokchetav and Ulutau massifs and on the Makbal Uplift is not typical of the Late Precambrian either. Finally, the boundaries of the ancient blocks are of fault nature and discordant with respect to the inner structure. It is therefore quite possible that there had existed a region of continuous continental crust in place of the future Kazakhstan–Tien Shan region by the beginning of the Late Precambrian. This ancient region must have constituted a single massif with the basement of the Eastern European Craton (in between, there had existed the Mugodzhary Early Precambrian) and Sinian Craton (via the Tarim Massif). It will be shown below that the same assumption may be true of the Altai–Sayan region and the Siberian Craton, adjacent on the east, and formulated for the Uralo–Mongolian Belt as a whole.

Proceeding from the assumption on the pre-Riphean age of the oldest metamorphic series of Central Kazakhstan and Northern Tien Shan as much more likely, we can

---

[8] V. M. SINITSYN seems to have been meant here (V. Kh.).

outline, following ZAITSEV & FILATOVA (1972), two principal stages in the development of this territory in the Early Precambrian, which may arbitrarily be dated as Late Archaean and Early Proterozoic.

7.10.2 The Late Archaean stage

is represented by the Zerenda series of the Kokchetav Block, the Bekturgan series of the Ulutau Block, the Aktyuz series of the Transilian Alatau, and probably also the Makbal series of the Tien Shan Kirghiz–Terskei zone. Initial composition of these formations, which are now represented by crystalline schists with garnet, micas, graphite, gneisses, amphibolites, quartzites, marble and which are generally metamorphized in the amphibolite facies (granulites and eclogites appear occasionally), must correspond to a carbonate-terrigenous formation with minor quantities of basic volcanics. Deposits of this character seem to be typical of a region with sufficiently well developed continental crust; however, eclogites, partly amphibolites, and also serpentinites and gabbro emerging in some places, in particular, in the Makbal Horst and possibly in Karatau and Kassan among the ancient Precambrian rocks may be regarded as protrusions of a primary oceanic floor.

Ubiquitous folding, regional metamorphism, and extensive metasomatic granitization correspond to the end of the Archaean (?). Complexes referred to the Lower Proterozoic overlie because of that unconformably the presumably Archaean complex throughout.

7.10.3 The Early Proterozoic (Karelian) stage

Metamorphic series, which are referred quite arbitrarily to the Lower Proterozoic in different parts of Central Kazakhstan, are of indubitably geosynclinal nature, however, they are not documented enough to enable the structural pattern of this stage to be reconstructed. Particularly complete and typical among these series is the Karsakpai series of Ulutau which fills a meridionally extending synclinorium of the same name. The Karsakpai Geosyncline (eugeosyncline) is a most typical example of suture jaspilite geosynclines which are so characteristic of the Early Proterozoic on practically all continents.

According to FILATOVA (1976), the sequence of the Karsakpai Geosyncline includes the following sequence of formations: 1) quartz-phyllite, 2) dacite-keratophyre-schist, 3) jaspilite-greenschist-spilite-basaltic, and 4) terrestrial dacite-rhyolitic.

The following interpretation may be given for this sequence: 1) formation of a flat-bottomed trough on a rising epi-Archaean peneplain; 2) formation of a trough-rift bounded by deep faults and of intracrustal magma chambers; 3) further widening and deepening of the rift, probably with partial oceanization of the crust; activity of subcrustal magma chambers under the conditions of high permeability; 4) compression instead of distension, decrease in permeability, and formation of new magma chambers. Total thickness of geosynclinal formations proper, according to L. I. FILATOVA, reaches 11 km, and the orogenic subaerial volcanics are more than 5 km thick. The closure of the

Karsakpai Geosyncline, which preceded the accumulation of these volcanic rocks, was accompanied by strong folding, regional metamorphism of the greenschist facies and a new phase of metasomatic (K-metasomatism) granitization, although much weaker than the preceding one. Formation of the Zhaunkar granites is associated with this phase in Ulutau. Probable analogues of the Karsakpai series in other parts of the region (Kokchetav and Chu blocks) exhibit manifestations of the same processes.

7.10.4 The Middle Proterozoic–Early Riphean stage

At least two epochs of folding, regional metamorphism, and granitization in Central Kazakhstan and Tien Shan produced a sufficiently thick and consolidated crust of continental type, and dry-land conditions obtained over vast territories. Against this background, the still high heat flow promoted vigorous magmatic activity which produced a thick (3 to 5 km) volcano-plutonic association of porphyroid-granitic composition, particularly well represented in the Ulutau Massif (the Maityube series and granite-gneisses), also known in the Sarysu–Teniz zone and on the Atasu–Mointy Uplift, in the latter in combination with a quartzite-phyllite formation. The granite-gneisses of this association in Ulutau have a radiometric age of 1,780 ± 50 m.y. and in the Atasu–Mointy region 1,400 m.y. (U/Pb method on zircon). A similar age of 1,275–1,070 m.y. in the Kirghizian Range was established (α Pb method on zircon) for the Karadzhilga–Kyzyl-tash granitoid pluton which intrudes both the Kirghizian and Ortatau series of the Lower Riphean (?) or Middle Proterozoic (KISELEV et al., 1974).

It follows from the above that the oldest basement of the Kazakhstan–Tien Shan region passed through the same evolutional stages as the basement of ancient cratons, ending with the stage of cratonization. Further history, however, was different, starting in places as early as the Middle Riphean to which in Northern Tien Shan and less distinctly on the remaining territory of the Kokchetav–Muyunkum Massif there corresponds an independent evolutional stage: Issedonian (after Yu. A. ZAITSEV) which is identical with the global Grenvillian stage.

7.10.5 The Middle Riphean (Grenvillian) stage

In this stage there formed along the southern margin of the Kokchetav–Muyunkum Massif a zone of subsidence which was, in all probability, of intracratonic character (intracratonic miogeosyncline?). Within this zone coinciding with the Kirghiz–Terskei zone of Caledonian Tien Shan, there accumulated in the Middle Riphean a rather thick quartzite-phyllite-carbonate formation resembling Riphean formations of the Southern Urals. At the boundary between the Middle and Late Riphean, it was deformed into folds and then overlapped, with an appreciable angular unconformity, by the Upper Riphean starting with a typical series of quartzose sandstones (the Dzhetysu series). A similar Kokchetav series occurs extensively also on the block of the same name, where it overlaps, with a sharp unconformity, the Early Precambrian basement, which enabled Yu. A. ZAITSEV to establish here the manifestation of the "Issedonian" phase of tectogenesis, with a subsequent setting in of tectonic conditions close to platform ones.

Riphean deformations $(R_{1-2}?)$[9] in the Atasu–Mointy Uplift–the northwestern relic of the Balkhash–Dzhungarian Median Massif (see below) seem to belong to the same phase.

7.10.6 The Baikalian stage (Late Riphean–Vendian)

The platform, more specifically quasiplatform conditions on the territory of the Kokchetav–Muyunkum Massif and, judging by the exposures within the Kyzyl Kum, on the Turgai–Syr Darya Massif prevailed during the remaining Late Precambrian and Cambrian. By the start of the Late Riphean, these two massifs must have constituted a single major continental block. However, the intermediate zone experienced regeneration of the geosynclinal conditions in the Late Riphean; a deep rift formed there, along which, as shown by the data of MAKARYCHEV (MAKARYCHEV & PAZILOVA, 1973; MAKARYCHEV, 1974; MAKARYCHEV & GES', 1981), the continental crust experienced rupture and was replaced by an oceanic crust. On both sides of this quasi-oceanic-type trough, along the margins of the microcontinents bounding it, there must have extended volcanic belts with acid magma eruptions, underlain by Benioff zones. The products of these eruptions make up the Koksu series in Ulutau (east of the trough) and the Bol'shoi Naryn series in the Dzhetymtau Range in the Median Tien Shan (south of the trough). At the end of the Riphean and beginning of the Vendian, downwarpings were superseded on both sides of the oceanic basin by upheavals and fold deformations; intrusion of granitoid plutons both in Karatau and in the Northern (Kirghiz Range) and Median Tien Shan (Kassan) is confined to the same border zones. The Vendian deposits overlying unconformably the Riphean ones are represented here by a molasse type formation. The latter formation begins by tillites or tilloids; red clastic rocks appear higher up. All these indications of active Baikalian tectogenesis attenuate in the northern direction along the western framing of the Kokchetav–Muyunkum Massif. Neither are they expressed in the massif proper or in the axial part of the basin; in the latter tholeiitic volcanics are superseded in the Vendian–Early Cambrian by andesite-basalts of calc-alkaline, island arc type.

7.10.7 The Caledonian stage (Vendian–Middle Devonian)

This stage was the principal one in the geosynclinal evolution of the entire Kazakhstan–Tien Shan region, with partial exception of the Dzhungaro–Balkhash system, a great part of which also experienced Caledonian tectogenesis.

The Caledonian development followed essentially different courses in the west and east of the region. In the west, the Turgai–Syr Darya and Kokchetav–Muyunkum massifs generally preserved their uplifted position. The Ishim–Talas strip between these massifs retained its mobility, despite partial Baikalian consolidation. However, as a result of the Baikalian consolidation, which was expressed in the regeneration of the granitic-metamorphic layer, the eugeosynclinal conditions of the Late Riphean were replaced here by miogeosynclinal in the Early Paleozoic. In the initial stage of evolution of the miogeosynclinal basin in the Late Vendian and Early Cambrian, subsidence proceeded at

---

[9] As mentioned before, they could well be pre-Riphean.

faster rates than sedimentation, so that there formed in the Karatau Range a deep trench with an accumulation of carbonate-siliceous-clayey (shale) sediments. Associated with the latter are known deposits of phosphate as well as vanadium-bearing shales; individual members are enriched with manganese and iron (KHOLODOV, 1973). The northern part of the Kokchetav–Muyunkum Massif had been cut through by the sublatitudinal Southern Kokchetav Trough in the Vendian (MINERVIN et al., 1971). This trough separated the Kokchetav and Ulutau blocks and had been an offshoot of the Yerementau–Chuili eugeosyncline until the Late Ordovician. This trough contains an ophiolitic complex and a Cambrian carbonaceous-siliceous-shale formation.

The conditions changed somewhat in the Middle Cambrian towards shallowing of the basin. Thick olistostromes of Middle Cambrian–Lower Ordovician age in the boundary zone of the Northern and Middle Tien Shan point to deformations in the neighbourhood in the south (MAKARYCHEV). But in the southeastern part of the geosyncline, starting from Karatau, a 3 to 4 km thick carbonate (limestone-dolomite) formation accumulated during the period from the Middle Cambrian to the Middle Ordovician. The sediments became thinner in the Greater Karatau, and there appeared at the base of the Ordovician a new ore horizon with Fe, Mn, Cu, and some other metals. The Upper Ordovician here lies unconformably and is of a molasse character, since it is composed of coarse clastics.

The Ordovician in the northern part of the miogeosyncline, in the framing of the Kokchetav and Ulutau blocks, as well as in the Southern Kokchetav Trough is of terrigenous composition and of a considerable thickness. In the Late Ordovician, there appeared volcanic material of andesitic composition (basaltic volcanics and jaspers were found in the Southern Kokchetav Trough in the Early Ordovician). The southeastern continuation of the Kokchetav–Muyunkum Massif (the Northern Kirghizian zone) underwent breakdown in the beginning of the Cambrian, and there formed the Chilik –Kemin eugeosynclinal trough which split this part of the massif into two blocks: Transilian in the north and Issykkul' in the south. At the end of the Cambrian, during the Salairian epoch of tectogenesis, this trough closed up after it was filled with volcanic rocks of the spilite-diabase formation, and there formed on both sides of it secondary troughs and volcanic arcs superposed on slopes of ancient blocks. In the Early and Middle Ordovician there deposited within these troughs a flyschoid terrigenous and an andesitic subaqueous-volcanic formation. At the end of the Middle and beginning of the Late Ordovician, the Northern Kirghizian zone was involved into upheaval, experienced fold and fault deformations and intrusion of major granite plutons. This was followed by a separation of the zone into uplifted and relatively downwarped areas. The latter became, in the Late Ordovician and Early Silurian, a site of accumulation of red continental molasses. The molasse troughs were partly inherited with respect to the earlier flysch troughs and partly superposed unconformably on older structures. In both cases the molasses form simple flat synclines in contrast to stronger though nonuniformly deformed older formations.

The principal events in this stage took place east of the Kokchetav–Muyunkum Massif. At the very end of the Precambrian or in the beginning of the Cambrian, there emerged vast areas with oceanic crust, to which the present Yerementau–Chuili and Chinghiz –Tarbagatai eugeosynclinal fold systems correspond (Fig. 47). The oceanic or quasioceanic character of these systems in the early stage of their evolution, in the Vendian–Cambrian (possibly, partly in the Ordovician) follows from the extensive

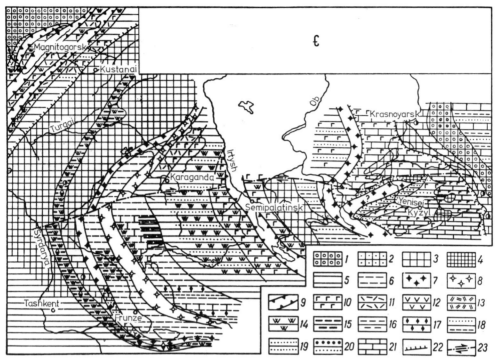

Fig. 47.   Paleotectonic diagram of the Kazakhstan–Altai region – Early Cambrian (after SEIDA-
LIN, without palinspastics). 1 = uplifts within ancient platforms; 2 = orogenic regions; 3 = median
massifs; 4 = geanticlinal uplifts; 5 = geosynclinal troughs; 6 = regions of regeneration of
geosynclinal subsidence on median massifs; 7 = eugeanticlines, analogues of mature island arcs; 8 =
the same, analogues of young island arcs; 9 = spreading zones in the Middle Paleozoic; Lithologic
assemblages and their main components: 10 = spilites and diabases; 11 = keratophyres; 12 =
andesites; 13 = rhyolites, 14 = cherts; 15 = carbonaceous shales; 16 = shales; 17 = quartz and
arkosic sandstones; 18 = terrigenous flysch; 19 = lower, marine molasse; 20 = upper, continental
molasse; 21 = carbonates; 22 = regional faults; 23 = transcurrent faults (analogues of transform
faults).

development of the ophiolitic association. Its fully crystalline components–ultrabasic
rocks and gabbro–presently occur within ophiolite belts along faults in cores of Caledonian anticlinoria (Zhalair–Naiman, Tekturmas, Maikain, Chinghiz, etc. belts), whereas
corresponding to the upper oceanic crust layers must be the siliceous-(jasper-micro-
quartzite)-basaltic (spilite, diabase) formation (more specifically association) known as
the Yerementau series in the northeastern part of the region and the Urtyndzhal series in
areas farther south. Silicites within this association replace basic volcanic rocks both
laterally in the direction of the basin peripheries and up the sequence, until volcanic series
are replaced completely by siliceous ones. Discussions are still under way as to whether
this formation is Upper Precambrian or Lower Paleozoic ("Precambrian Stratigraphy
…", 1971). The finds of Archeocyatha in the Chinghiz and articulateless brachiopods
and radiolarians in several other regions support the second opinion and point most
definitely to the Cambrian, mainly Lower Cambrian age of the formation. At the same

time some investigators assume its Ordovician (O_{1+2}) age (Tekturmas and Northern Balkhash anticlinoria). In V. S. ZVONTSOV's (1973) opinion there formed a similar siliceous-basaltic association in the Ordovician and in one area (Akzhal–Agadyr northwest of Lake Balkhash) also in the Silurian. It occupied a limited territory and was subordinate with respect to other formations.

L. P. ZONENSHAIN ("Tectonics of the Uralo–Mongolian …", 1974), using these somewhat contradictory data, arrived at a conclusion on the age "sliding" of the ophiolitic complex in a general direction from west (northwest) to east (southeast): from the Middle Riphean in Yerementau, Upper Riphean and Vendian in the framing of the Kokchetav Massif (i. e., farther east rather than west), Vendian in Akchetau, Lower Cambrian in Chinghiz to the Lower and Middle Ordovician in Tekturmas, and again Cambrian in the Kenterlau (Northern Balkhash area) Anticlinorium. These datings, alongside the datings of areas farther east were used by ZONENSHAIN to substantiate the concepts of the consecutive spreading of the region with oceanic crust in the Uralo–Mongolian (Central Asian) belt during the Late Precambrian and Paleozoic. Irrespective of the reliability of the picture presented by ZONENSHAIN it is quite likely that oceanic crust originated at the end of the Precambrian and further expanded at the beginning of the Cambrian or a little later; this was also a conclusion arrived at by several geologists in Kazakhstan (BESPALOV, 1971); ANTONYUK et al., 1977). Tectonic history of Central Kazakhstan has recently been elucidated from this point of view in an article by a group of Kazakhstan and Moscow geologists (ANTONYUK et al., 1977). In contradiction (probably apparent) with this conclusion are the data on the presence of ancient metamorphic rocks similar to the basement of the Kokchetav Massiv, at the base of the ophiolite complexes, more specifically siliceous-basaltic association. We already mentioned above that this comparison is sufficiently convincing only for the westernmost salients of the metamorphic basement (the Ishkeol'mes Anticlinorium), whereas Riphean age of this basement may be assumed for the area farther east. Apparently, the siliceous-basaltic association in the marginal parts of the eugeosyncline could overlie the still preserved but thinned out continental crust of the adjacent microcontinents. Or else, the ophiolitic complexes could be thrust over their continental framing. The Kokchetav –Muyunkum Median Massif formed such framing in the west and southwest, and the Balkhash–Dzhungarian Massif in the southeast.

The Balkhash–Dzhungarian Massif underwent considerable reworking in the Early and Middle Paleozoic: its best preserved relic can be observed in the Atasu–Mointy area, northwest of Lake Balkhash, where a moderately deformed 1.5–2.5 km thick terrigenous-carbonate (shelf) Cambrian–Ordovician formation overlies the Precambrian fold basement. These formations differ sharply from deposits of the adjacent Atasu–Tekturmas zone of the same age (ZVONTSOV, 1973) in terms of the facies nature, thickness and intensity of deformations, and, finally, in the absence of any appreciable unconformities. The southeastern continuation of this massif may be looked for in the southern part of the Dzhungarian Alatau (after AFONICHEV, 1967), and then at the base of the Dzhungarian Basin. The shallow occurrence of the Early Precambrian basement in the Northern Balkhash region is indicated, according to BESPALOV (1971), by the presence of large gneiss blocks in the ejections of Late Paleozoic volcanoes.

An appreciable change in the tectonic conditions within the Yerementau–Chuili and Chinghiz–Tarbagatai geosynclines started at the end of the Early Cambrian and finished

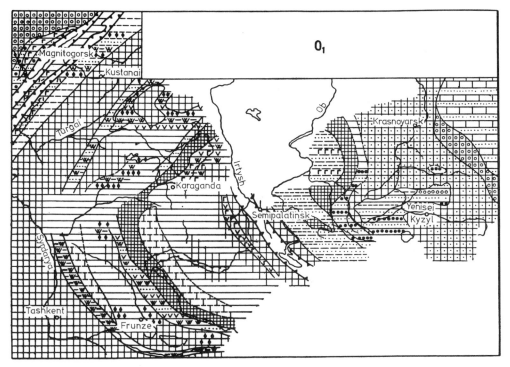

Fig. 48. Paleotectonic diagram of the Kazakhstan–Altai region – Early Ordovician (after O. A. SEIDALIN). Symbols as in Fig. 47.

by the beginning of the Ordovician. It was expressed in the replacement of basaltic volcanism by the andesite-basaltic (Fig. 47) one and in the appearance of graywackes – products of erosion of newly formed island arcs. The system of island arcs in place of cores of present anticlinoria took shape fully or almost fully at the boundary between the Cambrian and Ordovician, in pre-Arenig time in the Salairian epoch of tectogenesis, when the conditions of distension were replaced almost throughout by those of compression, judging by the extensive development of fold and probably also overthrust-nappe deformations.

In the Early and Middle Ordovician, environments of the late geosynclinal evolutional stage prevailed almost throughout in both eugeosynclines (Fig. 48). Terrigenous-volcanic series consisting of graywackes, shales, not infrequently in the flyschoid alternation, as well as of andesite-basaltic (sometimes basaltic) volcanics accumulated between island arcs within troughs of the type of inland or marginal seas. Cherts or limestones occur locally. These series are very thick (4 to 5 km or more). The paleotectonics and formations of this stage were characterized in great detail by V. S. ZVONTSOV ("Tectonics and Formations...", 1971) for the region of Tselinograd–Temirtau–Pavlograd. Subsidence in the Ordovician spread over a great part of the Kokchetav–Muyunkum Massif, being, however, weaker here, except in the newly formed transverse Southern Kokchetav Trough. Thus, partial inversion had found expression in the Kazakhstan–Northern Tien Shan geosynclinal region by the beginning of the Ordovician: there formed geanticlinal uplifts in axial zones of geosynclinal systems: island arcs with ophiolitic basement, and median massifs experienced partial subsidence.

At the end of the Middle and beginning of the Late Ordovician, the entire area of Central Kazakhstan and Northern Tien Shan, rather than only its eastern part, entered the epoch of upheavals and fold-overthrust deformations, which corresponded to the global Taconic epoch of Caledonian tectogenesis. The preliminary phase of movements here took place at the verge between the Llandeillian and Caradocian and the major phase at the boundary between the Middle and Late Caradocian.

The Taconic movements were the strongest on both sides and on the southeastern end of the Kokchetav–Muyunkum Massif, i. e., in the Ishim–Talas geosynclinal system, in the Stepnyak–Betpakdala zone of the Yerementau–Chuili system, and in the Northern Kirghizian zone of Northern Tien Shan. All these zones and the median massif separating them became an arena of vigorous granitoid plutonism. There emerged a huge bow-shaped granitoid (granodiorites, granites) belt extending from Petropavlovsk and Omsk to Northern Tien Shan and including major plutons such as Zerenda and Krykkuduk (in the north), Alma Ata and Terskei (in the south). In the north, the belt of Taconic granitoids continues into the basement of the Western Siberian Megasyneclise, and in the southeast it goes far into the Chinese Tien Shan. It is quite obvious that the Taconic granitization and accompanying metamorphism were responsible for the stabilization of the geosynclinal periphery of the Kokchetav–Muyunkum Massif and its following expansion at the expense of this periphery.

The Upper Ordovician coarse terrigenous shallow-water marine or lagoonal deposits, not infrequently variegated in contrast to the older green formations, constitute the lower (marine) Caledonian molasse. Even though they overlie unconformably the underlying layers and form superimposed troughs, these troughs are in general conformity with the Taconic fold structure.

At the end of the Ordovician, sedimentation practically ceased throughout the periphery of the Kokchetav–Muyunkum Massif, and it is from this particular time that one can speak of its transformation into a larger consolidation region: the **Kazakhstan – Tien Shan epi-Caledonian Median Massif** which was identified by ARKHANGEL'SKY (1941) and then by BOGDANOV (1965). This massif had not included, however, either the eastern, main zone of the Yerementau–Chuili Geosyncline or the Chinghiz–Tarbagatai Geosyncline. Taconic uplifts and deformations affected anticlines (geanticlines) of these zones. There also appeared minor intrusions of diorites and granodiorites, however, in troughs, the Upper Ordovician deposits are overlain conformably by Silurian sediments of similar composition and facies appearance; these sediments belong to the same lower molasse formation; their thickness ranges from 1.5–2 to 3.4 km. In places, they contain volcanic rocks: lavas and tuffs of intermediate, less frequently basic composition, as well as limestones, not infrequently of reef origin. The sediments become increasingly coarse clastic and deeper red. The general much more limited occurrence of Silurian (Fig. 49) in comparison with Ordovician deposits (no Silurian rocks were found in Central Dzhungaria) is indicative of the consequences of Taconic tectogenesis.

Transition from the Silurian to Devonian was essentially different in different zones of the region, and its character has been used for the division of the region into the Caledonides and Hercynides (see Fig. 44). Wherever the Silurian is present in the Yerementau–Chuili and Chinghiz–Tarbagatai systems, it is separated from the Devo-

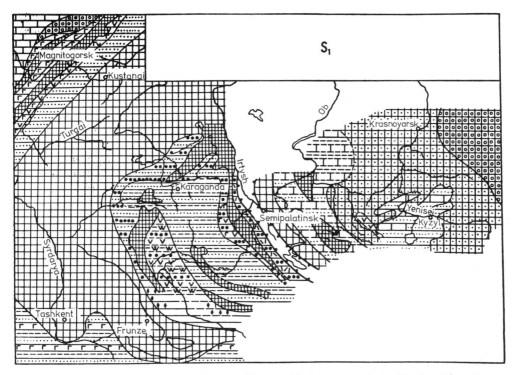

Fig. 49. Paleotectonic diagram of the Kazakhstan–Altai region – Early Silurian (after O. A. SEIDALIN). Symbols as in Fig. 47.

nian by a hiatus and unconformity (Fig. 50). The hiatus and the accompanying unconformity which is indicative of the major epoch of Caledonian tectogenesis become more extensive from east to west, in the direction of the central part of the Kazakhstan–Tien Shan Median Massif. Granitoid intrusions (granodiorite-granite formation) in Chinghiz – Tarbagatai and in the Chuili zone coincided with the same epoch. The Late Silurian and Early Devonian movements caused the region of Caledonian cratonization to expand appreciably further at the expense of the eastern zone of the Yerementau–Chuili system, the Chinghiz–Tarbagatai system, and Southern Dzhungaria.

Unlike the situation in these zones, the Silurian shallow-water marine basin (a marginal or inland sea) within the Dzhungaro–Balkhash residual geosyncline was inherited by the Early and Middle Devonian ones; there continued in it accumulation of a gray (greenish due to secondary alterations) terrigenous (sandy-clayey) formation which is quite typically represented in the Nura and Northern Dzhungarian troughs, where it reaches many kilometers in thickness. At the same time, there must have existed an uplift in place of the Northern Balkhash Anticlinorium. The Atasu–Mointy Block of the Balkhash –Dzhungarian Massif also remained uplifted (absolutely or relatively).

In between these two regions with opposite evolutional tendencies, there occurs a zone where the Upper Ordovician–Silurian molasses are superseded conformably by Lower Devonian ones, and the latter are overlain unconformably by Middle–Upper Devonian

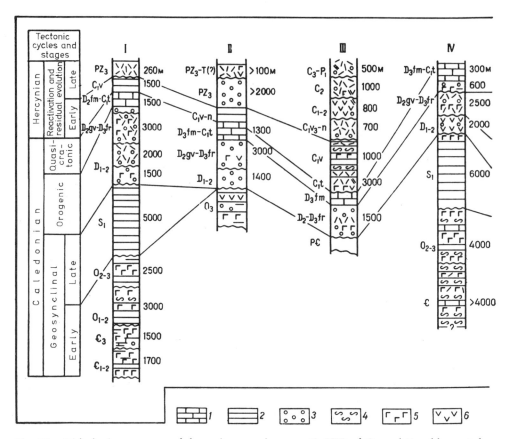

Fig. 50.   Lithologic sequences of the main tectonic zones (I–VII) of Central Kazakhstan (after TIKHOMIROV, 1975, simplified). I = Sarysu–Teniz zone; II = Teniz–Korzhunkul' syncline; III = southwestern flank of the Tokrau Basin; IV = Chinghiz Range, N part; V = N part of the Dzhungaro–Balkhash region, Karasor Trough; VI = northeast part of the Tokrau Basin; VII =

molasses. This zone includes the Selety and Shiderty depressions in the north of the Yerementau–Chuili system as well as the eastern part of the Sarysu–Teniz water divide (after GOLUBOVSKY, 1979); all these regions are situated outside (north or west) the **Devonian marginal volcanic belt** which originated, according to BOGDANOV (1965), in the border region between areas of complete Caledonian consolidation and Hercynian regeneration.

The main region of downwarping and sedimentation shifted at the boundary between the Silurian and Devonian to within the northwestern and northeastern periphery of the Balkhash–Dzhungarian Median Massif. The Devonian marginal volcanic belt which encircles in a horseshoe-like manner the Balkhash–Dzhungarian Massif and its peripheral troughs and which extends in conformity with the latter, is confined to the same boundary, crucial in the evolution of the region. The volcanic belt continued developing during the Early, Middle, and beginning of the Late Devonian; the andesite-basalt effusions were replaced by trachyrhyolite, rhyolite (prevailing), and dacite ones; at

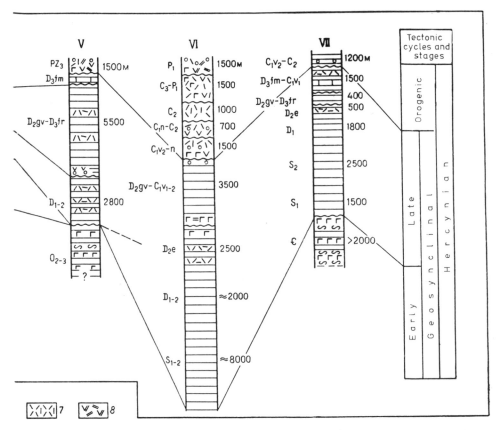

Balkhash Anticlinorium. 1 = carbonates; 2 = marine terrigenous sediments; 3 = continental clastic sediments; 4 = siliceous rocks; 5 = basalts, andesite-basalts; 6 = andesites; 7 = rhyolites; 8 = trachytes.

the end there again appear basalts. Rhyolites (and granitoids) predominate in the external part of the belt cross-section and andesite-basalts in the inner part (CHETVERIKOVA, 1970).

Toward the central part of the Kazakhstan–Tien Shan epi-Caledonian massif, the volcanic rocks are replaced by red continental, partly volcanic molasse; the accumulation of this (Upper Caledonian) molasse started in the Givetian, reached culmination in the Frasnian, and terminated in the Early Famennian. The greatest thicknesses (4 to 5 km) are confined to the Sarysu–Teniz zone and the Greater Karatau Trough (ZAITSEV, 1972). It is of significance that the deposition of the Devonian molasse was preceded by uplifting and folding in the region south of Karaganda, within the northwestern margin of the Hercynian geosyncline (or transitional zone between the Caledonides and Hercynides), which corresponds to the region of the **Tel'besian phase of tectogenesis,** identified by Usov (1936) in the Altai–Sayan region. Also coinciding with this phase seems to be the maximum of Devonian granitization expressed practically within the same area as the

Late Ordovician granitization, including Chinghiz and Tarbagatai, where Devonian granitoids play a more significant role than the Late Ordovician ones. However, the greater part of granitoids of this age gravitate toward the volcanic belt.

The Devonian, Givetian–Frasnian molasse occupies a dual position with respect to the Caledonian and Hercynian stages of evolution in the region. On the one hand, it completes the development of the Caledonides and is of the same age as that of the upper part of the classical Caledonian molasses of the Atlantic Belt, which suggests that it, unlike the Upper Ordovician–Silurian–Lower Devonian marine lower molasse, may be regarded as continental upper molasse. On the other hand, as pointed out by ZAITSEV (1972, p. 106), "the absence of any connections between the volcano-terrigenous formation and Caledonian structural zones, the prolonged hiatus prior to its accumulation, which often corresponds to the entire Silurian period, do not make it possible to regard this formation as a Caledonian molasse . . .". To this we will add that the given formation lies in several regions inconformably over the Caledonian complex proper and is connected through a gradual transition with overlying Hercynian formations (Fig. 51). This circumstance is stressed by MAZAROVICH (1976) who is among those denying the attribution to the Caledonian cycle of the Kazakhstan Devonian molasse. It is, however, obvious that this is a typical molasse associated in time with the Tel'besian, i. e. Acadian epoch of tectogenesis, if one uses global nomenclature. On a global scope, the Acadian epoch belongs to the Caledonian rather than to the Hercynian era of tectogenesis, so it is more logical to consider the Devonian molasses of Kazakhstan and Tien Shan to be Late Caledonian molasses. The fact that structurally they are part of the Hercynian complex, being separated by an unconformity from the Caledonian complex proper, should not surprise one, since major folding always precedes the upper molasse.

7.10.8 The Hercynian geosynclinal stage
(second half of the Devonian–first half of the Early Carboniferous)

The formation of molasses had terminated by the Famennian age (acid volcanism was under way in places), the tectonic relief underwent levelling, and there occurred general sea transgression covering both the Caledonides and Hercynides. In the Famennian, Tournaisian, and Early Viséan there deposited a carbonate (or terrigenous-carbonate) formation, mainly shallow-water; in some areas (Karaganda, Uspensk, and Akzhal – Aksoran troughs) it is replaced by a relatively deep-water siliceous-shale formation. Maximum thickness of this formation in the region of Caledonian consolidation was observed in the marginal Karatau–Naryn zone and in the superimposed Sarysu–Teniz graben-trough, which (especially the first) may be regarded as Hercynian "carbonate" miogeosynclines. For the remaining region of Caledonian consolidation, the Famennian–Viséan carbonate formation must be regarded as quasiplatform, since it is followed by the Upper Paleozoic secondary orogenic complex. This formation has the same character within the residual Balkhash–Dzhungarian Massif, at least in its western (Atasu) part, constituting the upper part of its sedimentary cover. In the Eastern Balkhash region, the Famennian–Viséan carbonate formation is replaced by the terrigenous-volcanic one (subaqueous volcanics of intermediate to acid composition), and in the Dzhungarian Alatau by the siliceous-terrigenous-tuffogene formation of considerable thickness. In fact, this latter formation alone (the Tastau series) alongside the siliceous-shale formation of the Karaganda region may be considered typical of the Hercynian geosyncline.

The tectonic conditions were entirely different during the same period of time in Northern Tien Shan. Here, there continued accumulating red continental molasses, indicating preservation of orogenic conditions inherited from the Caledonian stage.

7.10.9 The Hercynian early orogenic stage
(end of Early Carboniferous–beginning of the Permian)

In the middle of the Viséan age (more specifically at the boundary between the Early and Middle Viséan), the Dzhungarian–Balkhash Geosyncline experienced the effect of the phase of tectogenesis, which was called Saurian (after the Saur Range) in the fold system adjacent on the east. The movements of the Saurian phase were particularly strong in the northwestern part of the Dzhungarian–Balkhash Geosyncline, between the Spassk Fault zone and the Northern Balkhash region. The Saurian upheavals led to the separation of the Dzhungarian–Balkhash Basin into individual partial troughs with accumulation of lower molasse, paralic coal-bearing in the Karaganda Basin, marine in the Sayak – Dzhungarian residual trough which extends from the northeastern Balkhash region to the northern part of the Dzhungarian Alatau. The coal-bearing molasse accumulated also in the Teniz Depression within the Kazakhstan–Tien Shan Massif. In the Karatau–Naryn zone, there prevailed marine conditions, however, there appeared, alongside limestones, clastic rocks to conglomerates, as well as tuffs–vestiges of volcanism in the Dzhungaro–Balkhash system. A new horseshoe-shaped volcanic belt – the **Balkhash–Ilian** one – was formed in the mentioned system. It occupies a more inner position than the Devonian belt and fringes the relic uplift of the median massif together with the Sayak–Dzhungarian marine trough superposed on its northern edge. The subaerial volcanic products of this belt belong to the basalt-andesite-rhyolite formation. Granitoids, whose origin is believed to be connected with the Saurian phase of tectogenesis, are encountered within the belt and in its outer framing. They were found in the Northwestern Balkhash region (the Tokrau batholith), in the Bayanaul region in the northeast of the Yeremen-tau–Chuili system, in Chinghiz–Tarbagatai; in the south within the northern foothills of Tien Shan, in the Alma Ata region.

The coexistence of the Sayak–Dzhungarian Trough filling with marine sediments and the volcanic belt west and north of it had continued from the Late Viséan to the end of the Carboniferous and beginning of the Permian (?), the marine basin reaching its greatest dimensions at the end of the Middle and beginning of the Late Carboniferous (SHCHERBA, 1973). Part of the investigators identify within this interval a folding phase comparable with the Sudetian ("Tectonics of the Uralo–Mongolian ...", 1974) or Asturian ones (BESPALOV, 1971) and associate several granitoid intrusions with this phase. Others do not attach much importance to it. The full closure of the Dzhungaro–Balkhash Geosyncline and shaping of its structure took place in the Early Permian as a result of the Sayak epoch of tectogenesis, which might have corresponded to the European Saalian epoch, although it may be a little older. It is during this epoch that the diorite-granodiorite complex of the same name (Sayak) formed.

7.10.10 The Hercynian late orogenic stage (Permian period)

The remaining part of the Permian period constitutes the late orogenic stage of the evolution of the Dzhungaro–Balkhash system. A volcano-plutonic belt which experienced a new displacement inside the system and reached maximum width continued developing gradually along its periphery during the same period of time. The effusions became sharply contrasting in composition: from basalts to rhyolites, and intrusive magmatites are represented by plutons of monzonites, alkaline granites and syenites, to nepheline syenites; these plutons occur mainly within the Caledonian framing of the Dzhungaro–Balkhash system: from Chinghiz–Tarbagatai to Northern Tien Shan. Permian intrusions of leucocratic granites tend to the same volcanic belt.

Within the epi-Caledonian Kazakhstan–Tien Shan Massif, in the Late Paleozoic, starting from the Middle Carboniferous, depressions such as the Teniz and Dzhezkazgan basins became independent, whereas the peripheral Karatau–Naryn and transverse Sarysu–Teniz zones were involved into rather strong deformations: fold-overthrust in the first zone and fold-block in the second. The Upper Paleozoic deposits of the Teniz, Dzhezkazgan, and Karaganda basins are red continental coarse clastic molasses with minor quantities of evaporites and cupriferous sandstones in the Dzhezkazgan Basin. ZAITSEV (1972 a) stressed that the material was supplied to the depressions of the Kazakhstan–Tien Shan Massif from outside: from the west (eastern Uralian zones) or east (the Caledonian framing of the Dzhungaro–Balkhash system), and, consequently, the Hercynian molasses here are allotigeneous; yet, the great thickness of these molasses (up to 2.5–3 km) indicate increased mobility, i. e., reactivation of the massif itself.

In the Northern and Middle Tien Shan, epi-Caledonian troughs continued developing during the Middle and Late Carboniferous (only Late in the Middle Tien Shan) and partly during the Permian. Like in the west of Central Kazakhstan, the corresponding deposits form red molasses, however, associated with them here is a volcanic formation of andesitic (trachyandesite-basalt) composition, mainly Permian. Faults bordering the depressions controlled the distribution of Late Hercynian plutons. Along the southern boundary of Median Tien Shan, there had developed, from the Middle Carboniferous to the Early Triassic, the **Kurama volcano-plutonic belt** particularly distinct in the Kurama Range. In the northwestern direction, it is followed into the northeastern Kyzyl Kum and further north, as suggested by BOGDANOV (1965), seems to merge with the Valer'yanov andesite belt of Transuralia. The latter, however, is a little older than Viséan – Namurian and just like older Lower Carboniferous elements of the Balkhash–Ilian Belt, has a character transitional from late geosynclinal to orogenic. The volcanic formation of the Kurama Belt has a rather complex andesite-dacite-rhyolite composition, with a general increase in the content of silica, yet with basalts appearing in the top portion of the sequence. A similar evolution is observed in the composition of the plutons: from granodioritic through granitic to alaskitic, with deviations to monzonites and syenites.

Horizontal displacements, mainly along major wrench faults of northwestern-southeastern direction became active in the region during the orogenic period of the Hercynian stage. The Chinghiz Deep Wrench Fault is of this type (SAMYGIN, 1974). It originated as far back as the Devonian, during the period of Caledonian orogenesis, was passive during the quasiplatform stage of the Late Devonian and Early Carboniferous, but became active again and spread from the region of Caledonian consolidation into the Dzhungaro–Balkhash system during the Hercynian orogenic period. This fault and the minor faults accompanying it, had repeatedly, especially in the Late Paleozoic, been magma vents as well as paths of circulation of metalliferous (W, Mo, Au, and Cu) solutions. The amplitude of the horizontal displacement along the Chinghiz Wrench Fault reaches 100 km, the vertical displacements do not exceed 5 km, and the extent is longer than 700 km.

In the northwest, the Chinghiz Fault links with the Central Kazakhstan Fault which extends in the meridional direction from the Irtysh River to Lake Balkhash and which constitutes part of a major planetary lineament passing in the north, under the cover of the Western Siberian Megasyneclise, into the Arctic Basin and in the south into the Indian Ocean. This fault originated also in the epoch of Caledonian orogenesis (Early–Middle Devonian), however, the displacements were mainly vertical. The Dzhungarian and Zhalair–Naiman wrench faults extend parallel to the Chinghiz Fault farther southwest. The Late Paleozoic age of the Dzhungarian Wrench Fault was proved by BUSH, DMITRIEVA & FILATOVA (1968); it clearly cuts across the latitudinal structures of Dzhungaria. The Zhalair–Naiman Fault is situated differently; it coincides with the southwestern boundary of the corresponding stretch of the Yerementau–Chuili system and is much older, having originated prior to the Cambrian.

The Talas–Ferghana Wrench Fault is another major feature in the region, in its southwestern periphery. Its transcurrent nature was first noted by OGNEV (1939) and substantiated in great detail by BURTMAN (1961) who concluded that this fault originated as a strike-slip only at the very end of the Paleozoic and in the Late Permian. SUVOROV (1968) later presented proofs showing that the Talas–Ferghana Fault had existed as early as the Silurian, though as a normal fault. Horizontal displacements along it first started in the Devonian, i.e., again in the period of Caledonian orogenesis, however, in the Carboniferous and Permian this fault was again mainly a normal fault with an additional strike-slip component. Dextral displacements became predominant since the end of the Paleozoic. According to V. S. BURTMAN, the horizontal amplitude of the Talas–Ferghana Fault reaches 200 km (the northeastern limb is displaced southeast), and the vertical amplitude does not exceed 2.5–3 km. The horizontal displacement along the fault attenuates in the northwestern direction, being counterbalanced by overthrusts of the same trend in its southwestern limb (the Chatkal arcs of the Median Tien Shan), and in the region of the Karatau Range its continuation is already a deep normal fault.

In conjunction with the dextral wrench faults of northwestern-southeastern direction – the Chinghiz Fault, Dzhungarian, Zhalair–Naiman, etc. – there developed in Central Kazakhstan in the Late Paleozoic overthrusts and nappes of west-southwest and east-northeast orientation: Spassk, Uspensk, etc. They were subsequently crossed by sinistral wrench faults of the same strike, such as the Tekturmas one (SUVOROV, 1968, 1973).

Late Paleozoic deformations and the associated magmatism completed the long history of geosynclinal and orogenic evolution of the Kazakhstan–Tien Shan region, and in the beginning of the Mesozoic it entered the platform stage of evolution.

7.10.11 The Cimmerian–Early Alpine platform stage (Triassic–Paleogene)

This stage did not abound in events. Practically throughout this stage the greater part of the region experienced weak and slightly differentiated upheaval, representing on the whole the Kazakh Shield of the young Siberian–Turanian Platform; its western and southern margins were more mobile in the first half of the stage during the Triassic and Jurassic. Several grabens – Ubogan, Baikonur, and others – originated in the Late Triassic and continued developing in the Early Jurassic. The largest structures of this type, which had developed for a long time, aare the near-fault Jurassic troughs – Karatau (Leont'ev) and Ferghana (Yarkend–Ferghana) with vigorous terrigenous sedimentation and fold deformations of the Late Jurassic, which indicate continued activity of the Talas–Ferghana Fault. The Alakol' Jurassic trough is associated with the activity of another fault: the Dzhungarian one. In the central part of the shield, only flat troughs were formed: Karaganda, Ilian, Dzharkent, and some others inherited from the regional Late Hercynian structure. All these troughs disappeared in the Early Cretaceous epoch, when the main part of the shield experienced somewhat stronger upheaval. In the Late Cretaceous, the boundaries of the Kazakh Shield became more distinct in the northeast, north, west, and southwest, weak subsidence with accumulation of continental sediments involved individual areas of the Chu–Sarysu and Ilian depressions, as well as a band along the upper reaches of the Ishim River. These conditions had prevailed until the Middle Oligocene when arch-block uplifting started in the south of the region, which led to the formation of a highland here: the northern margin of the Central Asian epiplatformal orogen.

7.10.12 The Late Alpine stage of orogenic reactivation (Neogene–Quaternary)

The upheaval wich shaped the Tien Shan and Dzhungarian mountain edifices started by the end of the Oligocene, however, they were particularly extensive in the Pliocene – Quaternary time. The amplitude of positive vertical movements during the neotectonic stage exceeded 8 km in Northern Tien Shan and 5 km in the Dzhungarian Alatau. Alongside the uplifts there developed piedmont and intermontane troughs which were, however, smaller in terms of the area and amplitude of absolute subsidence. The piedmont troughs included the cis-Kirghizian (Southern Chu and Frunze) with Cenozoic fill of about 3 km in thickness, the Ilian with the Alma Ata and Dzharkent depressions, where the Mesozoic and Cenozoic thicknesses are 3 to 4 km, Southern cis-Balkhashian (1 to 1.5 km thickness), and Alakol'. They alternate echelon-like owing to the fact that spurs – Karatau, Chu–Ilian, Dzhungarian, and Chinghiz–Tarbagatai – branch off in the northwestern direction from the main region of recent upheavals. These "offshoots" of the epi-platformal orogen penetrate the body of the platform along the major Late Hercynian faults since the structural pattern of recent orogen is in good conformity with the Late Hercynian pattern preserved also in the intermediate platform stage. It is, therefore, no wonder that the Neogene–Quaternary troughs became superimposed on the Jurassic (Jurassic–Cretaceous) ones and the latter on the Middle Paleozoic troughs, with a certain displacement. This refers also to intermontane troughs, the largest of which are the Issykkul', Tekes (in the same strip), and Naryn troughs. The absolute subsidence amplitude of these troughs reaches 3–6 km (the Issykkul' Depression). As pointed out above, the troughs usually link with uplifts along faults along which near-fault fold deformations of the young cover are observed. The high seismic activity indicates continued vigorous tectonic processes in the south of the region, and repeated leveling points to further expansion of mountain building in the northern direction.

7.11 Structural and evolutional features of the region

The Paleozoic fold region of Central Kazakhstan, which is situated inside the Uralo–Okhotsk Geosynclinal belt, is considered in Soviet literature to be a typical example of a mosaically built geosynclinal region. Its structure, which had taken shape by the end of the Paleozoic, involves more or less similarly expressed elements of various orientations: meridional and latitudinal, northwestern and northeastern. This mosaic pattern has not infrequently been regarded a result of unconformable superposition of the Hercynian structure on the Caledonian one; in KASSIN's (1960) diagrams, this intersection is particularly distinct east of the Central Kazakhstan Fault identified later and north of the latitude of Dzhezkazgan. At the same time, both the Caledonian and Hercynian fold arcs and the troughs which produced them separated by rigid blocks, possess, in KASSIN's opinion, sufficiently variable strikes: the latter mainly latitudinal, the former meridional in the north to latitudinal in the south.

It followed from KASSIN's concept that the Hercynian stage in the evolution of Central Kazakhstan must be distinctly separate from the Caledonian stage. However, a different point of view was later expressed by SHATSKY (1938), who headed a large team from the USSR Academy of Sciences: "No, so to speak, drastic changes in the general folding pattern took place at the boundary between the Early Paleozoic and Devonian, i. e., between the Caledonian and Hercynian periods of folding within the Kazakh folded land: the Hercynian folding had inherited its general trends from Caledonian deformations of Kazakhstan." This concept which implied logically uselessness or at least arbitrariness of the Caledonides and Hercynides being identified in Central Kazakhstan, was developed further by SHATSKY's coworkers in this field, especially by MARKOVA (1964) who maintained the opinion that Kazakhstan has a single Paleozoic structure, the Caledonian folding being incomplete on the planetary scope.

Yet, in 1954, BESPALOV introduced a concept of the Dzhungarian–Balkhash Hercynian province, which he opposes to the remaining – Caledonian – Central Kazakhstan. BOGDANOV (1955) gave new life to ARKHANGEL'SKY's (1941) concept of the epi-Caledonian Kazakhstan–Tien Shan Median Massif, having, however excluded from it the Chinghiz–Tarbagatai zone and having drawn the boundary between the Caledonides and Hercynides along the Devonian volcanic belt. MARKOVA, even though she stressed the unity of the Paleozoic megacycle, proposed a scheme (a very good one, after all) of zoning of the region showing Early Caledonides, Late Caledonides, and Hercynides[10]. It is now common practice to separate Central Kazakhstan into regions of Caledonian and Hercynian consolidation, although boundaries between them are drawn differently.

The following can be stated in an attempt at objective assessment of the general structural and evolutional pattern of the Kazakhstan–Tien Shan region.

---

[10] Note the inner contradiction in MARKOVA's inferences on this problem ... "Three *distinct* tectonic zones have been identified on the territory of Central Kazakhstan, which have been *arbitrarily* (underlined by V. Kh.) described by us as Early Caledonian, Late Caledonian, and Hercynian" (MARKOVA, 1964, p. 137).

First of all, this region is situated at the general southeastern and southern turn of the Uralo–Okhotsk Belt, on an area where meridional Uralian trends give way to northwestern latitudinal trends of Central Asia. Faults directed along the bisecting line towards the angle formed by the meridional and latitudinal segments of the belt, i. e., in the northeastern direction appear regularly on this bend. A major fault zone of northeastern trend – the Bel'tau one – has been revealed in the Northern Kyzyl Kum, in the area of the Bel'tau Mountains, where coinciding with it is the end of the Bel'tau–Kurama Late Paleozoic volcanic belt. The southern end of the Ulutau Uplift is situated on its northeastern continuation; further on, it seems to cross the basement of the Dzhezkazgan Basin and continues along the "latitudinal" stretch of the Central Kazakhstan Devonian volcanic belt; more or less parallel to it are structures of the Southern Karaganda zone characterized by a trend generally anomalous for the Central Kazakhstan Paleozoic massif. It will be shown later that the principal meridional, northwestern, and latitudinal trends played different roles in the history of the region. Geosynclinal conditions died out from north and northwest to southeast and south, and the role of each of these trends changed accordingly.

The complex structural pattern of the Kazakhstan–Tien Shan region seems to be due further to the fact that a large continental crust block was preserved here during the formation of the Uralo–Okhotsk Belt; at the end of the Precambrian this block split into three smaller blocks (median massifs): Turgai–Syr Darya, Kokchetav–Muyunkum, and Balkhash–Dzhungarian. The shape of this Kazakhstan–Tien Shan Block and the directions of the faults bounding it were responsible for the predominance of meridional trends along its western periphery, latitudinal along the southern and northwestern along the northeastern periphery. At the end of the Caledonian stage, the Kokchetav – Muyunkum Massif, which had occupied a central position, became a center of crystallization of the regenerated and newly formed continental crust.

It is essential that, in contrast to the Hercynian stage, the geosynclinal development in the Caledonian stage encompassed practically the entire region, and it is only in individual areas of the Kokchetav–Muyunkum Massif that its expression was sharply weakened. The "oceanic" characteristics of the Kazakhstan–Tien Shan region were expressed most prominently in the Caledonian stage, more specifically at the beginning of the Cambrian. However, even during this climax there could hardly exist on the territory vast, thousand kilometers wide, areas with fully oceanic crust; these were rather smaller (up to 1,000 km) basins separated by microcontinents, but situated in the central part of the Uralo–Okhotsk paleo-ocean. As early as the Middle Cambrian and, in the opinion of ANTONYUK et al. (1977), even Vendian, there started a vigorous process of filling this space with the newly formed continental crust, which intensified at the end of the Cambrian and especially at the end of the Ordovician. Taconic deformations affected not only the region of subsequent epi-Caledonian consolidation, but also the territory of the future Dzhungaro–Balkhash system; the opinion of the "throughgoing" development of this system during the Early and Middle Paleozoic (AFONICHEV, 1967) does not seem sufficiently well grounded (KOSHKIN, 1971), even though the intensity of Caledonian deformations here must have been minimal.

The Caledonian structural pattern of the region must have been much more simple than the Hercynian one. The Caledonian "arcs" encircled the Kokchetav–Muyunkum Massif on the west and south, the Early (and partly also Late) Caledonides filled the space between them and the Balkhash–Dzhungarian Massif and also framed this latter massif on the northeast, linking with the former at its northern termination.

The powerful Late Ordovician granitoid magmatism in the west and south of the region was connected, in all probability, with the activity of a Benioff zone which encircled the Kokchetav–Muyunkum Massif on the west and south and plunged under this massif. Another such zone is expected to exist along the northeastern framing of the Chinghiz–Tarbagatai island arc. As early as the Silurian the geosynclinal process shifted sharply towards east and south, however, full separation of the Dzhungaro–Balkhash system took place in the second half of the Devonian, by the period of Tel'besian deformations. These deformations occurred along the northwestern periphery of the Balkhash–Dzhungarian Massif and were due to the collision between this massif and the northeastern promontory of the Early Caledonian consolidation region. The northwestern edge of the Balkhash–Dzhungarian Massif must have been determined by the Bel'tau–Karaganda diagonal fault. The Tel'besian deformations produced the Southern Karaganda fold zone, discordant somewhat with respect to the Caledonides in the west, but parallel to their Maikain branch in the east. The Tel'besian deformations here were subsequently reinforced by Hercynian overthrusts. The Hercynian Dzhungaro–Balkhash system cannot be compared with Caledonian geosynclines either in the dimensions or in the intensity of expression of the geosynclinal process. It is undoubtedly residual[11] with respect to the latter, as was first noted by BORSUK (1964), and is a typical brachygeosyncline (after PEIVE & SINITSYN, 1950) in the paleotectonic aspect and a basin of the type of an inland sea (for example, of the contemporary Black Sea) in the paleogeographic respect with a suboceanic if not subcontinental crust. That the crust was closer to oceanic is indicated by ophiolite outcrops along the Dzhungarian Fault. In the southeast, this rather unusual geosyncline linked with the authentic eugeosynclinal Irtysh–Southern Mongolian system.

Predominant in the Hercynian structure are northwestern trends which changed to latitudinal in the southeast and were interrupted by the Karaganda transform in the northwest. The formation of the specific Sarysu–Teniz zone of block-fold deformations is connected with northwest-trending faults. This zone is comparable with the Karatau–Talas Miogeosyncline, adjacent on the west, in terms of the thickness of the D_3–C_1 terrigenous-carbonate formation, but differs from it sharply in tectonic style. This particular zone furnishes a most striking example of unconformable superposition of Hercynian structures over Caledonian ones, however, the former are of Alpine type and the latter of Saxon type. The Chinghiz–Tarbagatai system is an opposite example of inherited Caledonian–Hercynian development; the Salairian movements initiated and the Taconian movements shaped individual anticlinoria, the synclinoria were deformed into folds at the end of the Silurian and beginning of the Devonian, while the general meganticlinorium structure took shape during the Hercynian stage. The deformations become less complex as one goes from one structural stage to another, from anticlinoria to synclinoria. The Sarysu–Teniz example refers to the ancient core of the Kazakhstan–Tien Shan region, the Chinghiz–Tarbagatai one to its periphery at the boundary with the Ob'–Zaisan Late Hercynian geosynclinal system.

---

[11] More specifically, residual-regenerated, provided the hiatus at the base of the Silurian is observable throughout its territory.

The volcanic belt in the Hercynian stage shifted consecutively towards the central part of the Dzhungaro–Balkhash system (and residual median massif?). As for the granitoid plutons, they fill practically the entire space of the system and occur also in Northern Tien Shan and Chinghiz–Tarbagatai. It seems more logical to associate all this magmatism with a single Benioff zone running in the latitudinal direction south of the entire Kazakhstan–Tien Shan and farther east of the Altai–Mongolian region (a continental margin of the Andean type) than with the existence of individual zones along the periphery of the Dzhungaro–Balkhash system. The irregular distribution of volcanic rocks must have been connected with permeability zones along the periphery of regions of earlier consolidation, surrounding the residual Dzhungaro–Balkhash Geosyncline.

The Chinghiz–Tarbagatai and Bel'tau–Kurama belts of intermediate and acid volcanics and granitoids, immediately adjacent to active Hercynian geosynclines may be of independent origin.

During the epoch of Alpine orogenesis, the upheavals which led to the regeneration of the mountain relief found expression mainly in the southern latitudinal strip of the region and spread only in a weakened form farther north along the northwest-trending faults (Zhalair–Naiman and Chinghiz). The northwestern structural orientation has also been observed in the neotectonic stage all over the territory, from Amu Darya to the Eastern Sayan.

It may be concluded in summary that the specific features of the Kazakhstan–Tien Shan structural pattern are due to three principal causes: first, its position at the bend of the Uralo–Okhotsk Belt from the meridional trend through southeastern to latitudinal; second, preservation of large blocks of older continental crust within it during the origination of the belt; third, consecutive alternation in time of the predominant roles of first meridional (the Caledonian stage), then northwestern (in the Hercynian stage), and finally, latitudinal (from the end of the Hercynian stage) structural trends.

8. Southern Tien Shan Hercynian geosynclinal fold system

8.1 Boundaries, principal structural subdivisions, geological coverage

The Southern Tien Shan Hercynian fold system extends from the lower reaches of Amu Darya River (the Sultanuizdag Range) across the Kyzyl Kum desert, where it has a northwest-southeast trend and farther on across the ranges of the Soviet Southern Tien Shan (Turkestan, Alai, Zeravshan, Gissar, Ferghana, and Kokshaal), passing in the east along the Kokshaal Range to the Chinese territory where it narrows down gradually and wedges out in this direction. Here, the system changes its trend to east-northeastern. In the southwest, in the lower reaches of Amu Darya River, in the Sultanuizdag tectonic node, the Southern Tien Shan system links with the Uralian one. The character of this linkage is discussed specifically at the end of this chapter. In the northeast, the Southern Tien Shan system is bounded by the Turgai–Syr Darya Median Massif whose eastern termination is buried under the young (Meso–Cenozoic) Ferghana Depression; an outlier of this (?) massif appears again in the Sarydzhaz Block in the southeast of Kirghizia. The **Kara Kum–Tadzhik Median Massif** provides a southern boundary of the Southern Tien Shan system on the Soviet territory; on the Chinese territory, the **Tarim Median Massif** plays a similar role. The system is separated from the two massifs by major faults: **Southern Ferghana** and **Major Gissar,** respectively.

Over its extent, measuring more than a fifteen hundred kilometers, the Southern Tien Shan system, like other fold systems, is divided naturally into segments which differ somewhat in their structure and evolution as well as in the character of the boundaries and which are separated by transverse and diagonal faults. Following D. P. REZVOY ("Tectonics of the Uralo–Mongolian . . .", 1974), we can identify three such segments: western or **Kyzyl Kum** having a southeastern trend and slightly involved into recent orogenesis; central or **Gissaro–Alai,** of sublatitudinal trend, which has experienced strong recent orogenesis; and eastern or **Ferghana–Kokshaal** of east-northeastern trend, also involved into neotectonic reactivation. The western segment is separated from the central one by the Western Tien Shan Fault of northeastern trend, and the central segment is separated from the eastern one by the southeastern stretch of the Talas–Ferghana Fault. The structural-facial zones of the central segment, especially the northern ones, experience an abrupt turn northwards or even north-northwestwards, as they approach the Talas–Ferghana Fault, in correspondence with the direction of the displacements along this fault. Owing to this, the continuation of the Southern Tien Shan zones can be followed at the eastern and northeastern closure of the Ferghana Basin (the Ferghana and Baubashata and adjacent ranges). The segments differ in the width, too: the western segment is more than 300 km wide, the central about 150 km, and the eastern segment is even narrower. They also differ in the combination of the structural-facial zones.

The Soviet part of the Southern Tien Shan system, which is larger, has been covered by extensive geological studies. Many years of studies conducted by SINITSYN (1957), REZVOY (1974), VINOGRADOV (1964), DOVZHIKOV (1977), and KUKHTIKOV (1968) have resulted in a certain concept of the Southern Tien Shan structure. In accordance with this concept, the Southern Tien Shan is clearly divided into several narrow longitudinal structural-facial zones separated by deep faults and exhibiting a practically independent evolution throughout the entire Paleozoic, i. e., during the entire geosynclinal (and epigeosynclinal-orogenic) history. D. P. REZVOY identified the following zones in the central segment, from north to south: 1) Karachatyr (foredeep); 2) Turkestan–Alai (Kichikalai in the east); 3) Zeravshan (or Turkestano–Zeravshan); 4) Zeravshan–Gissar (or Zeravshan–Alai, eastern Alai in the east); 5) Southern Gissar volcano-plutonic belt. Each of these zones is about 25–30 km wide. The Karachatyr zone is separated from the Turkestan–Alai one by the **Southern Ferghana Deep Fault,** the Turkestan–Alai from the Zeravshan by the **Turkestan Fault,** the Zeravshan from the Zeravshan–Gissar by the **Zeravshan Fault;** the Zeravshan–Gissar zone is bounded in the south by the **Northern Gissar** (Major Gissar) **Fault** beyond which there starts the **Southern Gissar volcano-plutonic belt.** The Major Gissar Fault constitutes only a short stretch of a large lineament which runs in the west across the Turanian Platform and the Caspian region to Europe and which continues in the east along the northern boundary of the Tarim Massif and the Sino–Korean Platform (the Sarmat–Turanian or Central Eurasiatic Lineament).

The inner structure of each of the above structural-facial zones is pictured in the light of the mentioned concept as an alternation of anticlinoria (frequently fan-like) and synclinoria. It has recently been conjectured that there exist in the region flat overthrusts of limited amplitudes. In what concerns the history of the Southern Tien Shan, the majority of the investigators were inclined to believe that the Southern Tien Shan Geosyncline had originated in the Early Paleozoic over a continental platform basement; the time of its origination remained open to discussion.

In contrast to this long established and seemingly well grounded concept of the Southern Tien Shan structure, there recently appeared a new concept of its much more complex, overthrust-nappe structure (PORSHNYAKOV, 1973) and origination over an oceanic rather than continental crust (BURTMAN, 1973; SCHULTZ JR., 1974; and MAKARYCHEV, 1978). The existence of nappes, at least in the Alai Range and in the mountain massif of Tamdytau in the Kyzyl Kum, is absolutely indubitable; however, there is no single model of the Southern Tien Shan nappe structure, yet; its various essentially different versions are now under discussion (BURTMAN, 1973; PORSHNYAKOV, 1973).

The above concepts proceed, as a matter of fact, from contrasting understandings of the conditions under which the general structure of the Southern Tien Shan formed, in particular, the specific bend of the constituent zones in the stretch of the Talas–Ferghana Fault. The first concept is based on the finding that the present structural pattern of the Median and Southern Tien Shan, which had taken shape by the beginning of the Mesozoic, is a direct reflection of its primary structural-facial zonality. The second concept assumes a more or less substantial rearrangement of this pattern as a result of overthrusting and displacement along the Talas–Ferghana Fault. MAKARYCHEV (1975) believes that overthrusting was limited only to the southern and eastern peripheries of the Ferghana Depression, and the nappes themselves moved northwards in the Southern

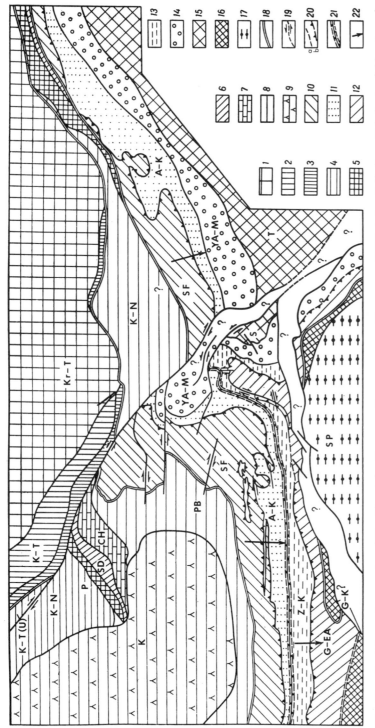

Fig. 51.   Pre-Mesozoic tectonics of Tien Shan (after Zubtsov, Porshnyakov & Yagovkin, 1974). 1–3 = Balkalian–Caledonian fold region of the Northern Tien Shan: 1 = Kirghiz–Terskei eugeosynclinal zone (Kr–T), 2 = Karatau–Talas miogeosynclinal zone (K–T), 3 = Kokdzhot–Uzunakhmat subzone (K–T/U); 4–9 = pre-Riphean Massif of Median Tien Shan, rejuvenated in the Caledonian and Hercynian epochs: 4–7 Chatkal–Naryn zone: 4 = Karatau–Naryn subzone (K–N), 5 = Pskem–Sarydzhaz subzone (P), 6 = Sandalash subzone (SD), and Chatkal subzone (Ch); 8–9 = Chatkal–Kurama zone: 8 = Pistali–Bosbutau subzone (PB), 9 = Kurama subzone (K); 10–11 = Hercynian fold region of the Southern Tien Shan: 10 = Southern Ferghana–Dzhangdzhir eugeosynclinal zone (SF), 11 = Alai–Kokshaal miogeosynclinal zone (KA–K); 12–13 = Zeravshan–Eastern Alai eugeosynclinal zone (G–VA), 13 = Zeravshan–Kalmakasui miogeosynclinal zone (Z–K); 14 = Yassy–Maidantag Hercynian miogeosynclinal zone (Ya–M); 15 = Tarim platform massif (T) and Suluterek Block (S); 16 = Gissar–Karakum zone of reactivated median massifs (G–K); 17 = Late Hercynian fold region of Northern Pamir (SP); 18 = boundaries of folded regions; 19 = boundaries of tectonic zones and subzones (a), including overthrust zones (b); 20 = horizontal displacements; 21 = Kul'gedzhili "bar" (zone of counter overthrusts); 22 = direction of displacement along overthrusts and wrench faults (author's addition – V. Kh.).

Ferghana and northwestwards in the Eastern Ferghana, i. e. in the direction of the eastern termination of the Turgai–Syr Darya Massif. The pre-Silurian ophiolitic complex exposed here along the Southern Ferghana Fault played a special role.

Another possible version of the Southern Tien Shan nappe structure was discussed by PORSHNYAKOV (1973). This version takes into account the fact that there exist three other ophiolitic zones in the north of Gissar–Alai, in addition to the Southern Ferghana ophiolitic zone proper passing through Kan (MAKARYCHEV & KURENKOV, 1974). If each of these zones is assumed to have been produced by a special eugeosynclinal trough, then the fill of these troughs must have been deformed in the process of general compression of the system and their fills must have been thrust in opposite directions over the broad geanticlines with largely limestone sequences, separating the troughs. This model of the Tien Shan nappe structure seems the least likely; PORSHNYAKOV, who had once been its supporter, presents convincing arguments against it (1973, pp. 163–164). Two other models deserve special attention, and it is difficult at present to decide between them.

The model of uni-linear movement of nappes from north to south is being advocated by BURTMAN (1973) and, in another version, by SCHULTZ, JR. ("Tectonics of the Uralo–Mongolian …", 1974). This scheme does not account for the Southern Gissar ophiolitic belt. The model of two-way counter over-thrusting from the north and from the south, in the direction from both marginal ophiolitic belts toward the center of the system was proposed by PORSHNYAKOV (1973). It can be readily seen that both models picture the structure of the northern half of the system practically in the same way, the differences concern only the southern half absent in the eastern segment (Fig. 51).

We will start our analysis of the Southern Tien Shan recent structure with its central segment which is the best exposed and studied one.

8.2 The Central (Gissar–Alai) segment of Southern Tien Shan

A marginal element of the Southern Tien Shan in the north is the Karachatyr Late Paleozoic rear trough superimposed on the southern margin of the Turgai–Syr Darya Median Massif. The Paleozoic formations of the Karachatyr zone are presently exposed in cores of Alpine horst-anticlines of the Ferghana Basin southern framing. The carbonate Middle Paleozoic rocks, which seem to constitute the cover of the massif, are overlain conformably by thick terrigenous marine Carboniferous strata (lower molasses) and lagoonal-continental red Lower Permian strata (upper molasses) deformed into large and quite simple folds.

The Karachatyr zone is superseded farther south, beyond a major fault – a marginal suture of the Southern Tien Shan system – by the **Southern Ferghana eugeosynclinal ophiolitic zone,** the first zone of the Southern Tien Shan proper in the north. Directly in the zone of the Southern Ferghana Fault, which had experienced repeated rejuvenation until the neotectonic epoch, the ophiolitic complex transformed into a typical serpentinite melange beautifully exposed in the Kan region, was studied in detail and described by MAKARYCHEV & KURENKOV (1974). These investigators have shown that the melange had formed here by the Namurian age (serpentinite clastoliths developed in the Namurian), but later this melange has been involved into vigorous movements again and, owing to this, it includes Jurassic and Cretaceous blocks which belong to the fill of the Ferghana Basin.

A relatively undisturbed cross-section of the ophiolitic complex was revealed by MAKARYCHEV (1978) farther south, in the Sartale region, where there occurs a sequence from serpentinized ultrabasic rocks and gabbro, through picrites, diabases, and spilites with interlayers of red jasper and siliceous shales to terrigenous-shale-volcanic formations with a fauna of Lower Silurian graptolites. The latter formations overlie transgressively the ophiolitic complex proper, suggesting its much older age. MAKARYCHEV believes that the Kan ophiolites are of Riphean, more specifically pre-Upper Riphean age, since blocks of green metamorphic schists aged 1,000 m.y. are present in the Kan melange, which may be regarded as an element of the ophiolitic complex and, at the same time, of the Ferghana Basin basement, i. e., of the Syr Darya Massif. In BURTMAN'S (1973) opinion, the greenschists of Kan type belong to the basement of the Ferghana Basin alone and were overthrust from the north on the Southern Ferghana ophiolitic complex. MAKARYCHEV'S (1978) opinion is that overthrusting in the southern framing of the Ferghana Depression proceeded from south to north and the Sartale sequence belongs to the nappes situated farther south and higher than the Kan melange. The Sartale ophiolites are assumed to be of Early Paleozoic age.

BURTMAN (1973) believes that the sequence of the Southern Ferghana Eugeosyncline lies in an allochthonous position throughout, making up one of the upper nappes of the Southern Tien Shan, and reconstructs the following primary sequence: 1) slaty, volcanic-terrigenous or siliceous-terrigenous Silurian formation (more than 1,000 m in thickness); 2) spilite-diabase and andesite-basaltic Devonian formation (500 to 3,000 m); 3) siliceous-carbonate-tuffaceous Early Carboniferous formation (up to 500 m).

REZVOY (1974) identifies an ophiolitic zone in the low piedmont region of the Turkestan and Alai ranges as a deep fault zone consisting of several "branches." This, rather than nappes, he maintains, accounts for the feature of the Southern Ferghana zone, such as "... frequent alternation of narrow bands of entirely different formational types and with absolutely dissimilar types of Middle Paleozoic sequences, which vary over short distances. This formational instability could result from frequent alternation of narrow, linearly extending sedimentation troughs and even narrower uplifts: cordilleras. The linear occurrence of foci of former ophiolitic activity and the presence of relatively large and also linearly extending ultrabasic bodies indicate the existence of deep faults here. Folded features are fragmentary in this tectonic zone; there predominate monoclines separated by dislocations of different orders" ("Geology of the USSR", Vol. 25, book 2, 1972, p. 226). It seems, however, that the nappe interpretation of all these characteristics of the zone is much more convincing than the deep-fault one. A subvertical deep fault originated (or became rejuvenated) here only in the Late Paleozoic (along the northern boundary of the zone) and was active in the neotectonic stage.

South of the Southern Ferghana zone, usually identified is the **Turkestan–Alai zone**[1] which occupies high foothill regions of the respective ranges. This zone is characterized by a complete sequence of the Middle Paleozoic with terrigenous (slate formation) Silurian rocks up to 2–2.5 km in thickness and carbonate Devonian, Lower and Middle Carboniferous (the "Alai" type of sequence, according to PORSHNYAKOV (1973), very

---

[1] Its eastern part has been identified as an independent Kichikalai zone by D. P. REZVOY.

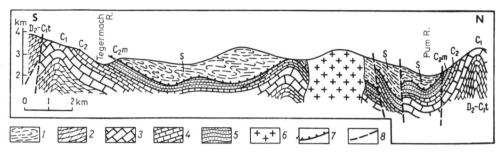

Fig. 52.   Profile across the Tegermach nappe on the right bank of Isfairam River (after ZAMALET-
DINOV, KLISHEVICH & YAGOVKIN, 1968). Allochthon: 1 = sandstones and shales, Silurian; autochthon: 2 = limestones and dolomites of the Alai series, 3 = Lower Carboniferous limestones, 4 =
Middle Carboniferous limestones; 5 = sandstones and shales of the Tolubai series, 6 = granodiorites, 7 = nappe base, 8 = post-nappe faults.

distinct in the relief), 1 to 3 km thick; total thickness of the Middle Paleozoic in the zone,
after BURTMAN (1973), ranges from 1 to 5 km; higher up there lies a flysch and a
Namurian–Late Carboniferous, 50 to 500 m thick, olistostrome formation. These
formations occur most frequently autochthonously or parautochthonously and form
large folds which make up an anticlinorium intruded by Middle Carboniferous
granodiorites along the axis as well as by less common bodies of syenites and monzonites.
BURTMAN (1973) refers this sequence to miogeosynclinal type (which nobody doubts)
and mainly autochthonous or parautochthonous, emerging from under the nappes in
tectonic windows. The lower of these nappes are composed mainly of terrigenous
Silurian rocks with graptolites; the **Tegermach nappe** has become widely known
(ZAMALETDINOV et al., 1968). Its size is 20 by 20 km, and it lies in the interfluve of rivers
Isfairam and Abshir on Middle Carboniferous limestones to the Lower Moscovian
substage (Fig. 52) and is intruded by Middle Carboniferous granodiorites. The surface of
overthrusting is quite parallel to the bedding of both Silurian and Carboniferous rocks,
and only the finds of graptolites in the former and foraminifers in the latter have made it
possible to establish the tectonic superposition of older over younger rocks. REZVOY
(1974), while admitting the existence of overthrusts here with an amplitude of up to 10–15
km, denies the phenomenon of regional nappe structure and accounts for the overthrusting by bilateral spreading apart of an anticlinal uplift composed of Silurian rocks. This
explanation is obviously quite artificial. Upper nappes, in particular, the Kirghizata one
in the east of the zone, are composed of volcano-terrigenous Silurian–Devonian rocks of
the eugeosynclinal zone, and in the uppermost nappes, there again occur Devonian–Carboniferous rocks which, according to BURTMAN (1973), are of a more northerly, Ferghana, origin. Troughs filled with Upper Paleozoic molasses are superimposed on this
complex nappe structure deformed into Late Hercynian syn- and antiforms and dislocated by subvertical faults.

    In the east, the Turkestan–Alai zone passes to the Eastern Alai one which is charac-
terized by a north-northeastern trend. The volcano-terrigenous Silurian–Carboniferous
series occurring synclinally in the central part of the zone (the Terekdavan synform) is

regarded by Burtman (1973) as an outlier of a large nappe originating from the Southern Ferghana eugeosyncline.

In the south, the Turkestan–Alai zone gives way to the Late Paleozoic **Surmetash**[2] **molasse trough** which extends in the east to the Alai Valley. Both the northern and southern boundaries of this synclinorium-like zone are of tectonic nature (overthrusts in the direction of the trough); particularly distinct is the southern fault which is considered by some investigators (Kukhtikov et al., 1968) to be an interzonal Turkestan deep fault.

The **Zeravshan (Turkestan–Zeravshan) zone,** which extends south of the Turkestan Fault, occupies the valley of a river of the same name as well as the facing slopes of the Turkestan and Zeravshan ranges. The Zeravshan Deep Fault forms its southern boundary. A thick (up to 3 km) series of a Silurian Llandoverian–Wenlock slate flyschoid formation plays a major role in the constitution of this zone; there emerge in places Cambrian–Ordovician strata, and the post-Wenlock carbonate Paleozoic rocks are exposed on the periphery of the zone. The terrigenous Silurian formation is folded into isoclinal-"Schuppen" folds which generally form, in the views of the supporters of the fixist model of the system structure, a fan-shaped anticlinorium, whereas, according to Burtman (1973), they form a complex nappe synform. The sequence of the zone is considered to be geanticlinal (starting obviously from the Ludlow) by all, however, Burtman (1973) believes that it had experienced considerable horizontal displacement southwards, whereas the "autochthonists" think that it had formed in situ as a result of inversion of the Cambrian–Silurian geosyncline (Dovzhikov et al., 1977). There stretches a chain of granitoid plutons in the axial part of the zone, which are accompanied by pegmatite fields; intrusions of Permian nepheline syenites are situated in the same zone.

The next zone farther south is the **Zeravshan–Gissar (Zeravshan–Alai) zone;** it is the southernmost zone of the Southern Tien Shan system proper; corresponding to it is the main part of the Zeravshan Range and the northern (also southern in the east) slope of the Gissar Range and the Karategin Range in the east. The sequence of the zone usually starts with the Silurian and includes the Devonian and Dinantian represented now in carbonate and now in terrigenous lithofacies, which was usually accounted for by alternation, starting from the Middle Devonian, of synsedimentary uplifts and troughs. The structure of the zone, quite complicated, was pictured as the Northern Gissar imbricate-fan-like anticlinorium accompanied on the north by a minor Zeravshan Anticlinorium. It is noteworthy that, first, the axial zones of practically all anticlinoria ever identified in the Southern Tien Shan have a **synclinal** structure and, secondly, no equivalent synclinoria were identified between them ("interanticlinorium" zones were distinguished instead). V. S. Burtman hypothetically characterizes the given zone as the one of tectonic overlapping of a miogeosynclinal autochthon by nappes of northern origin. The nappe structure of the zone has recently been confirmed by M. G. Leonov (personal commun.) (Fig. 53). At the same time, Zubtsov, Porshnyakov & Yagovkin (1974) believe that there predominate in the Zeravshan–Gissar zone imbricate overthrusts from the south, and that the tectonic suture separating this and the preceding zone is a specific zone where overthrusts of opposite directions come into contact (see Fig. 51) and whose eastern

---

[2] As it is broadly understood by Sinitsyn (1957) and Porshnyakov (1973), not by Rezvoy (1974) who identifies only the eastern part of this zone under this name.

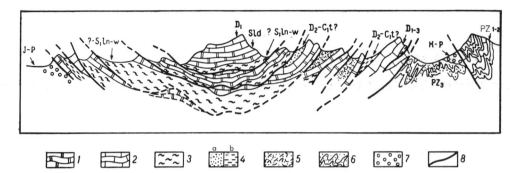

Fig. 53.   Schematic cross-section through the Zeravshan–Gissar zone in the region of the Gamza – Chimtarga Massif (after M. G. Leonov). 1 = dolomites and limestones of the Upper Silurian; 2 = Devonian limestones; 3 = quartz-sericite-chlorite schists (Yagnob suite); 4 = sandstones (a) and shales of the Middle Devonian–Tournaisian (b); 5 = terrigenous rocks (Pushnevat series); 6 = Upper Paleozoic flysch; 7 = Jurassic or Cretaceous–Paleogene; 8 = faults, basal surfaces of the nappes inclusive.

elevated part contains a small Suluterek Block and the Tarim Median Massif separated from it by the Talas–Ferghana Fault and in the west–basement salients of the Central Kyzyl Kum. In the Zeravshan–Gissar zone, there occur small plutons of Middle Carboniferous granodiorites – echoes of magmatism which found vigorous expression farther south.

The Zeravshan–Gissar zone together with the entire Southern Tien Shan system are bounded on the south by a major fault zone to which the **Southern Gissar volcano-plutonic belt** is confined. The Middle Paleozoic sequence in the zone of this belt is strongly reduced and metamorphosed, there appear serpentinized ultrabasic rocks, the Lower Carboniferous is represented by a spilite-diabase formation, the top of the Lower Carboniferous and the Middle Carboniferous by an andesite-basalt formation, the Middle–Upper Carboniferous by a flysch, the Lower Permian by a subaerial-volcanic formation of acid composition, the Upper Permian to Lower Triassic by red continental molasse. The multiphase Gissar granitoid batholith intruded into this zone in the Early – Middle Carboniferous. South of the belt, in the so-called Baisun zone of the southwestern Gissar spur as well as in Karategin, the Early Precambrian, Archaean basement of the **Kara Kum–Tadzhik Median Massif** is exposed at the surface. It is composed of crystalline schists and gneisses up to 3 billion years old, which are absolutely alien to the Southern Tien Shan.

8.3 The western (Kyzyl Kum) segment of Southern Tien Shan and its relationship with the Urals

The Kyzyl Kum differs from Gissar–Alai in three principal features: northwestern trend, width about twice comparing to the later, and salients of the Precambrian metamorphic basement, overlain directly by Devonian strata. The folded system expands at the expense of central zones with a corresponding northeastern bend of the northern zones.

The metamorphic basement outcropping in island mountains of Central Kyzyl Kum (Bukantau, Tamdytau, Auminzatau, Bel'tau, etc.) and traversed by drilling in between is composed of amphibolites, amphibolic schists, in places garnet-bearing, with interlayers of gneisses, crystalline schists, and quartzites; higher up there lie greenstone-altered volcanics of andesite-basalt composition, as well as various greenschists, quartzites, sandstones, siltstones with interlayers and lenses of dolomites and limestones. According to radiometric evidence, these formations are older than 800–700 m.y. (a dating of 1,300 m.y. is available for the lower amphibolites) and probably belong to the Middle or to the lower portion of the Upper Proterozoic, although Uzbekistan geologists ("The Precambrian of the Median and Southern Tien Shan", 1975) are inclined to consider them to be Early Precambrian. The Late Riphean age has been established more reliably (according to oncholites and other problematic plant remains and using radiometric evidence) for the overlying anchimetamorphic gray series of siltstones and micaceous shales with quartzite interlayers and lenses of dolomites and limestones. A terrigenous series with an appreciable content of coarser clastic rocks, such as sandstones, grits, and (occasionally at the base) conglomerates has also been referred arbitrarily to the Vendian. But in the last years a Lower Paleozoic fauna was found in the upper part of these metamorphic rocks.

This ancient complex with a small conglomerate member at the base is overlain unconformably by a carbonate formation of the Devonian, Lower and the lower part of the Middle Carboniferous, starting now with the Lower Devonian (in Kul'dzhuktau and Northern Nuratau even Silurian) and now Eifelian; its thickness ranges from 700 to 2,500 m. This carbonate formation is replaced by the lower molasses at the level of the Middle Moscovian stage.

The sequence, when regarded from a traditional viewpoint, belongs mainly to the Southern Bukantau and Auminza–Nuratau zones[3] which lie on the western continuation of the Turkestan–Alai zone of the Southern Tien Shan central segment. BURTMAN (1973) considers this sequence to be autochthonous and miogeosynclinal. It is overlain by an overthrust sheet packet in which a nappe composed of terrigenous-volcanic Silurian strata and overlain unconformably by Upper Silurian–Lower Devonian limestones occupies the lowermost position; higher up there lie, also unconformably, Viséan limestones and, finally, a Namurian–Middle Carboniferous clastic series. The sequence of this nappe, which has been called Bukanian in the Bukantau and Tamdytau mountains, is believed by V. S. BURTMAN to be geanticlinal. The second, higher-up, nappe is composed of Middle Paleozoic basic volcanic rocks and jaspers and occurs in the north of the Bukantau and Nuratau. It is underlain and in places overlain by ultrabasic rocks and gabbro which now make up individual slices and form serpentinite melange. An even higher-up lying nappe consists of slices which are composed of Riphean (?) greenschists, weaker metamorphized terrigenous and igneous Vendian–Lower Cambrian rocks (with Archeocyata bioherms) and also of ultrabasic rocks.

The Kyzyl Kum nappe structure is particularly well expressed in the Tamdytau Mountains (Fig. 54), where it was first noted by SABDYUSHEV & USMANOV (1971) and then

---

[3] Sometimes the Tamdytau zone is identified in between (DOVZHIKOV, ZUBTSOV & ARGUTINA, 1968), which is considered to be allochthonous by SABDYUSHEV & USMANOV (1971) as well as by SCHULTZ, JR. (1974).

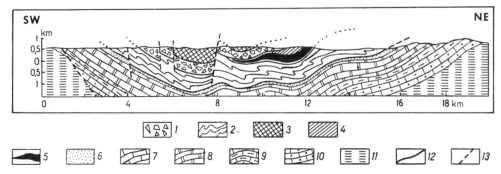

Fig. 54. Profile through Tamdytau Mountains in Kyzyl Kum (after BURTMAN, 1973). Allochthon: 1 = olistostrome (presumably Carboniferous); 2 = Silurian and Devonian terrigenous rocks; 3 = green metamorphic schists of the Precambrian or Early Paleozoic; 4 = gabbroids; 5 = ultrabasic rocks. Autochthon: 6 = Moscovian stage flysch; 7 = carbonate rocks of the Moscovian, Bashkirian, Namurian, and Viséan stages; 8 = the same for the Tournaisian and Famennian stages; 9 = the same for the Frasnian and Givetian stages; 10 = the same for the Eifelian stage and Lower Devonian; 11 = the same for the Early Paleozoic and Precambrian; 12 = basal surfaces of the nappes; 13 = late faults.

studied in detail by BURTMAN (1973). SABDYUSHEV & USMANOV (1971) are of the opinion that the fill of the ophiolitic trough here was pressed out on both sides of its axis, which is an interpretation similar to the earlier version proposed by PORSHNYAKOV (1973) for the central segment of the Southern Tien Shan and differing from the more recent and broader concept advanced by BURTMAN (1973).

The southernmost (southeastern) part of the Kyzyl Kum – the Zirabulak–Ziaetdin Mountains – is a direct continuation of the central segment Zeravshan–Gissar zone. A little farther north (the Southern Nurata Range), there is a wedging out termination of the Turkestan–Zeravshan zone with its thick terrigenous Silurian deposits. The structure of these zones still remains undeciphered from the new positions, yet, in general, this is the region of predominant occurrence of a miogeosynclinal carbonate formation. Several discordant granitoid plutons, mainly of Late Paleozoic age, were detected in the Kyzyl Kum. In the southwestern framing of the segment, along the so-called **Bukhara Terrace,** which is a buried continuation of the Major Gissar Fault, the intrusive bodies become more numerous and extend parallel to the fault. Traversed by drilling in this region are spilites and diabases, probably Lower Carboniferous, as well as volcanics of basic to intermediate composition, most probably Upper Paleozoic, i. e. almost an entire sequence of rocks characteristic of the Southern Gissar volcano-plutonic belt.

The Kyzyl Kum segment of the Southern Tien Shan system is bounded on the north by the **Northern Bukantau molasse trough,** which must be a direct continuation of the Southern Ferghana Karachatyr Trough. The trough is filled with coarse gray molasse which, however, contains interlayers of limestones with Middle–Upper Carboniferous fauna; Upper Paleozoic flora was found in the terrigenous rocks. The thickness exceeds 3 km in places; the molasse overlies unconformably the rocks up to the Moscovian stage (C_2), but is itself quite strongly deformed, so that there are even overturned layers and recumbent folds.

In the northwest of the Kyzyl Kum, the main part of the Southern Tien Shan system is truncated by the **Ural–Tien Shan wrench-fault** of northeastern trend. Frontal linkage of the Valer'yanov volcanic belt with the younger Bel'tau-Kurama one takes place in the zone of this fault, and the Bel'tau ancient block wedges into the region of their junction, as suggested by geophysical evidence (GARETSKY, KIRYUKHIN, in: "Tectonics of the Uralo–Mongolian ...", 1974). In the Sultanuizdag Range, an extreme northwestern promontory of the Kyzyl Kum basement, the miogeosynclinal Devonian carbonates typical of the Southern Tien Shan, which are tectonically overlapped by ophiolites on the northeast and which change gently their northwestern trend to meridional (Uralian), come into contact with Devonian volcanics, graywackes, and limestones resembling Uralian ones. This takes place along a meridionally trending fault in the west. The Northern Ust'yurt Massif lies on the continuation of the Uralo–Tobol Anticlinorium zone, therefore, we deal here only with the southern continuation of the Eastern Uralian Synclinorium and of the fault separating this zone from the Eastern Uralian Uplift.

Thus, almost directly south and southeast of the Aral Sea, the southern continuation of the Eastern Uralian zones comes into direct contact with the termination of the Southern Tien Shan system. Quite obviously, the two systems – the Uralian and the Tien Shan – used to belong to a single geosynclinal basin underlain by zones both with continental and oceanic crust and similar in the time of completion of their geosynclinal evolution and epigeosynclinal orogenesis and in the strongly expressed orogenic granitoid plutonism and the accompanying metallogeny; quite similar is probably also the nappe style of the structure with a general movement of the nappes westwards in the Urals, southwest and south in the Southern Tien Shan. Yet, there exist considerable differences between the two systems, which is perfectly natural, if one takes into account the noticeable differences between individual segments of each one of these systems, either Uralian or Tien Shan. The principal difference is in the stronger expressed oceanic characteristics of the Uralian system, i. e., much more extensive occurrence of the ophiolitic association as well as of geosynclinal volcanics in general, island arc ones inclusive. This produced the "phemic" appearance of the Urals in contrast to the "salic" character of the Tien Shan. This is why there is practically no mineralization in the Tien Shan, which is typical of ophiolitic complexes. It is essential, at the same time, that the Southern Tien Shan corresponds only to the easternmost Uralian zones, where the features generally typical of the Urals in general, most completely expressed in the Tagil and Magnitogorsk zones, start attenuating.

The appreciable differences between the Urals, even its eastern part, and the Southern Tien Shan make futile all attempts at drawing lines of individual structural-facial zones from one system to another and identifying zones "common" for both systems. Such attempts will fail because of the nappe structure and longitudinal segmentation of each of the systems being compared. It will be shown below how substantial the differences are, for example, between the central and eastern segments of the Southern Tien Shan itself.

8.4 The eastern segment of Southern Tien Shan

As already mentioned, structural elements of the central Gissar–Alai segment of the Southern Tien Shan within Eastern Ferghana, on approaches to the Talas–Ferghana

Fault, bend sharply northwards or even northwestwards, forming the famous **Eastern Ferghana sigmoid** (MUSHKETOV, 1934), with a very complex inner structure. In the extreme southeast of the segment, in the southwestern limb of the Talas–Ferghana Fault, there appears the **Suluterek salient** of the ancient crystalline basement: the former western termination of the Tarim Median Massif separated by this fault.

In the northeastern limb of the Talas–Ferghana Fault, the northern boundary of the Southern Tien Shan system is shifted south for nearly 200 km, which practically cancels the effect of the northerly bend of the Gissar–Alai zones at the opposite limb of the fault. This boundary in the segment in question is determined by the **Atbashy–Inyl'chek Fault** which is an analogue of the Southern Ferghana Fault. North of the Atbashy–Inyl'-chek Fault there extent the Caledono–Hercynides of the Karatau–Naryn zone of the Median Tien Shan and the Sarydzhaz salient of the ancient basement. The Southern Tien Shan system is thrust along this fault over the Median Tien Shan, just like the Median Tien Shan is thrust on the Northern along the "Nikolaev line". In particular, according to the data of KNAUF (1981), the northwestern flank of the Southern Tien Shan system is thrust over the Median Tien Shan structures at an angle of about 45°; this overthrust is accompanied by a serpentinite melange. However, the Atbashy gneiss block on the northern slope of the Atbashy Range must be an outlier of the basement nappe of the Sarydzhaz type, which could only be of northern origin.

The northwestern periphery of the Ferghana–Kokshaal segment – the **Atbashy and Inyl'chek zones** – are composed of a Silurian and Devonian terrigenous-carbonate series resembling formations of the Gissar–Alai segment central zone. BURTMAN (1973) refers them to the geanticlinal type. The southern limb of the Atbashy–Inyl'chek Fault contains outcrops of rocks of the eugeosynclinal, ophiolitic complex, including ultrabasic rocks, gabbro, diabases, spilites, keratophyres, and siliceous shales. The serpentinite melange mentioned above occurs along the fault. The structure of the main part of the Atbashy and Inyl'chek zones is usually considered to be of anticlinorial nature, and it is thrust over zones adjacent on the north and south.

In the central zones of the Kokshaal segment (**Aksai and Dzhangdzhir**), especially widespread are the Silurian slate formation and the Devonian–Lower Carboniferous carbonate formation (in places, in D_{2-3}, with subordinate subalkaline andesite-basalt volcanics), which reach great thicknesses (the carbonate formation alone is thicker than 5 km). However, alongside these typically miogeosynclinal sediments there exist carbonate-siliceous-terrigenous deposits ($S_2^2 - D_2^1$) which formed, according to KHRISTOV (1970), under the conditions of uncompensated subsidence. BURTMAN (1973) regards these deposits as leptogeosynclinal; he believes that they are thrust here over miogeosynclinal rocks which occur autochthonously.

Structurally speaking, this zone is assumed to be a synclinorium in the north and fan-like anticlinorium with an axial depression (see above) in the south. In the east, this zone comes into immediate contact with the Sarydzhaz Block of the Median Tien Shan along a fault which dips steeply southwards. In the southern part of the zone there prevails southern virgation, and even from the point of view of the "fixists" there exist numerous overthrusts with amplitudes to 8–10 km (KNAUF, 1981).

The southernmost band of the Kokshaal segment, including the **Maidantag and Kokshaal zone** proper and separated from the central part by major faults, is characterized by the presence of the Middle Paleozoic carbonate-siliceous-terrigenous and Upper Paleozoic (C_{2+3}) flyschoid terrigenous or carbonate-terrigenous formations up to 6–8 km in thickness. The Upper Paleozoic to Upper Carboniferous deposits together with Middle Paleozoic ones are involved in a very complicated south-vergent imbricate-folded structure; nappes may well be present here. In the southwestern limb of the Talas–Ferghana Fault, the continuation of the Maidantag terrigenous sequence could be detected in the **Yassy zone** east of the Ferghana Basin.

Thus, the folded-overthrust structure of the Kokshaal segment formed discontinuously from north to south, and the northern zones are, therefore, regarded as Early Hercynian (deformations starting from the Middle Devonian), central zones as Middle Hercynian, and southern as Late Hercynian. The northern and central zones contain granitoid intrusions of Middle Carboniferous (318–314 m.y.) and Permian–Carboniferous (300–270 m.y.) age, the southern zones Middle–Late Carboniferous (310–305 m.y.) and Permian–Carboniferous granitoid intrusions.

In the south, on the Chinese territory, the flyschoid terrigenous Carboniferous rocks of the Maidantag zone are superseded along a fault by carbonate Middle–Upper Paleozoic deposits ($C_1v–P_1$) of relatively small thickness (1,200 m), which makes up large coffer-box folds. Farther south, beyond another fault zone, the sediments of the same stratigraphic interval in the Kel'pin Mountains thin out to 250 m, and the content of terrigenous material increases: among the limestones there appear interlayers of marls and sandstones. This **Kel'pin block-folded zone** is regarded as a marginal uplift of the Tarim Massif and thus lies outside the Southern Tien Shan geosynclinal folded system.

The latter thins out gradually in the eastern direction, and the northern zones pinch out consecutively (the Median Tien Shan zones pinches out earlier, on the Soviet territory). The zones farther south are directly adjacent to the Baikalian metamorphic nucleus of the Northern Tien Shan, until, finally, beyond 90° E, the Southern Tien Shan system wedges out completely, and the Northern Tien Shan system links with the structures of Bei-Shan, a specific bridge between the Tarim Massif and the Sino–Korean Platform ("Some Problems ...", 1964).

It follows from the above that the Kokshaal segment differs from the Gissar–Alai one in many characteristics which become partly expressed already in Eastern Ferghana. These characteristics include the absence of analogues of the Gissar volcano-plutonic belt (its possible analogues appear only in the south of Bei-Shan), a distinct separation in the south of the segment of a zone with a predominantly terrigenous sequence of the entire Middle and Late Paleozoic (the Yassy zone of Eastern Ferghana is its equivalent) and with the post-Carboniferous age of the main deformations, as well as the predominance of northern vergence of overthrusts at the boundary with the Median Tien Shan (just like at the boundary between the Median and Northern Tien Shan). Some investigators (MAKARYCHEV & KURENKOV, 1974) believe that the displacements are of the same direction in the south of Ferghana as well as in Tamdytau (SABDYUSHEV & USMANOV, 1971), however, an opposite conclusion follows from the analysis performed by BURT-MAN (1973) and others of the vergence in the drag and flow folds within the body of nappes and of facies variations within their boundaries. By analogy with the Alps and partly with the Carpathians, it may be assumed that here too the phenomenon of retrocharriage took place in the stage of orogenesis.

8.5 Recent structure of Southern Tien Shan. The Ferghana Intermontane Basin

The present Tien Shan was shaped as a result of vigorous mountain-building movements which superseded in the Oligocene the prolonged domination of the quiescent platform conditions similar in principle to those presently typical of the Turanian Platform. As S. S. SCHULTZ pointed out for the first time in 1948, the present Tien Shan relief is due to the deformations which had the character of large-radius bends – megafolds. This conclusion was later confirmed by KOSTENKO (1961) based on a detailed structural-geomorphological analysis, however, with a significant correction maintaining that the limbs of these megafolds are frequently complicated by faults. The role of faults under the conditions of the Southern Tien Shan composed largely of practically nonmetamorphized Middle- and Upper Paleozoic folded series is generally minor, with the exception of its peripheral zones. This makes the Southern Tien Shan so different from the Median and especially Northern Tien Shan where, as pointed out above, recent movements were mainly or, as believed by some investigators (for example PATALAKHA & CHABDAROV, 1976), even exclusively of block nature, which must be due to the much higher degree of "rigidity" and isotropy of the folded basement, both Precambrian and Lower Paleozoic. The recent and Mesozoic–Early Paleogene structure in the Southern Tien Shan is markedly inherited from the Late Hercynian structure; in particular, the recent relative depressions occupied by longitudinal valleys, are inherited from the Late Paleozoic intermontane depressions.

Recent structure of the Southern Tien Shan is quite simple (Fig. 55). It is represented by three major (10 to 40 km in width and up to 500 km in length) parallel meganticlines (more specifically horst-meganticlines) which are expressed in the relief respectively by the Turkestan and Alai ranges (first), Zeravshan (second), and Gissar (third) ranges separated by narrower megasynclines (graben-megasynclines) – Zeravshan and Yagnob – to which valleys of the rivers of the same name are confined.

The depression bottoms lie at heights greater than 1 km, which points to a general predominance of uplifts. Total width of the Gissar–Alai horst-meganticline system is about 150 km. In the south, the Gissar–Alai uplift system is bounded by the Afghan – Tadzhik intermontane depression, in the north, by the Ferghana intermontane basin.

The Ferghana Basin separates the Southern Tien Shan from the Middle Tien Shan; it extends in a general west-northwest – east-southeast direction, reaching about 250 km in length and up to 100 km in width. The eastern part of the basin is the widest. It pinches out gradually westwards and links via a rather narrow corridor with the Tashkent – Golodnosteppian Trough of the Turanian Platform. In its general shape, which was most completely delineated by geophysical studies (TAL'-VIRSKY & ZUNNUNOV, 1972), the Ferghana Basin is a complex graben-megasyncline separated from the adjacent uplift systems of the Median Tien Shan (the Kurama Range, the Baubashata Mountains), Southern Tien Shan (the Turkestan and Alai ranges) and from the intermediate Ferghana Range, which extends along the diagonal Talas–Ferghana Fault, by major flexure-fault zones. The displacement amplitude along the largest of these faults, the Northern Ferghana one, which forms a boundary with the Kurama Uplift, reaches 5–7 km. The surfaces of the boundary faults are inclined away from the basin, which suggests their upthrust-overthrust nature and a certain overthrusting of the mountain framing on the basin.

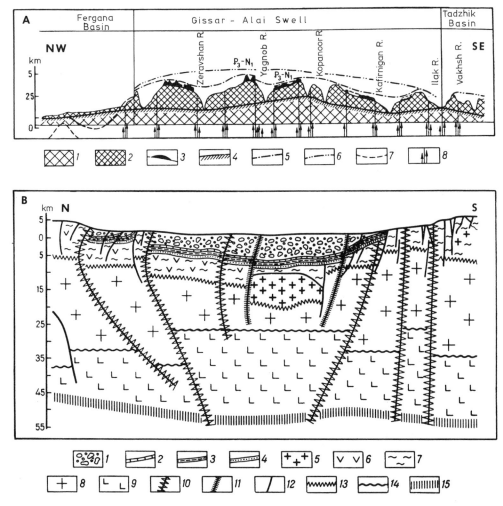

Fig. 55. Geological profiles through the Gissar–Alai Mountain range, after Kostenko (1961) (A) and through the Ferghana Basin, after Saidalieva & Akhmedov (1975) (B). Profile A: 1 = mountainous massif from sea level to the present topographic surface; 2 = general dissection of the mountain range by linear denudation; 3 = the same, by planar denudation; 4 = line of reference, showing retardation of regressive erosion in valleys of major mountain rivers; 5 = reconstruction of top surfaces of 1st-order megafolds in denudation region; 6 = the same for 2nd-order megafolds; 7 = reconstruction of Cenozoic downwarp in the accumulation region (from the end of the Oligocene to recent epoch); 8 = block displacement along fault. – Profile B: 1 = Neogene–Quaternary deposits; 2 = Paleogene; 3 = Cretaceous; 4 = Jurassic; 5–7 = Paleozoic: 5 = intrusions, 6 = volcanics, 7 = sedimentary formations; 8 = "granitic" layer; 9 = "basaltic" layer; 10–12 = faults: 10 = deep, 11 = regional, 12 = local; 13 = base of the Paleozoic; 14 = Conrad discontinuity; 15 = Mohorovičić discontinuity.

The northern and southern marginal zones and the Central Graben are identified in the structure of the Ferghana Basin. The graben is separated from these zone by faults with amplitudes up to 3–5 km (the Samgar Fault at the boundary with the northern slope of the depression). The basement in the Central Graben lies at the depth of 10–12 km; four to five km of these are filled with Mesozoic deposits, and up to 7.5 km are occupied by Cenozoic rocks. The marginal zones have a complex structure and consist of marginal troughs on the periphery, major horst-anticlinal uplifts of the Paleozoic or older basement, identified as "barrier ridges" and terraces facing the Central Graben, in the innermost part. The structure of the basin sedimentary cover is complicated by numerous folds which are the steepest and the most distinct in the relief directly along the periphery of the Central Graben. There are relatively few folds in the graben itself. They are flat-dipping, and some of them occur only within Cenozoic molasses; they owe their origin to the clay or salt (Middle Miocene salt) diapirism. The greater part of the Ferghana folds are suprafault and occur over basement faults mainly in marginal zones. There also are larger anticlinal uplifts which correspond to the bends in the Paleozoic basement. Jurassic, Cretaceous, and Paleogene deposits in the anticlinal folds contain oil and gas traps which are, however, not large. It was shown by SYTDYKOV (1977) that the pattern of the fold structure of the Ferghana Basin could be successfully explained by the influence of dextral displacement along the Talas–Ferghana Wrench Fault, according to the Moody and Hill model.

The basement of the Ferghana Basin is heterogeneous which is seen particularly well from the materials of magnetic exploration. The western part of the basin, together with the adjacent Kurama Range, is superimposed on the eastern termination of the Turgai–Syr Darya Median Massif; its eastern part is superimposed on the Eastern Ferghana Hercynides associated with the geosynclinal system of the Southern Tien Shan, partly overlapping the formations of the Median Tien Shan; the southern periphery of the basin belongs to the Southern Tien Shan.

The concepts of the Ferghana Basin deep structure have recently undergone serious alterations. It was not so long ago that the basin was believed to have concordant bends of the basement surface, of the Conrad and Mohorovičić surfaces and, therefore, thicker (more than 50 km), due to the sedimentary layer, crust (in comparison with adjacent

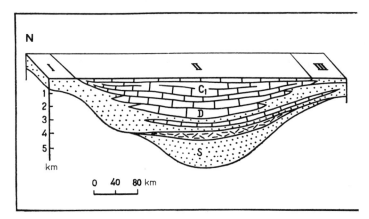

Fig. 56.   Diagram showing primary (Early Carboniferous) tectonic zonation of the Hercynian geosynclinal system of the Southern Tien Shan (after BURTMAN, 1973). Kirghizian continent: I = Kirghizian Caledonides, II = northern miogeosynclinal zone, III = northern leptogeosynclinal zone; IV = Turkestan oceanic structure; Alai continent: V–VI = southern leptogeosynclinal zone,

mountain structures). It was found later that the crust beneath Ferghana is no more than 50 km thick, and that about 40 km are occupied by a consolidated crust. Within adjacent mountain structures, especially Southern Tien Shan, the crust, which is here almost devoid of the sedimentary layer, thickens up to 50–55 km.

The region of vigorous upheavals of the Southern and Median Tien Shan is truncated in the west by a young flexure-fault zone of southwestern-northeastern trend, which separates it from the Kyzyl Kum region much less affected by recent orogenesis. This latter region is part of the Turanian Plate. The fault was called Western Tien Shan by REZVOY (1974). Similar flexure-fault zones of lesser size and undulations of the same strike were revealed parallel to it in Tien Shan itself, especially in the south.

8.6 Main stages of evolution

8.6.1 Pre-geosynclinal and early geosynclinal (Late Riphean–Early Carboniferous) history

The sharp differences in the interpretation of the Southern Tien Shan structure naturally affect interpretation of its tectonic evolution. Besides, the scarcity of information on the oldest formations of the system and on the conditions of their occurrence (for example, the Cambrian rocks) aggravate the situation further, leading to an even greater variety of opinions than is the case of Central Kazakhstan and Northern Tien Shan. If we still try to sum up the principal points of view, they will reduce to the following. The majority of the geologists (KOROLEV, 1981; KNAUF, 1981; AKHMEDZHANOV, ABDULLAYEV & BORISOV, 1979) are of the opinion that the Southern Tien Shan Geosyncline originated in the Early Paleozoic on a continental crust formed as early as the Middle Precambrian. SINITSYN (1957) was the first to have observed that no real geosynclinal conditions had existed in the Southern Tien Shan until the Ordovician, that they set in starting from the Silurian.

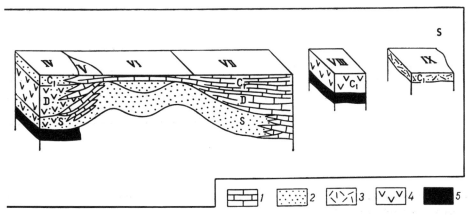

VII = southern miogeosynclinal zone; VIII = Gissar oceanic structure; IX = Tadzhik continent. 1 = carbonate sediments; 2 = terrigenous sediments; 3 = acid volcanics; 4 = basic and intermediate volcanics; 5 = ultrabasic rocks and gabbroids of the oceanic basement.

An opposite opinion is most consistently advocated by MAKARYCHEV (1978) who believes that in the Early Riphean the entire region of the Middle and Southern Tien Shan, including the Syr Darya Massif, had an oceanic crust and that continental crust formed in its place consecutively, from north to south. This process reached the Southern Gissar in the Middle Carboniferous and was thus completed in the given region. The first opinion is based on the existence of metamorphic salients referred (arbitrarily) to the Lower Precambrian and the second opinion on the presence of the ophiolitic association which is overlain unconformably by the Lower Silurian in Southern Ferghana.

An intermediate point of view was advanced by BURTMAN (1973). This investigator identifies three continental blocks in the region (Fig. 56): **Kirghizian,** including the Median Tien Shan and Syr Darya Massif in the north, **Alai** encompassing the Central Kyzyl Kum (where the basement is exposed at the surface) and central zones of the Southern Tien Shan, and, finally, **Tadzhik** in the south. The first two blocks are separated by the **Turkestan eugeosynclinal basin** with oceanic crust, which originated not later than the Silurian. Leptogeosynclinal zones (continental slopes and rises, V. Kh.) extended along its boundaries with adjacent continental blocks in the Middle Paleozoic, and in place of the Alai Block there used to be a miogeosyncline in the Devonian, Early and Middle Carboniferous. This miogeosyncline was directly adjacent to the Tarim Massif in the east. Further south, between the Alai and Tadzhik "continents" there originated in the Early Carboniferous (or earlier) yet another basin with oceanic crust: the **Gissar Basin;** the Alai and Tadzhik continents had been an integral whole prior to its formation.

We will also mention the viewpoint of SCHULTZ, JR. (1973) who considered probably complete allochthony of all structural-facial zones of the Southern Tien Shan. This point of view makes senseless the identification of the Alai continental block; there remain only the Kirghizian and Tadzhik blocks, according to BURTMAN's (1973) nomenclature. Yet, as believed by ZUBTSOV et al. (1974), there is a buried "bar" in the axial zone of the Southern Tien Shan, which used to link the southeastern projection of the Eastern European Platform with the Tarim Massif and in whose direction the nappes moved both from the north and south. The concept of BURTMAN (1973) seems the most likely one in general. In particular, he must be right in assuming that the ancient, mainly Precambrian continent extends southwards to the Southern Ferghana Suture (although, the ophiolites, whose presence is indicated by geophysical data, may be allochthonous here). Farther south, in the beginning of the Paleozoic or at the end of the Precambrian (MAKARYCHEV, 1978, seems to be right here), there formed an eugeosynclinal basin with its own ophiolitic association in the process of moving apart of the once continuous ancient continental crust. BURTMAN (1973) assumes this basin to be more than 1,000 km wide; it could even be wider.

A barrier reef the products of whose collapse were later buried in the form of olistoliths must have extended in the Vendian–Cambrian along the northern periphery of the basin. By the beginning of the Silurian there formed a volcanic island arc along the southern periphery of the basin. A deep-water trough (trench) formed in front of this arc. Leptogeosynclinal (according to V. S. BURTMAN) carbonate-siliceous-volcanic forma-tions of Middle Paleozoic age accumulated in it. In the rear of this arc, i. e., farther south, there deposited in the Silurian a thick flyschoid-slaty graptolite formation so typical of the Southern Tien Shan. Repeated (the first time prior to the Silurian, i. e., in the Taconic epoch), though relatively weak compression caused the deposits of this formation to rise above sea level at the end of the Silurian, and then they were overlapped transgressively by shallow-water carbonate Devonian rocks. The same Late Caledonian movements must have resulted in the transgressive occurrence of Devonian rocks directly on the Upper Precambrian or Lower Paleozoic ones in the Central Kyzyl Kum.

The events at the end of the Silurian–beginning of the Devonian did not, however, affect the structural pattern any significantly. Volcanic activity of island arc type continued during the Devonian and Early Carboniferous (andesitic magma); south of it there lies a band of thin siliceous sediments ("leptogeosyncline"), yet farther south there formed thick shelf carbonate deposits (up to 3 km).

8.6.2 The early orogenic stage (end of the Early to Middle Carboniferous)

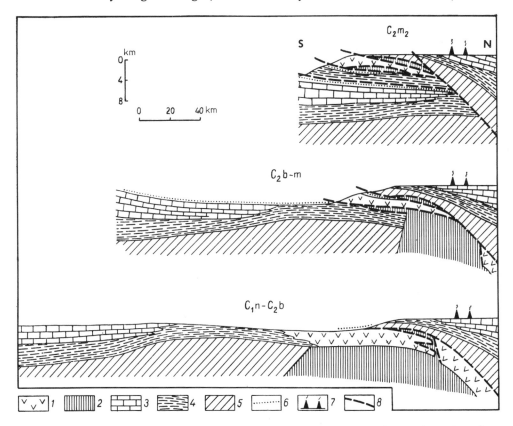

Fig. 57. Diagram showing the development of the nappe structure in the Southern Tien Shan Hercynides (after BURTMAN, 1973). Rocks of the Turkestan oceanic structure: 1 = volcanic rocks, 2 = ultrabasic rocks and gabbroids of oceanic basement. Rocks of the Alai and Kirghizian continents: 3 = Devonian and Early Carboniferous carbonate sediments; 4 = Silurian, Devonian, and Early Carboniferous terrigenous sediments; 5 = Early Paleozoic and Precambrian rocks; 6 = flysch and olistostrome deposits of the epoch of nappe formation; 7 = andesitic and more acid volcanism; 8 = surface of overthrusting.

By the end of the Early Carboniferous, this relatively quiescent development of the Southern Tien Shan geosynclinal system was interrupted by major horizontal movements which led to overthrusting of structural-facial complexes farther north on those farther south (Fig. 57). These movements reached maximum intensity in the Moscovian age. The overthrusting was, according to BURTMAN (1973), a result of the general convergence of the Kirghizian and Alai continental blocks, with a corresponding closure of the Turkestan eugeosynclinal basin, intermediate between them. The northern leptogeosynclinal complex overlapped in the process of overthrusting the eugeosynclinal one, the eugeosynclinal complex the southern leptogeosynclinal complex, and the latter the miogeosynclinal complex. BURTMAN (1973) has estimated the total amplitude of reduction of the geosynclinal system during the overthrusting epoch at 150–300 km for different areas, and the amplitude of individual overthrusts within 100 km, in places probably greater[4]. There accumulated olistostromal and flysch formations in front of the nappes; as one goes farther south, they become younger, from the Namurian to Moscovian age. A vigorous outburst of andesite-rhyolite volcanism took place in the rear of the overthrusting Kirghizian "continent", probably associated with the Benioff zone which plunges under this continental block. The continental crust spread apart alongside overthrusting in the main part of the Southern Tien Shan and partly, prior to it, along the southern boundary of the system. This produced the Southern Gissar Rift with a newly formed ophiolitic association which separated the Alai continental block from the Tadzhik one.

8.6.3. The late orogenic stage (Late Carboniferous – Middle Triassic)

In the Late Carboniferous, the Southern Tien Shan entered the orogenic stage of development proper. Its onset was accompanied by a large-fold deformation of the nappes and their dislocation by subvertical faults, mainly of upthrust, partly overthrust type. Here, northern vergence must have prevailed, directed away from the Southern Gissar Suture, where subduction probably took place along a Benioff zone inclined southwards under the Tadzhik microcontinent. Volcanic activity, which was particularly vigorous in the Early Permian, found expression along the northern margin of the microcontinent. Gissar granitic batholith emerged concurrently with this volcanism in the Late Paleozoic, within the same volcano-plutonic belt. Less clear are the formational conditions of the Late Paleozoic granitic plutons of the Southern Tien Shan proper: they may be products of remobilization of the sial basement of the Alai Megablock overlain by a thick pile of nappes.

The north-vergent movements within the southern zone of the Southern Tien Shan in the beginning of the stage in question reworked and suppressed the primary south-vergent movements. This is probably why PORSHNYAKOV (1973) and some other investigators assumed northerly direction of the nappe movement in the south of the Southern Tien Shan. The same reworking (retro charriage) took place in the zone of the northern

---

[4] MAKARYCHEV (1978) believes that the occurrence of nappes is confined to the northern zone of the Southern Tien Shan and that they are oriented from south to north, in the direction of Ferghana, rather than from north to south.

boundary of the Southern Tien Shan system, especially in its Kokshaal segment at the boundary with the Median Tien Shan.

As early as the beginning of the orogenic period a considerable part of the Southern Tien Shan rose as an uplifted dry land. There formed the Bukantau–Karachatyr rear trough along the northern framing of the system. Intermontane depressions originated within the system proper, the Surmetash–Gul'cha Depression being the largest. In the south, there also formed a molasse trough adjacent to the Tadzhik Massif. This happened with a delay, mainly in the Late Permian–Early Triassic. The Kokshaal segment (the southern part of it) lagged behind the Kyzyl Kum and Gissaro–Alai in its development; the unconformity at the base of the molasse complex is indistinct here, and the molasse making up the southern zones of the segment is mainly thin and of marine origin, and it experienced strong deformations.

The Upper Paleozoic molasse complex filling the orogenic troughs usually overlies unconformably the geosynclinal formations up to Middle Carboniferous inclusive and is in turn divided into two parts by an unconformity. The stratigraphic level of this unconformity is not accurately known, but it is to it that the above deformation phase (Saalian?) is confined as well as the intrusion of granitoid plutons (the diorite-granodiorite formation) of the Alai complexe, which are assumed to be Upper Permian–post-Karachatyr, even though they must be Early Carboniferous, in accordance with the radiometric data. The last considerable marine transgression into the region of the Karachatyr Trough of the Southern Tien Shan northern margin took place during the Karachatyr age of the Early Permian. This transgression concluded the formation of the lower, continental-marine and gray-colored molasse corresponding to the early orogenic stage of the system development. The upper molasse is wholly continental, largely red and more coarse, and its age ranges from the upper part of the Lower Permian to the Lower Triassic in places.

Concluding deformations of the Southern Tien Shan system, expressed rather unusually, were confined to the end of the Permian–Early or probably Middle Triassic. According to BURTMAN (1973), this phase is characterized by horizontal flexures, faults, and folds. It is during this phase that the principal activity of the Talas–Ferghana Fault was expressed, and a major Ferghana horizontal flexure ("sigmoid") formed in its southwestern limb.

The formation of the Alai complex of alkaline and nepheline syenites with the radiometrically determined age of 220–210 m.y. in the Gissar–Alai segment must refer to the same period of time. The geosynclinal-primary orogenic history of the Southern Tien Shan ended at this border line.

8.6.4 The platform stage (Late Triassic–Early Paleogene)

By the beginning of the Mesozoic, the Southern Tien Shan fold system had experienced general levelling. Tectonic movements, mainly in major fault zones which had existed in the preceding Hercynian stage, occurred against the background of the peneplain produced thus at the end of the Triassic–beginning of the Jurassic. Near-fault one-sided grabens formed along some of these faults. The Eastern Ferghana Trough in the southwestern limb of the Talas–Ferghana Fault was the largest of them; the continental coal-bearing – terrigenous Jurassic deposits there reach 2.5 km in thickness. At the end of the Jurassic, the fill of the trough experienced fold-fault deformations, in places quite appreciable. Another major trough – the Southern Ferghana one – extends along the fault of the same name which is part of the northern boundary of the Southern Tien Shan system; Lower Jurassic deposits, also continental lie in it. Yet another band of Jurassic troughs, of much smaller size, extends along the link zone of the Alai and Turkestan ranges. We will finally mention the zone of Early Jurassic downwarps with sediments up to 2 km in thickness along the southern boundary of the Southern Tien Shan, which continues in the east into the present Alai Valley between Tien Shan and Pamirs and farther on into the Kuchar and Yarkend troughs, at the foothills of the Eastern Tien Shan and Northwestern Kuen Lun, respectively. Andesite-basaltic volcanism found expression along the southern boundary of this trough.

At the boundary of the Jurassic and Cretaceous, there occurred a noticeable rearrangement of the structural pattern and general revival of the ascending movements, which was accompanied by fold deformations, relatively weak, with the exception of the Eastern Ferghana Fault mentioned above. All these events are echoes of compression processes expressed in the Pamirs – Karakorum system.

The **Ferghana Basin** took shape close to present at the beginning of the Cretaceous period. In the Early Cretaceous epoch it was occupied by a semiclosed basin which periodically linked with the marine basin of the Alai Trough in the east and which had a less stable connection with the Afghan – Tadzhik marine basin in the south. The depression was filled with a red clastic formation.

In the Early Cretaceous epoch, the Ferghana Basin experienced further subsidence with accumulation of a red sandy-clayey formation, mainly under the conditions of a closed brackish basin. It was only at the beginning of the Turonian and at the end of the Campanian – Maastrichtian that the sea invaded this region and organic limestones started depositing. The maximum thickness of the Upper Cretaceous deposits is 600 m. The Southern Tien Shan on the whole, as before, remained the region of uplifting, and individual longitudinal zones against its background were characterized by relative subsidence.

Similar environments prevailed in the region in question also in the Paleogene, up to the Oligocene. The Ferghana Basin in the beginning of the Paleocene used to be a strongly mineralized lagoon with settling sulfates, but later the water salinity decreased to normal. Subsequently, during the Eocene, the connection of the Ferghana Gulf via the Alai Strait with the Tadzhik Gulf and the Turanian Sea became now more and now less free, and the water salinity in it varied accordingly. Clays predominate in the sequence of the accumulating sediments; marls and limestones are of minor significance; the thickness amounts to hundreds of meters. These conditions kept on to the Early Oligocene.

8.6.5 Stage of orogenic reactivation (Oligocene – Quaternary)

At the end of the Early Oligocene, the Southern Tien Shan together with the remaining Tien Shan and Pamirs started experiencing intensifying arch-block upheavals. The sea was ultimately forced out from the Ferghana Basin, and the region was the site of accumulation of red fine clastic continental molasse, sometimes thicker than 1,700 m, during the remaining Oligocene and Early Miocene. In the Middle Miocene, temporary weakening of the upheaval of the Ferghana mountain framing and (or) aridization of the climate contributed to the increase of the salinity of the Ferghana Lake and deposition of evaporites which played a certain role in the formation of the fold structure of the basin.

The subsequent history of the region during the Neogene–Quaternary is characterized by an intensifying upheaval of the Southern Tien Shan and expansion of the territory occupied by this uplift at the expense of the adjacent depressions. The Southern Tien Shan upheaval was differentiated, and the longitudinal depressions – Zeravshan and Yagnob – inherited partly from the Late Paleozoic and more certainly from the Mesozoic and Early Paleogene were zones of relative subsidence. In the Early Pliocene, the Ferghana Basin ceased being a lake reservoir, and there formed the valley of the Syr Darya River. Along the periphery of Ferghana there emerged concentric zones of anticlinal uplifts, which were expressed in the relief as "adyr" uplifts (adyr = low foothills bordering the Ferghana Depression). However, there continued accumulating increasingly more coarse continental molasse on the alluvial plain of the central part of the depression. Total thickness of these molasses reached here 7.5 km during the interval from the Late Miocene to Quaternary.

8.7 Conclusions

In conclusion of the analysis of the structure and history of the Southern Tien Shan it is necessary to discuss again its position in the general structure of Eurasia. The Southern Tien Shan, just like Tien Shan in general, belongs, at least in its main part (see below), to the Uralo–Okhotsk (Uralo–Mongolian) folded geosynclinal belt. The Tien Shan system occupies in this belt, similar to the Urals, a marginal position, as was already pointed out by M. V. MURATOV and D. P. REZVOY (in: "The Uralo–Mongolian Folded Belt", 1974). At the same time, unlike the Uralian system, it does not border directly on the ancient platform and belongs to the sublatitudinal Central Asiatic part of the belt, parallel to and neighboring with another major geosynclinal fold belt of Eurasia: Mediterranean, extending similarly. This determined the formulation of the following problems: 1) does the Southern Tien Shan lie on the continuation of the Urals or is it the continuation of European Hercynides (Rhenides, after STILLE, 1928), as it follows from its trend; 2) if the Southern Tien Shan does belong to the "Uralides", does it belong to them wholly and where the boundary between the Uralides and Rhenides passes: inside the Southern Tien Shan or south of it; 3) what lies south of this boundary: the geosynclinal belt (system) directly or the zone of median massifs or even just a platform. These problems have been debated both in the Russian and non-Russian literature since the time of SUESS (1885); they interested ARKHANGEL'SKY (1941), MUSHKETOV (1934), VYALOV (1937), KOBER (1928), ARGAND (1922), STAUB (1928), and others (see review in the work by AKHMED-ZHANOV & BORISOV, 1975). There are much more factual data for their solution now, yet they are insufficient for the unambiguous interpretation. These data are as follows.

The Southern Tien Shan in the northwest is followed continuously to the Sultanuizdag Range in the lower reaches of the Amu Darya, however, it is truncated within this range by a major meridional fault which extends here from the Aral Sea and probably from the Urals. In the south, this fault (Urusai Fault, GARETSKY et al., 1972) abuts against an even larger Donets–Mangyshlak–Gissar (Sarmat–Turanian) Fault of global significance, along which there stretches via the Central Ustyurt a band of folded geosynclinal Middle Paleozoic, which preserves the Tien Shan northwestern trend and continues via the Caspian Sea to cis-Caucasia, northern Crimea lowlands and probably Northern Dobrogea; in cis-Caucasia and Crimea, this band has a distinctly miogeosynclinal character, but in the Transcaspia, strong linear positive magnetic anomalies are associated with it. These anomalies can hardly be accounted for by anything than a buried ophiolitic belt. In the southeast, this belt merges directly with the ophiolitic belt exposed in the Southern Gissar tectonic zone of Tien Shan.

Over the entire extent from the Crimea lowlands and cis-Caucasia to the Soviet Central Asia, the geosynclinal system in question is bounded on the south by a zone of median massifs (Karabogaz and Kara Kum–Tadzhik in the Transcaspia) whose basement, similarly to that in Central and Western Europe, includes both Baikalian and pre-Baikalian elements. The former are usually in the greenschist facies of metamorphism and the latter in the amphibolite or even granulite facies. The role of the pre-Baikalian or, more specifically, Early Precambrian complex increases distinctly eastwards, so that there appear in Karategin rocks aged up to three billion years.

This zone of median massifs is by no means part of the Uralo–Okhotsk Belt, and, as pointed out in the literature long ago, it lies in the same zone as the Tarim Promontory of the Sinian Craton which constitutes the southern boundary of the Central Asiatic part of this belt throughout its extent, with the exception of the Western Tien Shan.

Consequently, controversies may concern only the fold system of the Caucasian–Tien Shan trend, which is followed from Northern Dobrogea to Amu Darya and Southern Gissar. In the west, it is intimately connected with the Mediterranean Belt, constituting its outermost system. Within Ustyurt, it is separated from the southern virgation of the Uralian system by the Northern Ustyurt Ancient Massif. However, east of Amu Darya, the continuation of this system – the Southern Gissar zone – is directly adjacent to the main body of the Southern Tien Shan. It is, however, significant that there are no analogues of it in the Eastern Tien Shan; moreover, ZUBTSOV, PORSHNYAKOV & YAGOV-KIN (1974) find its continuation in Western Kuen Lun, i. e., southwest of the Tarim Massif. This is a specific and controversial problem which should be touched upon in connection with the analysis of the Pamiro–Kuen Lun system. Yet, it is disputable whether the entire system belongs to the Uralo–Okhotsk Belt, and it cannot be ruled out that its boundary with the Mediterranean Belt must be drawn farther north, although not so far inside the Southern Tien Shan as assumed by E. I. ZUBTSOV and his colleagues.

Still, taking into account the fact that the Donets–Caspian system, which links closely with the one just discussed in Eastern cis-Caucasia, lies on the southwestern continuation of the Southern Emba branch of the Urals and also the fact that the Uralian and Tien Shan elements interwine intricately in the Aral Sea, the entire region of Northern cis-Caucasia, Northern Caucasus, Northern Ustyurt, and Aral may be more rightly regarded as the region of junction of the Uralo–Okhotsk and Mediterranean belts, so that no hard and fast line of demarcation should be sought between them. The Mediterranean Belt has a longer evolutional history than the Uralo–Okhotsk one, and therefore it may be expected a priori that the structures of the former overlapped and reworked the structures of the latter in the course of geological time, which could also be the case in the southern part of Gissar–Alai.

References

General and Manuals

KORONOVSKY, N. V.: A Brief Course in Regional Geology of the USSR. – Moscow Univ. Moscow, 378 pp. (1976), 2nd ed., 1984.

LAZ'KO E. M.: Regional Geology of the USSR, Vols. I–II. – Nedra, Moscow, 334 + 464 pp. (1975).

NALIVKIN, D. V.: Geology of the USSR. – AN SSSR, Moscow–Leningrad, 813 pp. (1962).

Atlas of Lithologic–Paleogeographic Maps of the USSR. Vols. I–III (1968); IV (1967).

Belorussian and Baltic Faults, – Minsk, 119 pp. (1974).

Deep Structure of the Urals. – Nauka, Moscow, 383 pp. (1968).

Directions of the gas fields exploration in the Lower Paleozoic and Upper Proterozoic deposits of the Siberian platform. – Vniigazprom, Moscow (1977).

The Earth's Crust of Uzbekistan. – FAN, Tashkent, 287 pp. (1974).

Geochronological Boundaries and Geological Evolution of the Baltic Shield. – Nauka, Leningrad, 193 pp. (1972).

Geological Formations of the Sedimentary Cover of the Russian Platform. – Nedra, Leningrad, 168 pp. (1981).

Geological Structure of the USSR. Vols. I–VI. – Nedra, Moscow (1968).

Geology and Metallogeny of the Dzhungarian Alatau. – Nauka, Alma Ata, 255 pp. (1966).

Geology and Mineral Resources in the Northeast of the European SSR and Northern Urals. – Syktyvkar, Vol. I, 408 pp. (1971); Vol. II, 530 pp. (1973).

Geology of Oil and Gas of the Siberian Platform. – Nedra, Moscow, 552 pp. (1981).

Geology and Oil and Gas Fields of Timan–Pechora Province. – Syktyvkar, 127 pp. (1975).

Geology of the USSR, Vol. 2, the Arkhangel'sk and Vologda Regions and Komi ASSR, pt. I. – Gosgeoltekhizdat, Moscow, 1079 pp. (1963).

– Vol. 12, Permian, Sverdlovsk, Chelyabinsk, and Kurgan Regions, pt. I, 1, 723 pp.; 2, 304 pp. (1969).

– Vol. 13, Bashkirian ASSR and Orenburg Region, pt. I, 1, – Nedra, Moscow, 655 pp. (1969).

– Vol. 20, Central Kazakhstan. Geological Description, 2. – Nedra, Moscow, 380 pp. (1972).

– Vol. 21, Western Kazakhstan, pt. I, 1, 879 pp.; 2, 343 pp. – Nedra, Moscow (1970).

– Vol. 25, The Kirghiz SSR. Geological Description, 1. – Nedra, Moscow, 314 pp. (1972).

– Vol. 25, The Kirghiz SSR. Geological Description, 2. – Nedra, Moscow, 735 pp. (1972).

– Vol. 37, Karelian ASSR, pt. I, Geological Description, 740 pp. (1960).

– Vol. 40, Southern Kazakhstan. Geological Description, 2. – Nedra, Moscow, 534 pp. (1971).

Main Problems of the Geological Structure of the Russian Platform. – Nauka, Leningrad, 120 pp. (1979).

The Oldest Granitoids of the USSR. – Nauka, Leningrad (1981).

Paleogeography of the USSR, Vols. I–IV. – Nedra, Moscow, I, 270 pp. (1974); II, 180 pp. (1975); III, 200 pp. (1975); IV, 204 pp. (1975).

The Precambrian of the Middle and Southern Tien Shan. – FAN, Tashkent, 314 pp. (1975).

Precambrian Stratigraphy in Kazakhstan and Tien Shan. – Moscow Uni. Press, 237 pp. (1971).

Problems of Geology of Central Kazakhstan. – 661 pp. (1971).

Problems of Precambrian Geology. – Naukova Dumka, Kiev (1971).

Problems of Precambrian Geology of the Baltic Shield and Russian Platform Cover. – Trudy VSEGEI, N. S. **175**, 231 (1971).

Stages of Tectonic Evolution of Karelian Precambrian. – Nauka, Leningrad, 174 pp. (1973).

Stratigraphy of the USSR. Early Precambrian. – Gosgeoltekhizdat, Moscow, 398 pp. (1963).

Stratigraphy of the USSR. Late Precambrian. – Gosgeoltekhizdat, Moscow, 716 pp. (1963).

Stratigraphy of the USSR. Cambrian System. – Nedra, Moscow, 596 pp. (1965).
Stratigraphy of the USSR. Silurian System. – Nedra, Moscow, 531 pp. (1965).
Stratigraphy of the USSR. Permian System. – Nedra, Moscow, 536 pp. (1966).
Stratigraphy of the USSR. Jurassic System. – Nedra, Moscow, 524 pp. (1972).
Stratigraphy of the USSR. Devonian System. – Nedra, Moscow, Vol. I; 579 pp.; Vol. II; 376 pp. (1973).
Stratigraphy of the USSR. Triassic System. – Nedra, Moscow, 557 pp. (1973).
Stratigraphy of the USSR. Paleogene System[*]. – Nedra, Leningrad, 524 pp. (1975).
Structure of the Basement of Platform Territories of the USSR. – Nauka, Moscow (1975).
Tectonics and Formations of Kazakhstan. – Nauka, Alma Ata, 213 pp. (1971).
Tectonics and Magmatism of the Southern Urals. – Nauka, Moscow, 290 pp. (1974).
Tectonics and Zones of Oil and Gas Accumulation in the Kama–Kinel' Trough System. – Nauka, Moscow, 214 pp. (1965).
Tectonics of the Basement of East-European and Siberian platforms. – Nauka, Moscow, 210 pp. (1978).
Tectonics of the Eastern European Craton and Its Surroundings. – Nauka, Moscow, 240 pp. (1975).
Tectonics of the Peribaltic Region. – Mokslas, Vilnius, 92 pp. (1979).
Tectonics of pre-Paleozoic and Paleozoic Series of Tien Shan. – Ilim, Frunze, 195 pp. (1970).
Tectonics of the Ukrainian Shield. – Naukova Dumka, Kiev, 300 pp. (1972).
Tectonics of the Uralo–Mongolian Fold Belt. – Nauka, Moscow, 181 pp. (1974).

Special references

ABDULIN, A. A.: Geology of the Mugodzhary Range. – Alma Ata, 391 pp. (1971).
AFONICHEV, N. A.: Main stages of the development of the Dzhungar–Balkash geosynclinal system (Early Paleozoic, Silurian, Devonian). – Sov. Geol., No. 2, 33–53; No. 3, 61–82 (1967).
AISENSHTADT, G. E. A.: Main regularities of the formation of salt domes and oil fields in the Pericaspian basin. – In: Uslovia obrazovaniya i osobennosti neftagazonosnosti solyano-kupolnykh struktur. – Kiev, p. 58–66 (1966).
AIZBERG, R. E., GARETSKY R. G. & SINICHKA A. M.: The Sarmatian–Turanian lineament of the Earth's crust. – In: Problems of Theoretical and Regional Tectonics. – Nauka, Moscow, pp. 41–51 (1971).
AKHMEDZHANOV, M. A: Tectonics of the Pre-Mesozoic formations of the Middle and Southern Tien Shan. – FAN, Tashkent (1977).
AKHMEDZHANOV, M. A., ABDULLAYEV, R. N. & BORISOV, O. M.: Lower Paleozoic of the Middle and Southern Tien Shan. – FAN, Tashkent (1979).
AKHMEDZHANOV, M. A. & BORISOV, O. M.: On the southwestern boundary of the Uralo–Mongolian folded belt. – In: Problems of Regional Geology and Petrology of Soviet Central Asia. – Tashkent, p. 4–37 (1975).
AKHMEDZHANOV, M. A. et al.: Precambrian of Middle Asia. – Nauka, Leningrad (1982).
ANTONYUK, R. M.: Oceanic crust of the eugeosynclinal region in the East of Central Kazakhstan. – In: Tektonika Uralo–Mongolskogo skladchatogo poyasa. – Nauka, Moscow, p. 67–74 (1974).
ANTONYUK, R. M., LYAPICHEV, G. F., MARKOVA, N. G., PAVLOVA, T. G., ROZEN, O. M., SAMYGIN, S. G., TOLMACHEVA, S. G., SHUFANOV, V. I. & SHCHERBA, I. G.: Crustal structure and evolution in Central Kazakhstan. – Geotektonika, No. 3, 71–82 (1977).
ARGAND, E.: La tectonique de l'Asie. – Bruxelles (1922).
ARKHANGEL'SKY, A. D.: On the problem of nappe structure of the Urals. – Bull. Moskovsk. Obshch. Isp. Prir., Otd. Geol. 10 (1), 105–111 (1932).
– Geological structure and geological history of the USSR. – GONTI, Moscow (1941).

[*] The volumes of "Stratigraphy of the USSR" dealing with the Ordovician, Carboniferous, Cretaceous and Neogene systems are in print.

ARKHANGEL'SKY, A. D., MIKHAILOV, A. A., FEDYNSKY, V. V. & LUSTIKH, E. N.: Geological significance of gravity anomalies in the USSR. – Izv. Akad. Nauk SSSR, Ser. Geol., No. 4 (1937).

ARKHANGEL'SKY, A. D., ROSE, N. V., KOLYUBAKIN, V. V., ORLOV, V. P. & PADAREVSKAYA, A. I.: Tectonics of the Precambrian basement of the East European platform according to the USSR general magnetic survey. – Izv. Akad. Nauk SSSR, Ser. Geogr. Geofiz., No. 2, 155–194 (1937).

AVDALOVICH, V. S.: Volcanism of Tunguska Syneclise and its Relation to Tectonics. – Thesis, Moscow (1974).

BABICHEV, E. A., BOGOYAVLENSKAYA, I. A., BULYGO, L. V., MAZAROVICH, O. A., MINERVIN, O. V. & ROZEN, O. M.: On the nature of the boundaries of the Kokchetav Precambrian Massif (Central Kazakhstan). – Vestn. MGU, Geol. Ser., No. 1, 76–91 (1968).

BACKLUND, H. G.: Die Umgrenzung der Svekofenniden. – Bull. Geol. Inst. Univ. Uppsala, No. 27 (1937).

BANDALETOV S. M.: The Silurian of Kazakhstan. – Nauka, Alma Ata (1969).

BASHILOV, V. D. & KAMINSKY, F. V.: Problems of tectonics and magmatism of Timan. – Sov. Geol., No. 6, 127–133 (1975).

BEKKER, Yu. R.: Old molasse of the Urals. – Trudy Vsesoyuzn. Nauchno–Issl. Geol. Inst., N. S., 143, 110–118 (1968).

– Precambrian of the Urals. – In: Problems of Precambrian Tectonics in Eurasia. – Vladivostok, 225 pp. (1974).

BELYAEVSKY, N. A.: Earth's Crust in the USSR. – Nedra, Moscow, 280 pp. (1974).

BERTHELSEN, A.: Himalayan Tectonics: a key to the understanding of Precambrian shield patterns. – In: Himalaya. – Sciences de la Terre; CNRS, Paris, p. 61–66 (1976).

BESPALOV, V. F.: Dzhungar–Balkhash Hercynian geological province. – In: Problems of the Geology of Asia. – Moscow, Vol. 1, p. 129–154 (1954).

– Geological Structure of the Kazakh SSR. – Nauka, Moscow, 126 pp. (1971).

– The system of tectonic nappes of Kazakhstan. – Geotektonika, No. 2, 78–94 (1980).

BGATOV, V. I. et al.: To the paleotectonics of the North-West of the Siberian platform in Ordovician time. – Trudy Sibirsk. Nauchno–Issl. Inst. Geol., Geofiz., Miner. Syrya, No. 98, 17–39 (1969).

– To the paleotectonics of the North-West of the Siberian platform in Silurian time. – Trudy Sibirsk. Nauchno–Issl. Inst. Geol., Geofiz., Miner. Syrya, No. 98, 82–94 (1969).

BLAIS, S., AUVRAY, B., CAPDEVILA, R. & HAMEURT, J.: Les séries komatiitiques et tholéiitiques des ceintures archéennes de roches vertes de Finlande orientale. – Bull. Soc. Géol. France, Sér. 7, 30 (5), 965–970 (1977).

BLOKHIN, A. A.: New data on the geological structure of the Southern Urals. – Bull. Moskovsk. Obshch. Isp. Prir., Otd. Geol. 10 (1), 193–207 (1932).

BOCHKAREV, V. S. & POGORELOV, B. S.: New data on the age of the folded basement of the central regions of the West Siberian platform. – Dokl. Akad. Nauk SSSR, 179 (3), 664–665 (1968).

BOGDANOV, A. A.: Tectonics of the Ishimbai Region in the Urals. – MOIP, 143 pp. (1947).

– On certain general problems of ancient craton tectonics (as exemplified by the Eastern European Craton). – Sov. Geol., No. 9, 3–28 (1964).

– Tectonic zonation of the Central Kazakhstan and Tien Shan Paleozoides. – Bull. Moskovsk. Obshch. Isp. Prir., Otd. Geol. 40 (5), 40–68; (6) 8–42 (1965).

– Tectonic subdivision of Precambrian formations of the Eastern European Craton basement. – Vestn. MGU, Geol. Ser., No. 1, 8–26 (1967).

BOGDANOV, A. A., ZAITSEV, Yu. A., MAZAROVICH, O. A., MAKSIMOV, A. A., TIKHOMIROV, V. G. & CHETVERIKOVA, N. P.: Tectonic zoning of the Paleozoic massif of Central Kazakhstan. – Vestn. MGU, Geol. Ser., No. 5, 8–20 (1963).

BONDAREV, V. I. et al.: Geological structure and history of evolution of the Polar Urals, Pai–Khoi, Novaya Zemlya and North Pechora depression. – In: Geologia Sovetskoy Arktiki. – Leningrad (1971).

– History of tectonic development of the Northern island of Novaya Zemlya. – In: Geologia i stratigrafia Novoy Zemli. – Leningrad, p. 5–17 (1979).

BOROVIKOV, L. I. & BORSUK, B. I.: Geological structure of Central and Southern Kazakhstan. – Mater. VSEGEI, N. S., No. 41, 498 pp. (1961).

BORSUK, B. I.: Main fold systems of the Kazakhstan fold region and their structure. – Trudy Vsesoyuzn. Nauchno-Isl. Geol. Inst., N. S., **3**, 69–85 (1964).

BORUKAYEV, R. A.: Pre-Paleozoic and Lower Paleozoic of the North-East Central Kazakhstan (Sary-Arka). – Gosgeoltekhizdat, Moscow (1955).

BRONGULEEV, V. V.: Main features of the structure and development of the Mid–Paleozoic structural stage of Central Karatau. – Izv. Akad. Nauk SSSR, Ser. Geol., No. 2, 15–41 (1957).

BUBNOFF, S. VON: Grundprobleme der Geologie. – 3. Ed., Akademie-Verlag, Berlin (1954).

BUKHARIN, A. K., GARKOVETS, V. G. & PIATKOV, K. K.: Main features of the tectonic structure of the Paleozoides of the western part of the Southern Tien Shan and their framing. – In: Tektonika Uralo–Mongolskogo skladchatogo poyasa. – Moscow, p. 107–116 (1974).

BULINA, L. V. & SPIZHARSKY, T. N.: Heterogeneity of the basement of the Siberian craton. – In: Tektonika Sibiri, Vol. 3, p. 54–61, Moscow (1970).

BURTMAN, V. S.: On the Talas–Ferghana wrench fault. – Izv. AN SSSR, Geol. Ser., No. 12, 37–50 (1961).

– The Talas–Ferghana and San Andreas faults. – In: Faults and Horizontal Movements of the Earth's Crust. – Trudy GIN AN SSSR, No. 80, 128–151, Moscow (1963).

– Geology and Mechanics of Nappes Formation. – Nedra, Moscow, 97 pp. (1973).

– Structural Evolution of Paleozoic Fold Systems. – Nauka, Moscow, 289 pp. (1976).

BURTMAN, V. S., MOLDAVANTSEV, YU. E., PERIFIL'EV, A. S. & SCHULTZ, S. S.jr.: Oceanic crust of the Urals and Tien Shan Variscides. – Sov. Geol., No. 3, 23–36 (1974).

BUSH, V. A., DMITRIEVA, V. K. & FILATOVA, N. I.: Structural position, evolutional history and structure of the Dzhungarian Fault. – Geotektonika, No. 3, 77–87 (1968).

CHERVYAKOVSKY, G. F.: Middle Paleozoic Volcanism at the Eastern Slope of the Urals. – Nauka, Moscow, 258 pp. (1972).

CHETVERIKOVA, N. P.: Tectonics of the zone of linkage of Caledonides and Variscides in Central Kazakhstan. – Bull. MOIP, Otd. Geol., **45** (6), 5–28 (1970).

CHIRVINSKAYA, M. V., ZABELLO, G. D. & SMEKALINA, L. V.: Structural characteristics of the basement of the Dnieper–Donets Depression. – In: Geophysical Studies in the Ukraine. – Tekhnika, Kiev, pp. 11–26 (1968).

DAMINOVA, A. M.: Geological regularities of the evolution of magmatism and of the connected endogenous mineralization of the Taimyr Peninsula. – In: Magmatism i svyazannye s nim poleznye iskopayemye. – Moscow, p. 684–688 (1960).

DASHKEVICH, N. N., MUSATOV, D. I. & YATSKOVICH, V. L.: Deep structure of the western part of the Siberian craton and some aspects of its historic development. – In: Tektonika Sibiri, Vol. 3, p. 180–188, Moscow (1970).

DEDEEV, V. A.: Tectonic map of Precambrian basement of the Russian platform. – Geotektonika, No. 3, 27–36 (1972).

DEDEEV, V. A., ZAPOL'NOV, A. K. & KRATS, K. O.: Comparative Tectonics of Mezen' and Pechora Syneclises. – Nauka, Leningrad, 78 pp. (1969).

DEDEEV, V. A., ZHURAVLEV, V. S. & ZAPOL'NOV, A. K.: Timan and Pechora Fold System. – In: Basement Structure of Cratonic Regions of USSR. – Nauka, Leningrad, 82–89 (1974).

DOBROKHOTOV, M. N.: Some problems of the Precambrian geology of the Krivoy Rog–Kremenchug structural-facial zone. – Izv. Akad. Nauk SSSR, Ser. Geol., No. 4, 16–34 (1969).

DOVZHIKOV, A. E.: Tectonics of the Southern Tien Shan. – Nedra, Moscow, 127 pp. (1977).

DOVZHIKOV, A. E, ZUBTSOV, E. I. & ARGUTINA, T. A: The Tien Shan fold system. – In: Geological Structure of the USSR, Vol. 2, Tectonics, p. 303–326. Moscow (1958).

DUBINSKY, A. YA.: Tectonic zonation of the south-eastern part of the Ukrainian shield. – Sov. Geol., No. 10, 70–82 (1978).

EGIAZAROV, B. KH.: Geological description of the Severnaya Zemlya archipelago. – Trudy Nauchno–Issl. Inst. Geol. Arktiki, **81**, 388–423 (1957).

ERMAKOV, YU. G.: Principal features of paleo- and neotectonic movements within the North Black Sea Depression. – Dokl. AN SSSR, **207** (4), 923–926 (1972).

ESKOLA, P.: On the petrology of Eastern Fennoscandia. – Fennia, No. 45 (1925).

– The problems of mantled gneiss domes. – Quart. J. Geol. Soc. Lond., **104** (1948).

FEDOROV, S. F. & KUTUKOV A. I.: Geological structure and oil and gas bearing of the Saratov –Volga region. – Izv. Akad. Nauk SSSR, Ser. Geol., No. 3, 50–73 (1950).

FILATOVA, L. I.: Principal concepts and problems of stratigraphy of the Central Kazakhstan metamorphic complex. – Vestn. MGU, Geol. Ser., No. 1, 19–33 (1976).

FOMENKO, K. E.: Deep structure of the Caspian Depression according to geological-geophysical data. – Bull. MOIP, Otd. Geol., 47 (5), 103–111 (1972).

FREDERIKS G. & EMELYANTSEV, T.: Tectonics of the Ufa Plateau. – Zapiski Rossiisk. Miner. Obshch. Ser. 2, 61, No. 1, 170–178 (1932).

FROLOVA, N. V.: On conditions of sedimentation in the Archaean era. – Trudy Irkutsk. Univ., 5 (2), 38–68 (1951).

FROLOVA, T. I. & BURIKOVA, I. A.: Geosynclinal Volcanism (as exemplified by the eastern slope of the Southern Urals). – Moscow Univ. Press, 264 pp. (1977).

FRUMKIN, I. M.: Lower Precambrian geological complexes of the Aldan shield. – In: Tektonika, stratigraphia i litologia osadochnykh formatsyi Yakutii. – Yakutsk, p. 73–85 (1968).

– Evolution of the strata of the Earth's crust in the Archaean geological history of the Aldan shield. – In: Tektonika Sibiri, Vol. 11, p. 100–106 (1973).

GAFAROV, R. A.: Tectonics of the basement and types of magnetic fields of ancient cratons of the Northern Hemisphere. – In: Deep Tectonics of Ancient Cratons of the Northern Hemisphere. – Nauka, Moscow, 389 pp. (1971).

GAFAROV, R. A., LEITES, A. M., FEDOROVSKY, V. S. et al.: Tectonic zonation of the basement of the Siberian platform and stages of the formation of its continental crust. – Geotektonika, No. 1, p. 43–57 (1978).

GAFAROV, V. A.: Problems of the tectonics of the basement of the East European platform. – In: Tektonika Vostochno–Evropeyskoy platformy i yeye obramleniya. – Nauka, Moscow, p. 136–144 (1975).

GARAN', M. I.: Proterozoic and Lower Paleozoic of the Southern Urals. – Trudy Gorno–Geol. Inst. UFAN SSSR, No. 32, 373 pp. (1959).

– Precambrian and Cambrian in the Urals. – In: Stratigrafia pozdnego dokembria i kembria. – Moscow, p. 43–54 (1960).

GARBAR, D. I.: Jotnian of the Southwestern Onega Region. – Author's Abstr. of Candidate Thesis, Leningrad, 18 pp. (1970).

GARETSKY, R. G., KIRYUKHIN, L. G. & PERFIL'EV, A. S.: Sultanuizdag and the problem of relationship between the Urals and Tien Shan. – Geotektonika, No. 6, 88–96 (1972).

GARRIS, M. A.: Stages of magmatism and metamorphism in the pre-Jurassic history of the Urals. – Nauka, Moscow (1977).

GETSEN, V. G.: Structure of Basement of Northern Timan and Kanin Peninsula. – Nauka, Leningrad, 144 pp. (1975).

GEYER, P.: Precambrian of Sweden. – In: Precambrian of Scandinavia. – New York–London –Sydney (1963).

GILYAROVA, M. A.: Stratigraphy, Structure, and Magmatism of the Precambrian in the Eastern Part of the Baltic Shield. – Nedra, Leningrad, 223 pp. (1974).

GLUKHOVSKY, M. Z.: On nuclear and protogeosynclinal stages of the Archaean evolution of the SW part of the Aldan shield. – In: Voprosy regionalnoy geologii SSSR. – Nedra, Moscow, p. 178–182 (1971).

– Ring structures of the SE Siberia and their possible origin. – Geotektonika, No. 4, 50–63 (1978).

GLUKHOVSKY, M. Z., MORALEV, V. M. & KUZ'MIN M. I.: Tectonics and petrogenesis of the Katarchaean complex of the Aldan shield in connection with the problem of protoophiolites. – Geotektonika, No. 6, 103–117 (1977).

GLUKHOVSKY, M. Z., STAVTSEV, A. L. & KOGAN, V. S.: Tectonics and magmatism of Vit-im–Okhotsk region. – Geotektonika, No. 4 (1972).

GLUKHOVSKY, M. Z. & STAVTSEV, A. L.: Tectonics and main evolutional stages of the Aldan Shield. – In: Tektonika fundamenta drevnikh platform. – Nauka, Moscow, p. 65–75 (1973).

GOLUBEVA, I. I.: To the paleotectonics of the North-West of the Siberian platform in the Late Paleozoic. – Trudy Vsesoyuzn. Nauchno–Issl. Geol.–Razvedochn. Inst., No. 373, 139–154 (1975).

GOLUBOVSKY, V. A.: Tectonic significance of the Karabatyr Mts section in connection with the problem of the formation of the Caledonides in Central Kazakhstan. – Geotektonika, No. 6 71–76 (1979).

GORBATSCHEV, R.: Aspects and problems of Precambrian geology in Western Sweden. – Sver. Geol. Unders., Ser. C, **650**, (7), 63 pp. (1971).

GORBUNOV, G. I., ZAITSEV, YU. S. & CHERNYSHEV, N. M.: Main features of the stratigraphy and magmatism of the Voronezh crystalline massif. – Sov. Geol., No. 10, 8–25 (1969).

GORBUNOV, G. I., ZAITSEV, YU. S., RASKATOV, G. I. & CHERNYSHEV, N. M.: Principal tectonic features of the Voronezh Crystalline Massif. – In: Tectonics of the Basement of Ancient Cratons. – Nauka, Moscow, p. 44–49 (1973).

GORLOV, N. V.: The Structure of the Belomorides (Northwestern White Sea Region). – Nauka, Leningrad, 111 pp. (1967).

GORSKY, I. I.: Urals–Novaya Zemlya Fold Region. – In: Geological Structure of the USSR, Vol. 3, Tectonics. – Gosgeoltekhizdat, Moscow, 384 pp. (1958).

HAMILTON, W.: The Uralides and the motion of the Russian and Siberian platforms. – Bull. Geol. Soc. Amer., **81**, 2553–2576 (1970).

HEISKANEN, K. I., GOLUBEV, A. I. & BONDAR', L. F.: Orogenic volcanism of Karelia. – Trudy Inst. Geol. Karelsk. Fil. AN SSSR, No. 36, 216 pp. (1977).

HOLTEDAHL, O. (ed.): Geology of Norway. – Norg. Geol. Unders., No. 208, 540 pp. (1960).

HU PING, WANG CHING-PIN, KAU CHEN-CHIA, FANG HSIAO-T'IA & LU CH'ING: Some problems of geotectonics of the Sinkiang District. – Acta Geol. Sin., **44** (2), 156–170 (1964). (Engl. transl. Internat. Geol. Rev., **10** (10), 1173–1188).

IGOLKINA, N. S., KIRIKOV, V. P. & KRIVSKAYA, T. YU.: Principal stages of formation of the sedimentary cover of the Russian Platform. – Sov. Geol., No. 11, 16–35 (1970).

IVANOV, S. N.: On Baikalides of the Urals and Siberia. – Geotektonika, No. 5, 47–63 (1981).

IVANOV, S. N., EFIMOV, A. A., MINKIN, L. M., PERFIL'EV, A. S., RUZHENTSEV, S. V. & SMIRNOV, G. A.: The Nature of the Uralian Geosyncline. – Dokl. AN SSSR, **206** (5), 7–25 (1972).

IVANOV, S. N., PERFIL'EV, A. S., EFIMOV, A. A., SMIRNOV, G. A., NECHEUKHIN, V. M. & FERSHTATER, G. B.: Fundamental features in the structure and evolution of the Urals. – Amer. J. Sci., **275-A**, 107–130 (1975).

KABANOV, YU. F.: On the nappe structure of the Karaganda Basin Southern Framing. – Bull. MOIP, Otd. Geol., **47** (5), 39–46 (1972).

KALYAYEV, G. G.: Tectonics of the iron ore bearing province of the Ukrainian shield. – In: Geologia i perspektivy metallonosnosti dokembria Byelorussii i smezhnykh rayonov. – Minsk, p. 267–268 (1965).

KALYAEV, G. I.: Tectonics of the Ukrainian Shield and its position in the structure of the Eastern European Craton. – In: Tectonics of the Basement of Ancient Cratons. – Nauka, Moscow, p. 50–60 (1973).

KAMALETDINOV, M. A.: Nappe Structures of the Urals. – Nauka, Moscow, 230 pp. (1974).

KAMALETDINOV, M. A., KAZANTSEVA, T. T. & KAZANTSEV, YU. V.: Principal Problems of Crust Formation in the Urals during the Paleozoic. – Ufa (1978).

KARPINSKY, A. P.: General character of oscillations of the Earth's crust in European Russia. – Izv. Akad. Nauk, No. 1 (1894).

– On the tectonics of the European Russia. – Izv. Akad. Nauk, **12** (1919).

– Remarks on the character of the dislocations of rocks in the southern half of the European Russia. – Gorny Zhurn., **3** (1883).

KASSIN, N. G.: Development of geological structures of Kazakhstan. – In: Osnovnye idei N. G. Kassina v geologii Kazakhstana. – Alma Ata, p. 29–73 (1960).

KAUTSKY, I.: Stratigraphische Grundzüge im westlichen Kambrosilur der skandinavischen Kaledoniden. – Geol. Fören. Stockh. Forh., **71** (2), 253–284 (1949).

KEILMAN, G. A.: Gneiss complexes of the Urals. – In: Metamorphicheskiye poyasa SSSR. – Leningrad, p. 227–233 (1971).

– Migmatite Complexes of Mobile Belts. – Nedra, Moscow, 200 pp. (1974).

KHACHATRIAN, R. O.: The Tectonical Development of the Volga-Kama Anteclise and Oil and Gas Occurrence. – Nauka, Moscow, 160 pp. (1979).

KHAIN, V. E.: Some general patterns of platform evolution. – Dokl. AN SSR, N. Ser. 31 (2), 265–268 (1951).

– On the structure of the Caspian basin and structural links between the Caucasus and the Transcaspia. – Geol. Nefti, No. 9, 11–18 (1958).

– The Norgido-Nigerian mobile belt and conditions of origin of Western Thetys. – Dokl. AN SSSR, **189** (6), 1340–1343 (1969).

KHARITONOV, L. YA.: The Structure and Stratigraphy of the Karelides of the Eastern Part of the Baltic Shield. – Nedra, Moscow (1966).

KHERASKOV, N. P. & PERFIL'EV, A. S.: Main peculiarities of the geosynclinal structures of the Urals. – Trudy Geol. Inst. AN SSSR, No. 92, 35–63 (1963).

KHOLODOV, V. N.: Sedimentary Ore Genesis and Metallogeny of Vanadium. – Nauka, Moscow (1973).

KIRKINSKAYA, V. I. & POLYAKOVA, G. A.: Vendian paleogeography of the Siberian platform and regularities in the distribution of the terrigenous reservoirs of oil and gas. – Trudy Sibirsk. Nauchno–Issl. Inst. Geol., Geofiz., Miner. Syrya, No. 261, 23–28 (1978).

KIRSANOV, N. V.: Akchagyl of the Volga region. – In: Stratigrafia neogena vostoka Evropeiskoy chasti SSSR. – Moscow, p. 22–45 (1971).

KISELEV, V. V. & KOROLEV, V. G.: Precambrian Tectonics of Central Asia and Central Kazakhstan. – Ilim, Frunze, 79 pp. (1972).

KISELEV, V. V., KOROLEV, V. G., KRASNOBAYEV, A. A., KRIVOLUTSKAYA, V. N., USMANOV, U.: On pre-Vendian age of some granitoid intrusions in the western part of the Kirghiz Range (Tien Shan). – Dokl. Akad. Nauk SSSR, **214**, 407–409 (1974).

KNAUF, V. I.: Some problems of the formation of the consolidated Earth's crust of the Tien Shan. – In: Sootnoshenie geologicheskykh protsesov v paleozoiskykh skladchatykh soorugeniyakh Sredney Azii. – Frunze, p. 78–87 (1981).

KOBER, L.: Der Bau der Erde. – Borntraeger, Berlin (1928).

KOROLEV, V. G.: Basic features of tectonics, sedimentation and magmatism in the Late Precambrian and Early Paleozoic history of Tien Shan. – In: Sootnoshenie geologicheskykh protsesov v paleozoiskykh skladchatyhk soorugeniyakh Sredney Azii. – Frunze, p. 180–189 (1981).

KORZHINSKY, D. S.: Petrology of the Archaean complex of the Aldan platform. – ONTI, Leningrad–Moscow (1936).

KOSHKIN, V. YA.: Central Kazakhstan Fault. – Geotektonika, No. 1, 52–67 (1969).

KOSTENKO, N. P.: Main features of the neotectonics of Gissaro–Alai, Pamirs and Tadzhik basin. – Sborn. Trudov Geol. Fak. Moskovsk. Univ., p. 111–141. Moscow (1961).

KOSYGIN, YU. A. (ed.): Precambrian Tectonics of Siberia. – SO AN SSSR, Novosibirsk (1964).

KOZLOVSKAYA, A. I., RASPOPOVA, M. G., GLADKY V. N., GUREVICH, B. L. & CHIRVINSKAYA, M. V.: On the structure of the pre-Riphean basement of the Ukraine and Moldavia. – Sov. Geol., No. 6, 3–14 (1971).

KRASNOV, I. I., LURYE, M. L. & MASAITIS, V. L.: Geology of the Siberian Craton. – Nedra, Moscow (1966).

KRASOVITSKAYA, R. S. & PAVLOVSKY, V. I.: Tectonic structure of the Voronezh massif according to geophysical data. – In: Materialy po geologii i poleznym iskopayemym tsentralnykh rayonov Evropeiskoy chasti SSSR, No. 6, 285–290 (1970).

KRATS, K. O.: Precambrian of Karelia and Kola Peninsula. – In: Geologicheskoye stroyenie SSSR, Vol. 1. – Gosgeoltekhizdat, Moscow (1958).

– Geology of the Karelian Karelides. – AN SSSR, Moscow–Leningrad, 210 pp. (1963).

KRATS, K. O. & LOBACH-ZHUCHENKO, S. B.: Isotopic geochronology and deep crustal structure. – Geotektonika, No. 2, 74–79 (1970).

KRATS, K. O. et al.: Geological-geophysical zonation and evolution of the regional metamorphism of the Early Precambrian rocks of the Russian platform basement. – In: Geologia metamorphicheskikh kompleksov, No. 6, 34–39, Sverdlovsk (1977).

– Geology and geophysics of the eastern part of the Baltic shield. – In: Geologia i poleznye iskopayemye drevnikh platform. – Nauka, Moscow, p. 5–12 (1984).

KRESTIN, E. M.: Komatiites of Late Archaean greenstone belts of the Voronezh crystalline massif. – Sov. Geol., No. 9, 84–97 (1980).

– Precambrian of the region of KMA (Kursk magnetic anomaly) and main peculiarities of its development. – Izv. Vuzov, Geol. i Razvedka, No. 3, 3–18 (1980).

KRICHEVSKY, G. N.: Peculiarities of structure of interdomal zones of the Pericaspian basin in relation with the exploration of great gas fields in the suprasalt complex. – In: Nauchno-tekhnicheskiy sbornik po geologii, razrabotke i transportu prirodnogo gaza. – Leningrad, p. 45–55 (1963).

KUKHTIKOV, M. M.: Tectonic Zonality and Major Patterns of Structure and Evolution of the Gissar–Alai Region in the Paleozoic. – Donish, Dushanbe, 298 pp. (1968).

KUNIN, N. YA.: Tectonics of the Middle Syr Darya and Chu-Sarysu Depressions According to Geophysical Data. – Nedra, Moscow, 263 pp. (1968).

KUZNETSOV, E. A.: On the tectonics of the eastern slope of Middle Urals. – Bull. Moskovsk Obshch. Isp. Prir., Otd. Geol., 10 (2) (1933).

– Evolution of views on the tectonics of the Urals from A. P. KARPINSKY to our days. – Izv. AN SSSR, Ser. Geol., No. 4, 637–653 (1937).

LAPINSKAYA, T. A.: On the stratigraphic subdivision of the Precambrian basement of the Volga-Urals region. – Trudy Moskovsk. Inst. Neftekhim. i Gazovoy Promyshlennosti im. Gubkina, 61, 55–62 (1966).

LAZAREV, YU. I.: The tectonic evolution of the Early Karelides of Karelia. – Geotektonika, No. 5, 50–64 (1973).

LEIPZIG, A. V. & MAZOR, YU. R.: Formations and formational series of Siberian Craton. – Bull. MOIP, Otd. Geol., No. 3 (1970).

LEITES, A. M. & FEDOROVSKY, V. S.: Tectonics of Western Aldan Shield (Olekma–Vitim Mountain Country). – Geotektonika, No. 2, 46–60 (1972).

– Main stages of formation of the continental crust of the South of the Siberian platform in the Early Precambrian. – Geotektonika, No. 1, 3–23 (1977).

– The structure and the main stages of the formation of the continental crust of Southern Siberia in the Early Precambrian. – Trudy Geol. Inst. Akad. Nauk SSSR, No. 321, 109–170 (1978).

LEITES, A. M., MURATOV, M. V. & FEDOROVSKY, V. S.: Paleoaulacogens and their place in the evolution of ancient cratons. – Dokl. AN SSSR, 191 (1970).

LOBACH-ZHUCHENKO, S. B., CHEKULAEV, V. P. & BAIKOVA, V. S.: Epochs and Types of Granitization in the Baltic Shield Precambrian. – Nauka, Leningrad, 207 pp. (1974).

LUCHININ, I. L.: Some new data on polycyclic evolution of the Uralian mobile belt. – Dokl. AN SSSR, 207 (4), 927–930 (1972).

LUNDEGÅRDH, P. H.: Neue Gesichtspunkte zum schwedischen Präkambrium. – Geol. Rdsch., 60 (4), 1392–1405 (1971).

MAGNUSSON, N.: The Precambrian history of Sweden. – Quart. J. Geol. Soc. Lond., 121, 1–30 (1965).

MAKARYCHEV, G. I.: The problem of formation of the crustal "granitic" layer as exemplified by the Western Tien Shan. – Geotektonika, No. 5, 3–18 (1974).

– Two types of sequence of the ophiolitic association in the Southern Tien Shan. – Dokl. AN SSSR, 220 (3), 676–679 (1975).

– Geosynclinal Process and Formation of Continental Crust in Tien-Shan. – Nauka, Moscow, 196 pp. (1978).

MAKARYCHEV, G. I. & GES', M. D.: Tectonic nature of the junction zone of the Northern and Median Tien Shan. – Geotektonika, No. 4, 57–72 (1981).

MAKARYCHEV, G. I. & KURENKOV, S. A.: The Paleozoic serpentinite melange of the Kan zone (Southern Tien Shan). – Bull. MOIP, Otd. Geol., 49 (4), 22–34 (1974).

MAKARYCHEV, G. I. & PAZILOVA, V. I.: Structure of the basement and early evolutional stages of the Karatau Geosyncline (Southern Kazakhstan). – Geotektonika, No. 6, 75–87 (1973).

MAKARYCHEV, G. I. & SHTREIS, N. A.: Tectonic position of the Southern Tien Shan ophiolites. – Dokl. AN SSSR, 210 (5), 1164–1166 (1973).

MAKHLAEV, L. V. & KOROBOVA, N. I.: Genetical Series of Taimyr Precambrian Granitoids. – Krasnoyarsk Publ. House, 158 pp. (1972).

MALICH, N. S.: Sedimentary Formations of the Siberian Craton. – Nedra, Moscow (1974).

MAMAEV, N. F.: Geological structure and evolutional history of the eastern slope of Southern Urals. – Trudy Inst. Geol. UFAN SSSR, No. 73, 170 pp. (1965).

MAMAEV, N. F. & CHERMENINOVA, I. V.: Lower Paleozoic and Precambrian in the Eastern Slope of the Urals. – Nauka, 100 pp. (1973).

MARKOVA, N. G.: Patterns of distribution of folded zones of different age, as exemplified by Central Kazakhstan. – In: Folded Regions of Eurasia. – Moscow, 376 pp. (1964).

MASAITIS, V. L.: Magmatic cycles of the Siberian platform. – In: Problemy svyazi tektoniki i magmatizma. – Nauka, Moscow, p. 201–212 (1969).

MASAITIS, V. L., MIKHAILOV, M. V. & SELIVANOVSKAYA, T. V.: Popigai Meteorite Crater. –
Nauka, Moscow (1976).
MASAITIS, V. L., YEGOROV, L. S., LEDNEVA, V. P. et al.: Proterozoic, Paleozoic and Mesozoic
magmatic complexes of the Siberian platform. – Trudy Vsesoyuzn. Nauchno–Issl. Geol. Inst.,
265, 5–16 (1977).
MASLENNIKOV, V. A.: Absolute geochronology of the Precambrian of the eastern part of the Baltic
shield. – In: Geologia i glubinnoye stroyenie vostochnoi chasti Baltiyskogo shchita. – Nauka,
Moscow–Leningrad (1968).
MATVIYEVSKAYA, N. D.: Geological structure of Timan–Pechora Province. – In: Formation Con-
ditions and Distribution Regularities of Oil and Gas Deposits. – Moscow, 333 pp. (1974).
MAZAROVICH, O. A.: Geology of Devonian Molasses. – Nedra, Moscow, 207 pp. (1976).
MICHOT, P.: Le segment orogénique fondamental du Rogaland méridional (Norvège). – In: Etages
Tectoniques. – Neuchâtel, Baconnière, 332 pp. (1967).
MIKHAILOV, A. E.: Tectonics of the Middle and Upper Paleozoic in the Western Part of Central
Kazakhstan. – Nauka, Moscow, 245 pp. (1969).
MILANOVSKY, E. E.: Rift Zones of the Continents. – Nedra, Moscow (1976).
MILOVSKY, A. V. & BARANOV, V. V.: On Precambrian folding and Precambrian structures in south-
eastern Urals. – Vestn. Moskovsk. Univ., Geol., No. 5, 29–41 (1971).
MINERVIN, O. V., BABICHEV, E. A. & ROZEN, O. M.: Pre-Ordovician siliceous-volcanic deposits
of the Kokchetav massif and its framing. – Mater. Geol. Tsentralnogo Kazakhstana, 10,
214–225, (1971).
MIRCHINK, M. F., KHACHATRIAN, R. O. et al.: Tectonics and Zones of Oil and Gas Accumulation
of the Kama–Kinel' System of Troughs. – Nauka, Moscow (1965).
MKRTCHIAN, O. M.: General Laws of the Distribution of the Structural Forms in the Eastern Part
of the Russian Platform. – Nauka, Moscow, 136 pp. (1980).
MORALEV, V. M. & GLUKHOVSKY, M. Z.: Early Precambrian basic and ultrabasic complexes of the
Aldan shield. – In: Dokembriy. – Nauka, Moscow, p. 92–97 (1980).
MURATOV, M. V.: Geosynclinal fold belts of Eurasia. – Geotektonika, No. 6, 3–18 (1965).
– Geosynclinal fold systems of the Precambrian and some peculiarities of their development. –
Geotektonika, No. 2, 47–73 (1970).
– (ed.): Explanatory Note for the Tectonic Map of the Ukrainian SSR and Moldavian SSR. – Kiev
(1972).
– Main structural elements of the continents, their relations and age. – MGC, 24 sessia, Dokl.
Sovetsk. Geol., Tektonika. – Nauka, Moscow, p. 5–17 (1972).
– Tectonics of the basement of the Eastern European Craton and history of its formation. – In:
Tectonics of the Basement of Ancient Cratons. – Nauka, Moscow, pp. 12–143 (1973).
MUSHKETOV, D. I.: Modern views on the tectonics of Middle Asia. – Zapiski Leningradsk.
Gornogo Inst., 8, 1–16 (1934).
NALIVKIN, D. V.: The Geological History of the Urals. – Sverdlovsk (1943).
– Paleogeography of the Uralian Geosyncline in the Paleozoic. – Izv. AN SSSR, Geol. Ser., No. 5,
3–11 (1972).
NALIVKIN, V. D.: Stratigraphy and Tectonics of the Ufa Plateau and Yurezan–Sylva Depression. –
Gostoptekhizdat, Leningrad–Moscow, 207 pp. (1949).
– On the morphological classification of platform structures. – Geol. Nefti i Gaza, No. 8 (1962).
NEIMAN-PERMYAKOVA, O. F.: On the problem of Silurian and Devonian deposits of the Western
slope of Middle Urals. – Trudy i Mater. Sverdlovsk. Gornogo Inst., No. 5 (1940).
NEVOLIN, N. V.: Pre-Vendian structure of the Russian platform. – Trudy Inst. Geol. Geophys. SO
AN SSSR, 543, 110–117 (1982).
NEVOLIN, N. V. et al.: Main features of the structure of the basement of the East European
platform. – In: Geologicheskiye resultaty prikladnoy geophysiki. – Nedra, Moscow, p. 88–91
(1968).
NEVOLIN, N. V., BOGDANOVA, S. V. & LAPINSKAYA, T. A.: Principal structural features of the
basement of the Eastern European Craton. – MGC, XXIII Sess., Dokl. Sov. Geol., Problema 5.
– Nauka, Moscow, p. 88–91 (1968).
NIKITIN, I. F.: The Ordovician of Kazakhstan. I. Stratigraphy. – Nauka, Alma Ata (1972).
– The Ordovician of Kazakhstan. II. Paleogeography, Paleotectonics. – Nauka, Alma Ata (1973).

NIKOLAEV, V. A.: On the main structural line of Tien Shan. – Zapiski Vserossiisk. Miner. Obshch. Ser. 2, 62, No. 2, 347–354 (1933).

NOVIKOVA, A. S.: Tectonics of the Basement of the East European Platform. – Nauka, Moscow (1971).

OFFMAN, P. E.: Tectonics and Volcanic Pipes of Central Part of Siberian Craton. – In: Tectonics of USSR, Vol 4. – AN SSSR (1959).

OGARINOV, I. S.: Structure and Zoning of the Earth's Crust in the Southern Urals. – Nauka, Moscow, 86 pp. (1973).

OGNEV, V. N.: Talas-Ferghana fault. – Izv. Akad. Nauk SSSR, Ser. Geol., No. 4, 71–79 (1939).

OKHOTNIKOV, V. N.: Western deep fault in the Polar Urals. – Sov. Geol., No. 9, 103–108 (1968).

– Structure and eastern boundary of the western part of the Polar Uralian Uplift. – Sov. Geol., No. 2, 80–92 (1973).

OKHOTNIKOV, V. N. & STREL'NIKOV, S. I.: Tectonic features of the northern part of the Polar Urals eastern slope . – Sov. Geol., No. 1, 129–136 (1974).

OSTROVSKY, M. I., ZOLOTOV, A. N., IVANOV, T. D. & SARKISOV, Yu. M.: The Riphean–Early Paleozoic stage of formation of the cover of central and northern areas of the Eastern European Craton. – Sov. Geol., No. 10, 87–98 (1975).

PALEI, P. N.: Principal tectonic features of the Baltic Shield. – In: Problems of Regional Tectonics of Eurasia. – AN SSSR, Moscow, p. 11–34 (1963).

– Oldest formations of platform type within the Baltic Shield. – In: Problems of Theoretical and Regional Tectonics. – Nauka, Moscow, p. 178–185 (1971).

PATALAKHA, E. I. & CHAVDAROV, N. M.: Geodynamics of the Kazakhstan segment of the Earth's crust on the neotectonic stage of development. – Izv. Akad. Nauk SSSR, Ser. Geol., No. 2, 1–11 (1976).

PAVLOV, A. P.: Samarskaya Luka and Zhiguli. – Trudy Geol. Komiteta, 2 (5) (1887).

PAVLOV, S. F.: The structural evolution of the Tungusska syneclise in Late Paleozoic. – Trudy Tomskogo Univ., 232, 166–176 (1974).

PAVLOVSKY, E. V.: Zones of Pericratonal subsidence – Craton structures of first order. – Izv. AN SSSR, Ser. Geol., No. 12 (1959).

– Early stages of evolution of Earth's crust. – Izv. AN SSSR, Ser. Geol., No. 5 (1970).

PAVLOVSKY, E. V. & MARKOV, M. S.: Some general problems of geotectonics (on the irreversible evolution of the Earth's crust). – Trudy Geol. Inst. Akad. Nauk SSSR, No. 93, 9–53 (1963).

PEIVE, A. V., SHTREIS, N. A., MOSSAKOVSKY, A. A., PERFIL'EV, A. S., RUZHENTSEV, S. V., BOG-DANOV, N. A., BURTMAN, V. S., KNIPPER, A. L., MAKARYCHEV, G. I., MARKOV, M. S. & SUVOROV, A. I.: Paleozoids of Eurasia and some problems of evolution of the geosynclinal process. – Sov. Geol., No. 12, 17–25 (1972).

PEIVE, A. V., SHTREIS, N. A., PERFIL'EV, A. S., POSPELOV, I. I., RUZHENTSEV, S. V. & SAMYGIN, S. G.: Structural position of ultrabasic rocks on the western slope of the Southern Urals. – In: Problems of Theoretical and Regional Tectonics. – Nauka, Moscow, p. 9–24 (1971).

PEIVE, A. V. & SINITSYN, V. M.: Some basic problems of the geosynclinal theory. – Izv. Akad. Nauk SSSR, Ser. Geol., No. 4, 28–52 (1950).

PEIVE, A. V. et al.: Tectonics of the Urals. Explanatory Note to the Tectonic Map of the Urals. – Nauka, Moscow, 120 pp. (1977).

PERFIL'EV, A. S.: Tectonic Features of Northern Urals. – Nauka, Moscow, 223 pp. (1968).

– Formation of the Earth's Crust of the Uralian Eugeosyncline. – Nauka, Moscow, 188 pp. (1979).

PLYUSNIN, K. P.: Technique of Studying Tectonic Structures of Fold Belts (as exemplified by the Urals). – Perm', 217 pp. (1971).

PODSOSOVA, L. L. (ed.): Magmatic Complexes in the Structure of the Tumen' Urals and their Metal-Bearing. – Tumen' (1981).

POGREBITZKY, Yu. E.: Paleotectonic Analysis of Taimyr Fold System. – Nedra, Leningrad, 248 pp. (1971).

POGREBITSKY, Yu. E. et al.: Fold systems and platform covers of the Arctic shelf of Middle Siberia. – In: Geologia Sovetskoy Arktiki, p. 36–37 (1971).

POLKANOV, A. A.: Geological description of the northern part of the gabbro-labradorite pluton of Volhyn'. – Trudy Leningradsk. Obshch. Yestestvoispytat., Otd. Geol. Miner., 67 (2), 187–204 (1938).

- Geology of Karelia and Kola Peninsula. – Soviet Geol., No. 28 (1947).
- Geology of the Hoglandian-Jotnian of the Baltic shield and the problem of the Precambrian hiatus. – Izv. Akad. Nauk SSSR, Ser. Geol., No. 1, 5–28 (1956).
POLKANOV, A. A. & GERLING, E. K.: Geochronology and geological evolution of the Baltic shield and its folded frame. – Trudy Labor. Geol. Dokembria AN SSSR, No. 12, 1961.
POPOV, V. I.: Nuclear Theory of the Evolution of the Earth's Crust. – Univ. Press, Tashkent (1960).
PORSHNYAKOV, G. S.: Hercynides of the Alai and Adjacent Regions of the Southern Tien Shan. – Leningrad Univ. Press, 216 pp. (1973).
PORTNYAGIN, E. A., GNUTENKO, N. A., KOVAL'CHUK, I. A., KUZEMKO, V. N. & PAVLOV, V. I.: Carboniferous volcanism and certain problems of tectonics of Gissar (Southern Tien Shan). – Bull. MOIP, Otd. Geol., 48 (2), 82–93 (1973).
PORTNYAGIN, E. A., KOSHLAKOV, G. V. & KUZNETSOV, E. S.: On the problem of relationships between deep Paleozoic structures of the Southern Tien Shan and buried Tadzhik-Afghan Massif. – Bull. MOIP, Otd. Geol., 49 (3), 18–23 (1974).
POSTNIKOVA, I. E.: Upper Precambrian of the Russian Platform and the Presence of Oil and Gas in it. – Nedra, Moscow (1976).
PRITULA, Yu. A., SAVINSKY, K. A., KOGAN, A. B. et al.: Structure of the Earth's crust of the Tunguska syneclise. – Sov. Geol., No. 10, 29–39 (1973).
PRONIN, A. A.: Main Features of the Tectonic Development of the Urals. Variscan Cycle. – Nauka, Moscow–Leningrad, 160 pp. (1965).
- Main Features of the Tectonic Development of the Urals. Caledonian Cycle. – Nauka, Leningrad, 265 pp. (1971).
PUCHKOV, V. N.: Structural Links between cis-Polar Urals and the Adjacent Part of the Eastern European Platform. – Nauka, Leningrad, 209 pp. (1975).
- Bathyal Complexes of the Passive Margins of the Geosynclinal Regions. – Nauka, Leningrad, 260 pp. (1979).

RAMBERG, I. B. & SMITHSON, S. B.: Gravity interpretation of Southern Oslo Graben and adjacent Precambrian rocks, Norway. – Tectonophysics, 2 (6), 419–460 (1971).
RAMSAY, W.: Über die präkambrischen Systeme im östlichen Teil von Fennoscandia. – Zbl. Miner., Geol., Pal., No. 2 (1907).
RAVICH, M. G.: Taimyr. – In: Geologicheskoye stroyeniye SSSR, Vol. 2 (1958).
RAVICH, M. G. & POGREBITSKY, Yu. E.: Stratigraphic scheme of the Precambrian of Taimyr. – Trudy Nauchno-Issl. Inst. Geol. Arktiki, 145, 13–27 (1965).
RAZNITSYN, V. A.: Tectonics of Southern Timan. – Nauka, Moscow–Leningrad, 151 pp. (1964).
- Tectonics of Central Timan. – Nauka, Leningrad, 221 pp. (1968).
REZANOV, I. A.: Peculiarities of the Structure and Development of the Mesozoides of the North East of the USSR. – Nauka, Moscow (1968).
REZVOY, D. P.: Southern Tien Shan – marginal geosynclinal system of the Ural-Mongolian Paleozoic mobile belt. – In: Tektonika Uralo-Mongolskogo skladchatogo poyasa, Moscow, p. 116–125 (1974).
ROMANOV, V. A.: Type Sequences of Southern Uralian Precambrian. – Nauka, Moscow, 132 pp. (1973).
RONOV, A. B.: History of Sedimentation and Oscillatory Movements in the European SSSR (according to data by the volumetric method). – AN SSSR, Moscow–Leningrad, 391 pp. (1949).
RONOV, A. B. & MIGDISOV, A. A.: Evolution of the chemical composition of shield rocks and sedimentary cover of the Russian and Northern American platforms. – Geokhimiya, No. 4, 403–438 (1970).
ROTAR', A. F.: Mashak Series (Riphean) in Southern Urals. – Sov. Geol., No. 4, 116–123 (1974).
ROZEN, O. M.: The Riphean of the Kokchetav Massif. – Izv. AN SSSR, Ser. Geol., No. 7, 102–114 (1971).
RUSAKOV, M. P.: On the problem of presence of overthrusts and normal faults in the eastern part of the Kirghiz steppes. – Izv. GGRU, 49, No. 2 (1930).
RUZHENTSEV, S. V.: Geology of the southern part of the Urals Sakmara zone (Kosistek Region). – In: Problems of Theoretical and Regional Tectonics. – Moscow, p. 25–41 (1971).

RYABENKO, V. A.: Principal Features of Tectonic Structure of the Ukrainian Shield. – Naukova Dumka, Kiev, 126 pp. (1970).

SABDYUSHEV, SH. SH, & USMANOV, R. R.: Tectonic nappes, melange, and the ancient oceanic crust in Tamdytau (Western Uzbekistan). – Geotektonika, No. 5, 27–36 (1971).

SAIDALIEVA, M. S. & AKHMEDOV, KH.: Intermontane Basins of the Western Tien Shan and their Oil and Gas Prospects. – FAN, Tashkent (1975).

SALOP, L. I.: Principal features of stratigraphy and tectonics of the Baltic Shield Precambrian. – Trudy VSEGEI, 175 (N. S.), 6–87 (1971 a).

– Two types of Precambrian structures – gneiss fold ovals and gneiss domes. – Bull. MOIP, Otd. Geol., 46, 5–3 (1971 b).

– General Stratigraphic Scale of Precambrian. – Nedra, Leningrad (1973).

SAPIR, M. KH.: Tectonics of Jurassic-Cretaceous deposits. – In: Geology and Petroleum and Gas Occurrence in the Yenisei–Khatanga Trough. – Leningrad (1974).

SAMYGIN, S. G.: The Chinghiz Fault and its Role in the Structure of Central Kazakhstan. –Nauka, Moscow, 207 pp. (1974).

SAMYGIN, S. G. & TRET'YAK, V. G: On structural relationship of the Chinghiz Fault with the basement of the Alakol' Depression and Dzhungarian Fault (Eastern Kazakhstan). – Geotektonika, No. 2, 89–94 (1971).

SAVINSKY, K. A.: On the borders of the Siberian craton. – Geol. and Geophys., No. 4, 58–69 (1971).

SAVINSKY, K. A., ALEKSANDROV, V. K., MORDOVSKAYA, T. K., OSHCHIPKOV, YU. S. & DANILOV, F. V.: On the Problem of Boundaries of Siberian Craton. – Geol. and Geophys., No. 4 (1971).

SAVINSKY, K. A. & TUGOLESOV, D. A.: Modern surface structure of the pre-Riphean basement of the Siberian craton. – Trudy Vostochno–Sibirsk. Nauchno–Issl. Inst. Geol., Geophys., Miner. Syrya, No. 2, 5–15 (1971).

SCHULTZ, S. S.: Analysis of Neotectonics and the Relief of Tien Shan. – Geografgiz, Moscow (1948).

SCHULTZ, S. S., JR.: Geological Structure of the Urals and Tien Shan junction Zone. – Nedra, Moscow, 206 pp. (1972).

– Formation of the continental crust of the Paleozoic fold belts and their modern structure (on the Tien Shan example). – In: Tektonika Uralo–Mongolskogo skladchatogo poyasa. – Moscow, p. 156–176 (1974).

SEDERHOLM, J.: On the geology of Fennoscandia. – Bull. Comm. Geol. Finl., No. 98 (1932).

SEMENENKO, N. P.: Geochronology of the East European craton and its framing. – Trudy XV sessii Komissii po opredeleniyu absolutnogo vozrasta geologicheskykh formatsii. – Nauka, Moscow, p. 80–84 (1970).

– Comparison between the Ukrainian and Baltic Shield Precambrian.–Geotektonika, No. 5, 93–98 (1972).

SEMENENKO, N. P., TKACHUK, L. G. & KLUSHIN V. I.: Galitsian region of Ripheides and Caledonides and its foredeep. – In: Materialy VI syezda Karpato–Balkanskoy geologicheskoy assosiatsii, Kiev, p. 225–231 (1965).

SERGEEVA, E. I.: Riphean sedimentation on Kola Peninsula. – Trudy Leningradskogo Obschch. Yestesvoyspytat., 73 (2), 76–80 (1973).

SHABLINSKAYA, N. V.: Faults according to geophysical data and their comparison with cosmic images on the example of the West Siberian and Turan platforms. – In: Glubinnoye stroyeniye vostochnykh nefteperspektivnykh territoriy SSSR po rezultatam kompleksnoy interpretatsii geologicheskikh i geophysicheskikh dannykh. – Leningrad, p. 87–92 (1979).

SHATSKY, N. S.: On tectonics of the Central Kazakhstan. –Izv. Akad. Nauk SSSR, Ser. Geol., No. 5–6, 737–769 (1938).

– Outlines of tectonics of the Volga-Uralian oil-bearing region and adjacent part of the Urals western slope. – MOIP, Moscow, 131 pp. (1945).

– Principal features of structure and evolution of the Eastern European Craton. – Izv. AN SSSR, Ser. Geol., No. 1, 5–62 (1946).

– On troughs of the Donets type. – Izbrannye Trudy, 2, 544–553, Moscow (1964).

SHATSKY, N. S. & BOGDANOV, A. A.: On the international tectonic map of Europe, scale 1:2.500.000. – Izv. Akad. Nauk SSSR, No. 4, 3–25 (1961).

SHCHERBA, I. G.: Hercynian Structure of the Northern Balkhash Region. – Nauka, 163 pp. (1973).

SHLYGIN, E. D. & SHLYGIN, A. E.: Certain principles of geotectonic zoning of Kazakhstan. – In: Fold Regions of Eurasia. – Moscow, 376 pp. (1964).

SHLYGIN, E. D.: On the new content of the idea of foldings intersection in Central Kazakhstan. – Izv. Akad. Nauk Kazakhsk. SSR, Ser. Geol., No. 3, 1–7 (1977).

SHTREIS, N. A.: Stratigraphy and tectonics of the greenstone belt of the Middle Urals. – In: Tektonika SSSR, Vol. 3, Moscow (1951).

SHURKIN, K. A.: Principal features of geological structure and evolution of the Eastern Part of the Baltic Shield. – In: Geology and Deep Structure of the Eastern Part of the Baltic Shield. – Nauka, Leningrad, p. 5–60 (1968).

SIEDLECKA, A.: Late Precambrian stratigraphy and structure of the north-eastern margin of the Fennoscandian Shield (East Finnmark–Timan region). – Norg. Geol. Unders., 29 (316), 313–348 (1975).

SIDORENKO, A. V., LUNEVA, O. I. & NEMOVA, T. V.: On the sedimentary genesis of the Kola Peninsula granulites. – Dokl. AN SSSR, 198 (5), 1182–1185 (1971).

SIMON, A. K.: The final stage of the development of the Proterozoic of the mobile zone of the Kola Peninsula. – In: Tektonika fundamenta drevnikh platform. – Nauka, Moscow, p. 95–103 (1973).

SIMONEN, A.: Das finnische Grundgebirge. – Geol. Rdsch., 60 (4), p. 1406–1421 (1971).

SINITSYN, N. M.: Scheme of the tectonics of Tien Shan. – Vestn. Leningradsk. Univ., Ser. Geol. Geogr., No. 12, 5–25 (1957).

SMIRNOV, G. A.: History of tectonic evolution of the Urals after the data of lithofacies studies. – Geotektonika, No. 2, 29–37 (1971).

SMIRNOV, G. A. & SMIRNOVA, T. A.: Materials of the Urals Paleogeography. Essay III. Famennian Age. – Trudy Gorno–Geol. Inst. UFAN SSSR, No. 60, 85 pp. (1961).

– Materials of the Urals Paleogeography. Essay IV. Tournaisian Age. – Inst. Geol. Geochem. UFAN SSSR, 205 pp. (1967).

SOBOLEV, I. D.: Tectonics and magmatism of the Urals. – In: Magmatism, Metamorphism, and Metallogeny of the Urals, Vol. 1. – Sverdlovsk, 552 pp. (1963).

SOBOLEV, N. D.: Amodecian (Great Donets) basin, its geological location and subdivision. – Trudy Neftyan. Konf., 1938. Izd. Akad. Nauk Ukrainsk. SSR (1939).

SOKOLOV, B. A. & LARCHENKOV, E. P.: Tectonic nature and tendencies of development of the Viluy syneclise. – Izv. Vuzov, Geol. i Razvedka, No. 6, 25–31 (1978).

SOKOLOV, V. A. et al.: Stages of the Tectonic Development of the Precambrian of Karelia. – Nauka, Leningrad (1973).

SOKOLOV, V. L.: Problems of Gas Fields in the North Caspian Depression. Scientific and Technical Review. – Moscow, 55 pp. (1970).

SOKOLOV, V. L. & KRICHEVSKY, G. N.: On peculiarities of the salt dome structure of the Peri-caspian basin in connection with exploration of oil and gas. – In: Uslovia obrazovaniya i osobennosti neftegazonosnosti solyano-kupolnykh struktur. – Kiev, p. 146–153 (1966).

SOLLOGUB, V. B.: Deep structure of the Earth's crust of the Dnieper–Donets aulacogen. – In: Stroyeniye zemnoy kory i verkhnei mantii Tsentralnoy i Vostochnoy Evropy. – Kiev, p. 158–169 (1978).

SOLLOGUB, V. B., CHEKUNOV, A. V., KALYUZHNAYA, L. T. et al.: Inner structure of the crystalline basement in the SE part of the Korosten' pluton according to seismic data. – Geophys. Sborn., No. 517, p. 122–130 (1963).

SOLLOGUB, V. B. & CHEKUNOV, A. V.: Results of deep seismic sounding: the Ukrainian SSR. – In: Crustal Structure of Central and Southeastern Europe. – Naukova Dumka, Kiev, 286 pp. (1971).

SOLLOGUB, V. B., LITVINENKO, I. V., CHEKUNOV, A. V. et al.: New DSS data on the crustal structure of the Baltic and Ukrainian shields. – Tectonophysics, 20 (1–4), 67–84 (1973).

SPIZHARSKY, T. N.: Siberian craton. – In: Geologicheskoye stroyeniye SSSR, Vol. 3, pp. 35–48 (1958).

– Siberian craton, its origin and evolution. – In: Voprosy sravnitelnoy tektoniki drevnikh platform. – Moscow, p. 122–134 (1964).

STAUB, R.: Der Bewegungsmechanismus der Erde. – Borntraeger, Berlin (1928).

STAVTSEV, A. L.: New principles of the elaboration of the tectonic terminology. – Sov. Geol., No. 4, 49–62 (1965).

STILLE, H.: Über europäisch-zentralasiatische Gebirgszusammenhänge. – Nachr. Ges. Wiss. Göttingen, Math.-Phys. Kl. (1928).

– Uralte Anlagen in der Tektonik Europas. – Z. Dt. Geol. Ges., 99, 1947 (1949).

– Das Verteilungsbild der assyntischen Faltungen. – Geologie, 4 (3), 219–222 (1955).

SUDOVIKOV, N. G.: Materials for geology of the south-western part of Kola Peninsula. – Trudy Leningrad. Geol. Tresta, No. 10 (1935).

SUDOVIKOV, N. G. et al.: Geology and Petrology of the Southern Framing of the Aldan Shield. – Nauka, Leningrad (1965).

SUDOVIKOV, N. G., GLEBOVITSKY, V. A., SERGEEV, A. S., PETROV, V. P. & KHARITONOV, A. L.: Geological Evolution of Deep Zones of Mobile Belts (Northern Ladoga Region). – Nauka, Leningrad, 227 pp. (1970).

SUESS, E.: Das Antlitz der Erde. – Vol. 2, Prag–Leipzig (1885).

SURKOV, V. S. & ZHERO, O. G.: The Basement and the Evolution of the Platform Cover of the West Siberian Platform. – Nedra, Moscow (1981).

SUVEIZDIS, P. I.: Principal features of structure and formation of the sedimentary cover in Southern Baltic. – In: Geophysical Studies in the Southern Baltic Region. – Vilnius, p. 41–52 (1972).

SUVOROV A. I.: Structural and Formational Patterns of Deep Faults. – Nauka, Moscow, 315 pp. (1968).

– Recent data on the structure of the Tekturmas zone (Central Kazakhstan). – Dokl. AN SSSR, 213 (3), 677–680 (1973).

SVETOV, A. P.: Platform Basaltic Volcanism of the Karelian Karelides. – Nauka, Leningrad, 208 pp. (1979).

SYCHEVA–MIKHAILOVA, A. M.: Some regularities of the mechanism of formation of salt structures in the Pericaspian basin. – Voprosy geonomii. – Moscow, p. 146–168 (1972).

SYTDYKOV, B. B.: Dynamical Connection of Tectonic Structures of the Ferghana Basin with the Talas–Ferghana Wrench Fault. – FAN, Tashkent (1977).

TAL'-VIRSKY, B. B. & ZUNNUNOV, F. KH.: Tectonic zoning of the Ferghana intermontane basin after geophysical evidence. – Geotektonika, No. 2, 119–123 (1972).

TALVIRSKY, D. B.: Tectonics of Yenisei–Khatanga Oil–Gas-Bearing Region after Geophysical Data. – Nedra, Moscow, 168 pp. (1976).

TALVIRSKY, D. B. et al.: Tectonics of the eastern Part of the Yenisei–Khatanga Trough and its oil and gas prospects. – Sov. Geol., No. 10, 118–123 (1974).

TARBAEV, B. I., MOKRUSHIN, I. M. & KONOVALOV, G. A.: Stages of tectonic development of Northeastern part of Pechora Syneclise in Paleozoic, and problem of oil and gas. – Geol. of Oil and Gas, No. 12, 26–30 (1973).

TATARINOV, P. M. & STARITSKY, YU. G.: Minerageny of the Siberian Craton. – Nedra, Moscow (1970).

TETYAEV, M. M.: The Alpine fold belt in the East of the USSR. – I Vostochno–Sibirsky krayevoy nauchno–issledovatelskiy syezd, No. 1, p. 23–35, Moscow–Irkutsk (1932).

– To the problem of the tectonics of the Northern Caucasus. – Problemy Sovetsk. Geol., 5 (10) (1935).

TIKHOMIROV, V. G.: Paleozoic Magmatism and Tectonics of Central Kazakhstan. – Nedra, Moscow, 148 pp. (1975).

TIKHOMIROV, V. V.: Similar features of the Upper Cretaceous history of the Russian platform and Lesser Caucasus. – In: Pamyati Akad. A. D. Arkhangelskogo, Moscow, Akad. Nauk, p. 409–421 (1951).

TRUNOV, V. P.: Main features of the paleogeographical evolution of the Siberian platform and its folded framing in the Late Precambrian. – Trudy Vsesoyuzn. Neftyan. Nauchno–Issl. Geol. –Razvedochn. Inst., No. 373, 21–41 (1975).

TSIRYUL'NIKOVA, M. YA. et al. (eds.): Geology and Deep Structure of the Eastern Part of the Baltic Shield. – Nauka, Leningrad, 196 pp. (1968).

TSZYU, Z. I.: Main features of tectonic development of Timan–Pechora Province. – In: Oil and Gas Geology of the Northeastern European Part of USSR, No. 1, 3–25, Moscow. (1964).

TUGARINOV, A. I. & BIBIKOVA, E. V.: The geochronology of the oldest formations of the Kola Peninsula. – Geokhimia, No. 9, 1275–1281 (1975).

TUGOLESOV, D. A.: Surface structure of the pre–Riphean basement of the Siberian craton. – Sov. Geol., No. 8, 50–65 (1970).

USOV, M. A.: Phases and cycles of tectogenesis of the West Siberian Region. – Tomsk (1936).

USPENSKAYA, N. YU.: The belt of major faults in the territory of the platform of the south of the European part of the USSR and Middle Asia. – Sov. Geol., No. 3, 88–96 (1961).

VAKAR, M. G. & EGIAZAROV, B. KH.: Main stages of the geological history of the Taimyr and Severnaya Zemlya. – Trudy Nauchno–Issl. Inst. Geol. Arktiki, 145, 153–163 (1965).

VAKHROMEYEV, I. S., KLEMIN, V. P. & SENCHENKO G. S.: Stages of the tectono-magmatic development of the Magnitogorsk megasynclinorium. – In: Tektonika i magmatizm Yuzhnogo Urala. – Nauka, Moscow, p. 90–101 (1974).

VALEEV, R. N., KLUBOV, V. A. & OSTROVSKY, M. I.: Comparative analysis of conditions of formation and spatial distribution of aulacogens of the Russian Platform. – Sov. Geol., No. 4, 58–67 (1969).

VARDANIANTS, L. A. & TIKHOMIROV, S. N.: Geological map of the surface of the crystalline basement of the East European platform in the borders of the USSR. – In: Geologia dokembria. – Leningrad, p. 36–40 (1968).

VARSANOFIEVA, V. A.: Geological structure of the Pechora region. – Trudy Ylychskogo Gosudarstevennogo Zapovednika, No. 1, 5–214, Moscow (1940).

VASIL'KOVSKY, N. P.: To the problem of the evolution of the Earth's crust. – In: Deformatsia porod i tektonika. – Moscow, p. 22–23 (1964).

VÄYRYNEN, H.: Suomen kallioperä sen synty ja geologinen kahitya. – Helsinki, Ottawa (1954).

VELIN, E.: The Svecofennian fold system in Northern Sweden. – Geotektonika, No. 5, 53–60 (1972).

VESELOVSKAYA, M. M.: Results of petrographic studies of the crystalline basement of the Russian Platform. – Izv. AN SSSR, Ser. Geol., No. 7, 32–53 (1963).

VIALOV, O. S. & VIALOVA, R. I.: Scheme of the tectonics of the Urals. – Sov. Geol., No. 12, 23–32 (1939).

VINOGRADOV, A. P., TUGARINOV, A. I., KNORRE, K. G. et al.: On the age of the crystalline basement of the Russian platform. – In: Opredeleniye absolutnogo vozrasta dochetvertichnykh geologicheskykh formatsii. – Nauka, Moscow, p. 132–148 (1960).

VINOGRADOV, P. D.: Main stages of the formation of the structure of the western part of the Paleozoic Tien Shan geosyncline (Central Tadjikistan). – In: Tektonika Pamira i Tien Shanya. – Nauka, Moscow, p. 192–207 (1964).

VOINOVSKY-KRIGER, K. G.: Essays on tectonics of the Lemva facies-structural zone (western slope of the Polar Urals). – Bull. MOIP, Otd. Geol., No. 4, 5–29 (1966).

VOLOBUYEV, M. I. et al.: The age of the basement and geosynclinal formations of the Grenvillides of the Yenisei Range. – In: Opredeleniye absolutnogo vozrasta rudnykh mestorozhdeniy i molodykh magmaticheskikh porod. – Nauka, Moscow, p. 39–47 (1976).

VOLOBUYEV, M. I., ZYKOV, S. I., STUPNIKOVA, N. I., STRIZHOV, V. P. & BELOV, V. P.: On the geochronology of the epigeosynclinal orogenic formations and the main geosynclinal complex of the Baikalian epoch in the Yenisei Range. – In: Opredeleniye absolutnogo vozrasta rudnykh mestorozhdeniy i molodykh magmaticheskikh porod. – Nauka, Moscow, p. 146–155 (1976).

VOROBYEV, I. V.: Upper Proterozoic Geosynclinal System of the Yenisei Range and the History of its Development. – Thesis, Moscow Univ. (1969).

WATSON, J.: Eo–Europa. The evolution of a craton. – In: Europe from Crust to Core. – J. Wiley & Sons, New York–Sydney–Toronto, p. 59–80 (1977).

YANSHIN, A. L.: Tectonics of the Kargala Mountains. – Bull. Moskovsk. Obshch. Isp. Prir., Otd. Geol., 10 (2), 308–345 (1932).

YAROSH, A. YA. & KASSIN, G. G.: Connection between the structure of the cis-Uralian Foredeep and the pre-Riphean basement on the territory of the Middle Urals. – Trudy Sverdlovsk. Gornogo Inst., No. 83, 26–31 (1972).

YEGIAZAROV, B. KH.: Severnaya Zemlya. – In: Geology of the USSR, Vol. 26, p. 237–323 (1970).

ZABIYAKA, A. I.: Structural-facial zonation of the Upper Proterozoic of the Taimyr geosynclinal region. – In: Geology and Mineral Resources of Krasnoyarsk Region. – Krasnoyarsk, 250 pp. (1971).

ZAITSEV, YU. A.: The historical-geological content of the "transitional" stage as exemplified by the Kazakhstan and Central Asia Paleozoides. – Geotektonika, No. 5, 99–114 (1972a).

– On the identification of the Middle Riphean Issedonian Folding in Kazakhstan. – Vestn. MGU, Geol. Ser., No. 4, 19–35 (1972b).

ZAITSEV, YU. A. & FILATOVA, L. I.: Stages of geological evolution of Kazakhstan in the Precambrian. – Vestn. MGU, Geol. Ser., No. 4, 19–35 (1972).

ZALALAYEV, R. SH. & BEZZUBTSEV, V. V.: On the Chelyuskin ultrabasic belt (Eastern Taimyr). – Geol. and Geophys., No. 12, 123–133 (1975).

ZAMALETDINOV, T. S., KLISHEVICH, V. L. & YAGOVKIN A. V.: The Tegermach Hercynian nappe in the Southern Tien Shan. – Geotektonika, No. 5, 86–92 (1968).

ZAPOL'NOV, A. K.: Tectonics of Bol'shezemelskaya Tundra. – Nauka, Leningrad, 120 pp. (1971).

ZAVARITSKY, A. N.: Some main problems of the geology of the Urals. – Izv. Akad. Nauk SSSR, Ser. Geol., No. 3 (1941).

ZHARKOV, M. A.: Cambrian saliferous formation of Siberian Craton. – Sov. Geol., No. 2 (1966).

ZHDANOV, V. V.: On two types of the Earth's crust without "granitic" layer in the North of the Baltic shield. – Sov. Geol., No. 5, 101–111 (1965).

ZHIVKOVICH, A. E.: The Structure and Bauxites Prospects of the Middle Part of the Ufa Amphitheater. – Thesis, Moscow Univ. (1980).

ZHURAVLEV, V. S.: Comparative Tectonics of the Pechora, Caspian, and North-Sea Exogonal Depressions of the Eastern European Craton. – Nauka, Moscow, 398 pp. (1972).

ZHURAVLEV, V. S. & GAFAROV, R. A.: Scheme of tectonics of the North-East of the Russian platform. – Dokl. Akad. Nauk SSSR, 128 (5), 1023–1025 (1959).

ZHURAVLEV, V. S., PERFIU'EV, A. S. & KHERASKOV, N. P.: Spatial and temporal relations of Uralides and Preuralides on the eastern boundary of the Russian platform. – Bull. Moskovsk. Obshch. Isp. Prir., Otd. Geol., 40 (5), 106–130 (1965).

ZNOSKO, K., KUBICKI, S. & RYKA, W.: Tectonics of the crystalline basement of the Eastern European Craton on the territory of Poland. – Geotektonika, No. 5, 79–92 (1972).

ZUBTSOV, E. I., PORSHNYAKOV, G. S. & YAGOVKIN, A. V.: A new scheme of the Tien Shan pre-Mesozoic tectonics. – Dokl. AN SSSR, 217 (5), 1153–1156 (1974).

ZVONTSOV, V. S.: Structure and development peculiarities of the Dzhungar–Balkhash median mass in the Caledonian epoch. – Trudy Inst. Geol. Nauk im. Satpayeva, 32, 41–48 (1973).

Author Index

Abdulin, A. A. 165
Abdullayev, R. N. 237
Afonichev, N. A. 206, 218
Aizberg, R. E. 43
Akhmedov, Kh. 235
Akhmedzhanov, M. A. 237, 243
Antonyuk, R. M. 189, 206, 218
Argand, E. 243
Argutina, T. A. 229
Arkhangel'sky, A. D. 35, 133, 200, 208, 217, 243

Babichev, E. A. 179
Backlund, O. 10
Baranov, V. I. 139
Bashilov, V. I. 69
Bekker, Yu. R. 162, 163
Berthelsen, A. 23
Bespalov, V. F. 206, 213
Bezzubtsev, V. V. 122
Bgatov, V. I. 110
Bibikova, E. V. 12
Blais, J. 49
Blokhin, A. A. 133
Bogdanov, A. A. 23, 26, 66, 70, 160, 175, 184, 185,
    188, 192, 200, 208, 210, 214, 217
Bogdanova, S. V. 35, 36
Bondarev, V. I. 110
Borisov, O. M. 237, 243
Borsuk, B. I. 219
Borukaev, R. A. 175
Bronguleev, V. V. 186
Bubnov, S. v. 65
Burkharin, A. K. 188
Bulina, N. K. 92
Burikova, I. A. 165, 173
Burtman, V. S. 149, 154, 215, 222, 224–227, 229,
    230, 232, 233, 236, 238–241
Bush, V. A. 196, 215

Chabdarov, N. M. 234
Chekunov, A. V. 27, 43, 44
Chermeninova, I. V. 163
Chernyshev, F. N. 133
Chetverikova, N. P. 192, 211
Chirvinskaya, M. V. 43

Daminova, A. M. 122
Dashkevich, N. N. 93
Dedeev, V. A. 36, 37, 68, 70

Dmitrieva, V. K. 215
Dovzhikov, A. E. 222, 227, 229
Dubinsky, A. 27

Egiazarov, B. Kh. 120, 159
Eisenstadt, G. A. 48
Emelyantsev, T. 133, 142, 147
Ermakov, Yu. G. 66
Eskola, P. 10, 16

Fedorov, S. F. 65
Fedorovsky, V. S. 14, 85, 87, 93, 102
Filatova, L. I. 178, 179, 185, 190, 200, 201
Filatova, N. I. 215
Fiveg, M. P. 169
Fotiadi, E. E. 36
Frederiks, G. N. 133, 142, 147
Frolova, N. V. 83
Frolova, T. I. 165, 173
Frumkin, I. M. 83

Gafarov, R. A. 8, 36, 37, 47, 66, 71, 92, 93
Galdobina, L. P. 17
Galitsky, V. V. 186, 187
Garan', M. I. 161, 162
Garbar, D. I. 21
Garetsky, R. G. 231, 243
Gar'kovets, V. G. 188
Gerling, E. K. 15
Ges', M. D. 185, 200, 203
Getsen, V. G. 71, 75
Geyer, P. 23
Gilyarova, M. A. 16, 17
Glukhovsky, M. Z. 82, 83, 86, 87, 102, 106, 115
Golubeva, I. I. 113
Golubovsky V. A. 210
Gorbunov, G. I. 34
Gorlov, N. V. 16, 17
Gorsky, I. I. 133

Hamilton, W. 173

Igolkina, N. S. 52, 54, 56, 61
Ivanov, S. N. 135, 161, 164, 167

Kabanov, Yu. F. 193
Kalyaev, G. I. 27, 29, 32
Kamaletdinov, M. A. 133, 136, 139, 142–149, 153,
    162, 166, 173

Kaminsky, F. V. 69
Karpinsky, A. P. 41, 43, 63, 133, 147
Kassin, N. G. 175, 199, 217
Kautsky, I. 65
Kazantsev, Yu. V. 173
Kazantseva, T. T. 173
Keilman, G. A. 150
Khachatrian, R. A. 57, 58
Khain, V. E. 43, 65
Kheraskov, N. P. 133, 162
Kholodov, V. N. 204
Khristov, E. V. 232
Kirikov, D. A. 52, 54, 56, 61
Kirkinskaya, V. I. 106, 108, 109
Kirsanov, N. V. 64
Kiryukhin, L. G. 231
Kiselev, V. V. 182, 184, 202
Klemin, V. P. 144
Klishevich, V. L. 226
Klitin, K. A. 96
Knauf, V. I. 178, 182, 232, 237
Kniazev, H. S. 110
Kober, L. 243
Konovalov, G. A. 73
Kopnin, V. I. 169
Korobova, N. I. 120
Korolev, V. G. 178, 182, 184, 200, 237
Korzhinsky, D. S. 83
Koshkin, V. Ya. 197, 218
Kostenko, N. P. 234, 235
Kozlovskaya, E. A. 32
Krasovitskaya, R. S. 34
Krats, K. O. 10, 16, 17, 23, 37, 49
Krestin, E. M. 34
Krichevsky, G. I. 44, 46
Krivskaya, T. Yu. 52, 54, 56, 61
Kukhtikov, M. M. 222, 227
Kunin, N. Ya. 187, 188
Kurenkov, S. A. 224, 233
Kutukov, A. I. 65
Kuznetsov, E. A. 133, 149

Lapinskaya, T. A. 35, 36
Leites, A. M. 14, 85, 87, 93, 102
Leonov, M. G. 227, 228
Lobach-Zhuchenko, S. B. 10
Luchinin, I. L. 166
Lundegårdh, P. H. 23

Magnusson, N. 23
Makarychev, G. I. 185, 199, 200, 203, 204, 222, 224,
    225, 233, 238, 240
Makhlaev, L. V. 120
Mamaev, N. F. 163
Markov, E. P. 110
Markov, L. G. 110

Markova, N. G. 217
Masaitis, V. L. 95, 102, 117
Maslennikov, V. A. 12
Matukhina, V. G. 110
Matviyevskaya, N. D. 69
Mazarovich, O. A. 212
Melkumov, A. A. 75
Milanovsky, E. E. 62
Milovsky, A. V. 139
Minervin, O. V. 204
Mirchink, M. F. 57, 58
Mokrushin, I. M. 73
Moralev, V. M. 83
Muratov, M. V. 1, 14, 22, 23, 43, 87, 199, 200, 243
Mushketov, D. I. 232, 243

Nalivkin, D. V. 133, 164
Nalivkin, V. D. 43, 147
Naumov, V. A. 97
Neiman-Permyakova, O. F. 133
Nevolin, N. V. 35, 36
Nikolaev, V. A. 175, 182, 184, 199, 200
Novikova, A. S. 16, 20

Offman, P. E. 95
Ogarinov, I. S. 137, 145
Ognev, V. N. 200, 215
Okhotnikov, V. N. 154
Orlov, I. V. 193
Ostrovsky, M. I. 52, 54

Patalakha, E. I. 234
Pavlova, T. G. 96
Pavlovsky, E. V. 83, 94, 96, 102
Pavlovsky, V. I. 34
Pazilova, V. I. 200, 203
Peive, A. V. 135, 200, 219
Perfil'ev, A. S. 133, 135, 151, 152, 154, 155, 157, 162,
    164
Plyusnin, K. P. 148, 170
Podsosova, L. L. 155
Pogrebitsky, Yu. E. 101, 106, 112, 119–122, 124,
    131, 159
Poliakova, G. A. 106, 108, 109
Polkanov, A. A. 10, 15, 33
Popov, V. I. 199
Porshnyakov, G. S. 222–225, 227, 230, 240, 244
Postel'nikov, E. S. 96
Postnikova, I. E. 51–54, 66
Pritula, Yu. A. 95
Pronin, A. A. 133, 163, 165, 170
Puchkov, V. N. 145, 151, 153, 165
Pyatkov, K. K. 188

Ragozina, A. L. 16
Ramsay, W. G. 10, 16

Ravich, M.G. 121, 122
Rezanov, I.A. 103
Rezvoy, D.P. 221, 222, 225–227, 237, 243
Romanovich, B.S. 159
Rose, T.A. 35
Rozen, O.M. 178
Rusakov, M.P. 193
Ruzhentsev, S.V. 135, 137, 144, 166

Sabdyushev, Sh.Sh. 229, 230, 233
Saidalieva, M.S. 235
Salop, L.I. 21–23, 37, 86, 102
Samygin, S.G. 196, 214
Sapir, M.Kh. 125, 129
Savinsky, K.A. 80, 92, 94
Schultz, S.S.Jr. 222, 224, 229, 238
Sederholm, J. 10
Seidalin, O.A. 205, 207, 209
Semenenko, N.P. 8, 33, 35
Senchenko, G.S. 137, 144, 145
Shablinskaya, N.V. 131
Shatsky, N.S. 8, 35, 41, 70, 71, 160, 175, 200, 217
Shcherba, I.G. 213
Shlygin, A.E. 178
Shlygin, E.D. 175, 178
Shtreis, N.A. 135, 166, 199, 200
Shurkin, K.A. 49
Sidorenko, A.V. 15
Siedlecka, A. 68
Simonen, A. 21
Sinitsyn, N.M. 200, 222, 227, 236
Sinitsyn, V.M. 200, 219
Smirnov, G.A. 71, 135, 165–168
Sobolev, D.N. 55
Sobolev, I.D. 133, 138, 150, 151, 170
Sokolov, V.A. 15, 17, 20
Sokolov, V.L. 44, 46, 48
Sollogub, V.B. 27, 33, 43, 44
Spizharsky, T.N. 92, 94, 116
Staub, R. 243
Stavtsev, A.L. 43, 82, 83, 87
Stenar', M.M. 15
Stille, H. 8, 70, 160
Strakhov, N.M. 151
Strel'nikov, S.I. 154
Suess, E. 81, 100
Sudovikov, N.G. 10, 86
Surkov, V.S. 131
Suveizdis, P.I. 38
Suvorov, A.I. 195, 199, 215
Svetov, A.P. 20
Sytdykov, B.B. 236

Tal'-Virsky, B.B. 234

Tarbaev, B.I. 73
Tetyaev, M.M. 55, 200
Tikhomirov, V.G. 181, 210
Tikhomirov, V.V. 35, 65
Tret'yakov, V.G. 196
Trunov, V.P. 103
Tsiryul'nikova, M.Ya. 10
Tszyu, G.I. 69
Tugarinov, A.I. 12, 35
Tugolesov, D.A. 94, 95

Urvantsev, N.N. 119, 122, 124
Usmanov, R.R. 229, 230, 233
Usov, M.A. 211
Uspenskaya, N.Yu. 43

Vakar, V.A. 120
Vakhromeev, I.S. 144
Vardanyants, L.A. 35
Varsanof'eva, V.A. 133
Vasil'kovsky, N.P. 199
Velin, E. 22, 23
Veselovskaya, M.M. 35
Vialova, R.I. 133
Vinogradov, A.P. 35
Vinogradov, P.D. 222
Voinovsky-Kriger, K.G. 153
Volobuyev, M.I. 101, 106, 119, 122, 123
Vorobyev, I.V. 101, 106
Vyalov, O.S. 133, 243

Watson, J. 22

Yagovkin, A.V. 223, 226, 227, 244
Yanshin, A.L. 133, 169

Zabiyaka, A.I. 120
Zaitsev, Yu.A. 175, 176, 178, 182,
    183, 185, 190, 200, 202, 211, 212, 214
Zalalayev, R.Sh. 122
Zamaletdinov, T.S. 226
Zapol'nov, A.K. 37, 70
Zavaritsky, A.N. 133
Zhdanov, V.V. 15
Zhero, O.G. 131
Zhivkovich, A.E. 170
Zhuravlev, V.S. 43, 63, 69–71, 75, 157, 163
Znosko, J. 37
Zonenshain, L.P. 206
Zubtsov, E.I. 223, 227, 229, 238, 244
Zunnunov, F.Kh. 234
Zvontsov, V.S. 195, 206, 207

Subject Index

Abraly Synclinorium 191
Adamovka–Mugodzhary Anticlinorium 146
Afghan–Tadzhik Basin, Depression 234, 242
African (African–Arabian) Craton 1
Akbastau Anticlinorium 191
Akchetau Anticlinorium 206
Aksai Zone 232
Aktau–Mointy Uplift 190
Aktyuz series 200, 201
Akzhal–Aksoran Synclinorium, Trough 121, 194
Alai continental Block 238–240
Alai granite, syenite complex 241
Alai–Kokshaal Zone 223
Alai Trough 242
Alakol' Depression 198, 216
Alakol' wrench Fault 196, 198
Alapaevsk–Techensk Synclinorium 150
Aldan Anteclise, Shield 79–82, 86, 87, 89, 91–94, 97
Aldan Block, Megablock 82, 88, 89
Aldan complex 83, 86, 89
Aldan Shield 14
Aldan–Timpton (Yengra) series 82, 83, 85, 86
Aldan–Uchur series 82, 86
Aleksandrovsk Synclinorium 146
Alma Ata Depression 198, 216
Alma Ata Granite Pluton 208
Alpine Foredeep 67
Altynsu Nappe 193
Anabar Anteclise, Massif, Salient 79, 81, 89, 91, 92, 94, 99
Anabar–Khatanga Saddle 125
Anabar Massif 124
Anabar–Mirny Fold Range 91
Anabar–Olenek Anteclise 94
Anabaro–Baikalian Litosphere Plate 91
Angara Block, Megablock 92, 93
Angara–Lena Anteclise 94
Angara–Lena Pericratonal Downwarp, Syneclise, Trough 79, 94, 96, 97
Angara–Lena Zone of dislocation 97
Angara Overthrust 98
Angara series 96
Angara Syneclise 100
Angara–Vilyui Trough 100, 101
Aralbai series 179
Arctic Belt 1
Arkhangel'sk Swell 39
Asha series 162
Ashchisu Synclinorium 190, 191

Astrakhan Swell 44
Atasu–Mointy Block, Uplift 192, 194, 197, 202, 203, 209
Atasu–Uspensk Synclinorium 194, 212
Atbashy Block 232
Atbashy–Inyl'chek Fault 232
Atbashy Zone 232
Auminza–Nuratau Zone 229
Auminza series 187
Ayat (Ayat–Irgiz) Synclinorium 134, 146
Azov Block, Massif 27, 32
Azov–Volyn tectono-magmatic epoch 32, 33

Baidaratsky (Kara) Trough 158
Baidauletov Fault 194
Baikal–Vilyui Suture 91
Baikalian marginal Suture 85
Baikalian Megablock 92
Baikonur Graben 216
Baikonur Synclinorium 179, 184
Balkhash–Dzhungarian Massif 190, 195, 197, 203, 206, 209, 210, 212, 218, 219
Balkhash–Ilian volcanic belt 197, 213, 214
Baltic–Moscovian Syneclise 53–55, 64, 65
Baltic-Podolian Lineament 40
Baltic Shield 9, 10, 26, 27, 32, 34, 35, 37–40, 48–50, 52–55, 59, 60, 63–65, 81
Baltic Syneclise 38, 58, 60, 62, 65, 69
Baltic–Ukrainian uplift Zone 38
Bardym Nappe 147, 148
Barents Sea Massif, Barentsia 5
Barentsian Plate 70
Bashkirian Anticlinorium 70, 134, 136, 137, 143, 145, 146, 160, 161, 163, 166, 167, 169
Batomga series, Zone 85, 88
Bayanaul Synclinorium 190
Bekturgan series 200, 201
Belaya Depression, Foredeep 134, 135, 141, 147, 149, 172
Belomorian Aulacogen 52
Belomorides 25
Belorussian Massif 63
Belorussian, Mazurian–Belorussian Anteclise 38, 39, 42, 55, 58, 60
Belorussian–Voronezh Anteclise, Uplift 64, 65
Bel'skaya series 96
Bel'tau–Karaganda Fault 219
Bend Anteclise 130
Berdyaush Pluton 50, 161

Berezov Trough 79, 97
Bergaul series 17
Besedin complex 34
Bessaz series 200
Bilimbaevo Saddle 148
Bilyakchan Trough 88
Biryusa Block 89, 93
Black Sea Basin, Depression 5
Black Sea Pericratonal Downwarp, Trough 27, 34, 60–62, 64
Bodaybo litosphere Plate 91
Bohus–Arendal granite complex 26, 51
Bol'shekinel' Arch 41
Bol'shezemelskaya Tundra Anteclise, Massif, Swell, Uplift 70, 73–75
Bol'shezemel'sky (Khoreiver) Massif 160
Borotala Synclinorium 196
Borovskoye Pluton 178
Bothnian–Ladoga Lineament 21
Botuoba Fold Range 91
Bragin Salient 42
Brest Trough 40
Bug complex 49
Bug–Podolian Block 32, 33, 49
Bukantau–Karachatyr Trough 241
Bukhara Terrace 230
Bulay series 96
Bureya Massif 4, 5
Burzyan series 143, 161
Buzuluk Trough 45, 59
Bystra series 74

Canadian Shield 81
Carpathian Geosyncline 60, 63
Carpathian Uplift 62
Caucasian Geosyncline 60, 63–65
Central Aldan Block 83
Central Asian Belt 1
Central European Basin 59
Central European Geosyncline 38, 64
Central Dzhungarian Anticlinorium 195, 196
Central (Ferghana) Graben 235, 236
Central Kazakhstan Fault 191, 195, 197, 215, 217
Central Kazakhstan Fold Region 174–176, 218
Central Kazakhstan Marginal Volcanic belt 189, 192, 210, 218
Central Kazakhstan Median Mass 164
Central Kola Block 10, 13, 14
Central Russian, Middle Russian Aulacogen 39, 52, 53, 63, 64
Central Siberian Anteclise (Nepa–Botuoba Uplift) 79, 94–96, 98, 99
Central Siberian Platform 81
Central Taimyr Zone, Trough 118, 124, 128, 130
Central Terskei Uplift 183
Central Turgai (Tyumen'–Livanov) Fault 141, 174

Central Uralian Anticlinorium, Uplift, Zone 134, 137, 145, 148, 161–165, 167, 169
Chara–Aldanian litosphere Plate 91
Chara Block, series 82, 83, 85, 86, 91, 93
Chara–Olekma Block, Zone 86, 88, 89
Chara Zone 174
Chatkal–Kurama Zone 223
Chatkal–Naryn Zone 183, 223
Chatkal Subzone 223
Chelkar Dome 44
Chelyabinsk–Suunduk Anticlinorium 146
Chelyuskin Median Massif 121, 122
Chernov Ridge 67, 134, 153, 165
Chernyshev Range, Ridge 67, 68, 73, 134, 152, 158, 170
Chetlassky Kamen Uplift 68
Chilik–Kemin Trough 183, 204
Chingiz Anticlinorium 174, 190, 205, 206, 210
Chingiz, Kalba–Chingiz Fault 174, 191, 196, 214, 215, 220
Chingiz–Tarbagatai Geosynclinal Fold System 174, 175, 191, 192, 197, 204, 206, 208, 209, 212–214, 217, 219, 220
Chu–Balkhash Anticlinorium 190, 211
Chu Basin, Depression 182, 184, 198
Chu (E Betpakdala) Block, Swell 180, 182, 202
Chu–Sarysu Depression 180, 216
Chudnov–Berdichev granites 33
Chuili–Ketmen Zone 195
Chuili Zone 209
Chukotka Massif 4
Chulaksai suite 163
Chul'man Depression 88
Chuja Anticlinorium, Uplift 85
Cis-Carpathian Foredeep Trough 4, 8
Cis-Kopetdagh Trough 5
Cis-Uralian Pericratonal Downwarp 53, 55
Coppermine River Traps 63
Cumberland Nappe 143

Dala Porphyries 23, 50
Dala Sandstones 23
Dala series 26
Dalslandian Tectonomagmatic Epoch 26, 51
Degtyarka–Ufalei Wrench Fault 149
Demurin Dome 29
Denisov Trough 68, 73, 76
Denmark–Poland Suture Zone 38
Dnieper–Donets Aulacogen, Depression 29, 40, 42–44
Dnieper granitoid complex 29
Dniester Pericratonal Downwarp, Trough 27, 34, 51, 62, 64
Dniester Trough 34
Dobrogean Geosyncline 64, 65
Don–Medveditsa Dislocations 40, 60

Donets Basin 43, 57, 58, 130
Donets–Caspian Fold System, Geosyncline 43, 44, 64
Donets–Caspian Fold System 244
Donets–Caspian Fold Zone 8, 55
Donets–Caspian Zone 170
Donets Fold System 40
Donets Fold System, Ridge 8, 27
Dzhangdzhir Zone 232
Dzharkainagach Anticlinorium 180, 184
Dzharkent Basin, Trough 198, 216
Dzhetysu series 202
Dzhezkazgan Basin, Depression 179–181, 214, 218
Dzhida Zone 3
Dzhugdzhur Pluton 87
Dzhungarian Basin 192
Dzhungarian Gates Graben 196
Dzhungarian Wrench Fault 196, 198, 215
Dzhungaro–Balkhash Geosyncline, Fold System 3, 174, 175, 191, 192, 195, 196, 203, 209, 213, 214, 217–220

Eastern Alai Zone 222, 226
Eastern Betpakdala Massif 178
Eastern European Craton, Platform 1, 6, 8, 27, 36, 37, 39, 42, 49, 50, 62, 64, 65, 67, 69, 73, 77, 81, 91, 96, 130, 132, 135, 147, 160, 162–164, 170, 172, 200, 238
Eastern Ferghana Fault 242
Eastern Ferghana Sigmoid 232, 241
Eastern Ferghana, Yarkend–Ferghana Graben 198, 216, 241
Eastern Finland Block 20
Eastern Ghats Belt 82
Eastern Russian Depression 64
Eastern Taimyr Depression 125
Eastern Ulutau Fault 179, 181
Eastern Uralian (Uralo–Tobol) Anticlinorium, Uplift 139, 140, 146, 150, 166–169, 171, 172, 231
Eastern Uralian Synclinorium 139, 140, 166–168, 170, 231
Ebeta Anticlinorium 134
Efimov series 178, 190
Eurasian (Central Eurasian) Platform 175, 198

Fennoscandia 9, 10
Ferghana Basin, Depression, Trough 5, 174, 221, 222, 224, 225, 233–237, 242, 243
Ferghana–Kokshaal Segment 221, 241
Fore-Baikalian Trough 80, 94, 98, 100
Fore-Patom Trough 96
Fore-Timan Trough 69, 74
Frunze (Southern Chu) Basin, Foredeep 182, 198, 216

Galician Geosynclinal System 8, 51, 63

Gamburtsev–Chernov Ridge 68, 73
Gamburtsev Ridge 67
Gimola series 17, 21, 49
Gissar–Alai Segment, Uplift 221, 224–228, 231, 232, 234, 241, 244
Gissar–Eastern Alai Zone 223
Gissar Eugeosynclinal, Oceanic basin 238
Gissar–Karakum Zone 223
Gissar (Southern Gissar) Volcano-Plutonic Belt 222, 228, 230, 233
Gondwana Supercraton 1
Gothian Cycle, Epoch of Diastrophism 22, 23, 26, 35, 50
Greater Karatau Trough 211
Greater (Bol'shoi) Naryn series 185, 203
Gremyakha–Vyrmes Pluton 14
Grenville Belt 62, 82
Guben Massif 161

Hindustan (Indian) Craton 81
Hoglandian series 25
Hyperborean Craton 1

Ichkeletau–Susamyr Fault 184
Ili Basin, Depression 195, 197, 198, 216
Imandra–Varzuga Trough, series 13, 14, 23, 25
Indol–Kuban Trough 4
Ingulets Arch 31
Ingulets granites 30
Innuitian Geosynclinal System 131
Inyl'chek Zone 232
Irendyk Zone 145
Irkineevo Suture 91
Irkutsk Syneclise 100
Ishim–Naryn Fold System 175
Ishim–Talas (Baikonur) Zone 175, 184, 185, 203, 208
Ishkeol'mes Anticlinorium 189, 190, 206
Issykkul' Basin, Depression, Trough 5, 198, 199, 216
Issykkul' (Kungei-Terskei) Block 182, 183, 204
Izhma–Omra complex 72, 74
Izhma–Pechora Basin, Depression 68–70, 72

Jotnian Sandstones 21, 26, 33

Kalevian 21
Kalmykkul' Synclinorium 179, 180, 184
Kaltasa Aulacogen 52
Kama Pericratonal Downwarp, Depression 51, 53
Kamennoborsk (Petrozavodsk) suite 20
Kandalaksha Graben 39
Kanin–Timan Ridge 67, 69, 70, 72, 74, 75, 77
Kara Massif, Swell 118, 120–124, 127, 128
Kara (Baidaratsky) Trough 158
Kara–Novaya Zemlya Uplift 130

Karabogaz Median Mass 244
Karabulak series 187
Karachatyr Zone (Foredeep) 222, 224, 230, 241
Karadzhilga–Kyzyltash Pluton 202
Karaganda Basin, Synclinorium 193, 198, 212–214, 216
Karasor Trough 210
Karataikha Basin, Foredeep, Trough 68, 153, 158
Karatau, Leont'ev Graben 198, 216
Karatau Nappe, Saddle, Salient 134, 135, 142, 143
Karatau–Naryn (Karatau–Chatkal) Zone 175, 183, 185–187, 212–214
Karatau–Naryn Zone, Subzone 223, 232
Karatau series 137, 161, 162
Karatau–Talas Fault („Nikolaev line") 183–185
Karatau–Talas Miogeosyncline 219
Karatau–Talas Zone 223
Karelian Epoch of Diastrophism, Tectonomagmatic Epoch 22, 50
Karelian Massif, Megablock 14, 18, 21, 23, 25
Karelides 37
Karmakul' Saddle, Synclinorium 158, 159
Karsakpai Fault 179
Karsakpai Geosyncline 201
Karsakpai series 179, 200–202
Karsakpai Synclinorium 179
Kassan Uplift 187
Katanga (Anteclise) Swell 95
Kazakh Shield 216
Kazakhstan–Tien Shan Geosynclinal Region 199
Kazakhstan–Tien Shan Median Massif 176, 188, 208, 209, 211, 214, 217, 218
Kazan Saddle 41
Kel'pi Zone 233
Kemin series 200
Kempendyai Trough 79, 101
Kempirsai Anticlinorium 146
Kerensk–Chembar Arch 41
Keweenawan traps 63
Keyvy granites 12
Keyvy series 12
Keyvy-tundra Block, Synclinorium 12
Khandyga Fold Range 91
Khanka Massif 4
Kharayelakh Depression 99
Kharbei Anticlinorium 72, 134, 154, 155
Khatanga Depression 100
Khibiny Pluton 14, 60
Khoreiver Depression, Trough 68, 70, 73, 76
Kichikalai Zone 222
Kil'din Trough 69
Kirensk Fold Range 91
Kirghiz–Terskei Zone 201, 202, 223
Kirghizata Nappe 226
Kirghizian Continental Block 238–240
Kirghizian series 200, 202

Kirov–Kazhim (Kazan–Sergiev) Aulacogen 41, 52, 58
Kirovograd Block 31, 32, 33
Kirovograd-Zhitomir granite complex 31
Kochechum Depression 99
Kodar–Kemen, Kodar–Kamenka Lopolith 87
Kodar–Udokan Block, Trough 87–89
Kokchetav Block, Massif 161, 178–180, 189, 190, 200, 202, 206
Kokchetav–Muyunkum Massif 1, 3, 175, 178, 181, 182, 184, 185, 202, 203, 206–208, 218, 219
Kokchetav series 189, 202
Kokdzhot–Uzunakhmat Subzone 223
Kokshaal Zone 232
Koksu series 203
Kola Megablock 10, 13, 14, 23, 25, 27, 29
Kola series 10, 12, 17, 49
Kolva Anticlinal Zone, Arch 68, 70, 73–77
Kolva (Middle Pechora) Saddle 135, 147
Kolyma Massif 4
Komi-Permian Ridge, Uplift 41
Konka–Verkhovtsevo series 29, 32, 34, 49
Konka–Yalyn Depression 27
Korosten Pluton 33
Korsun–Novomirgorod Pluton 32
Kos'va–Chusovaya Saddle 147
Kotel'nich (Kotel'nich–Sysol) Uplift 41
Kotuy Trough 78
Kozhim Uplift 134, 152, 153
Kraka Nappe 143–145, 148
Kremenchug–Krivoy Rog Synclinorium, Trough, Zone 31, 32, 34
Krestets (Krestsy) Aulacogen 52, 53
Krivoy Rog Basin 30
Krivoy Rog Fault 31
Krivoy Rog series 29, 31, 34
Krykkuduk granite complex 189, 208
Ksenofontovo Saddle 135
Kuchar Trough 242
Kumpu Formation 21
Kura Trough 5
Kuraili (Ayaguz) Depression 198
Kurama, Bel'tau–Kurama Volcanic Belt 188, 214, 218, 220, 231
Kurama Subzone, Uplift 223, 234
Kureika Depression 99
Kursk series 34, 35
Kustanai Saddle 133
Kustanai Synclinorium 188
Kustanai, Tyumen'–Kustanai Synclinorium, Zone 141, 146, 169
Kyutungdy Graben, Trough 79, 95
Kyzyl Synclinorium, Zone 145, 146
Kyzylkum Segment 221, 228–231, 241

Laborov Swell, Uplift 155

Ladoga Aulacogen, series 18, 50, 52
Late Kareliam granites 19
Latvian Saddle 38, 40, 53
Lemva Synclinorium, Zone 134, 137, 145, 152–154, 165, 167
Lena-Anabar Trough 125, 126
Lenivaya series 121, 127
Linda Suture 91
Linn granite 23
Loknov Promontory, Uplift 40, 53, 54
Longdor (Anticlinorium) Uplift 85
Lopian Salient 73
Lopian series 17, 49
Lovozero Pluton 14, 60
Lower Amga Fold Range 91
Lower Angarian litosphere Plate 91
Lower Kotuy Fold Range 91

Magnitogorsk Nappe, Segment, Synclinorium, Zone 134, 138, 143, 146, 166–168, 170, 172, 173, 231
Maidantag Zone 232
Maikain Anticlinorium 190, 205, 219
Maikop series 62
Maikube Basin 198
Maili–Dzhair Anticlinorium 196
Maimakan–Batomga Block 82, 85, 86
Maimecha–Kotuy magmatic complex 129
Main Uralian Deep Fault 151, 172, 173
Maityube series 179, 202
Major Gissar Fault 221, 230
Major Karatau Fault 186, 187, 198
Major Novaya Zemlya Fault 159
Makbal Horst 180, 182, 200, 201
Makbal series 200, 201
Maksyutov metamorphic complex 145
Malozemel'sky (Kolguev) Massif 160
Malun–Idre Trough 23
Mama–Bodaibo Synclinorium, Trough 85
Mama–Vitim Fault 85
Mara Depression 99
Mar'evka Synclinorium 184
Mashak suite 161
Mediterranean Geosynclinal Belt 1, 4, 5, 55, 60, 66, 244
Mezen' Syneclise 38, 39, 55, 58, 60
Mid-Siberian Anteclise 94
Mid-Siberian Platform 94
Middle Dnieper (Zaporozhye) Block 28, 31, 32
Middle Uralian (Kvarkusha–Kamennogorsk, Vishera–Chusovaya) Anticlinorium 139, 147–149
Middle Uralian (Sylva–Solikamsk) Foredeep 135, 142, 147, 149, 171
Mikashevichi Salient 42
Mikhailov series 34
Minyar series 70

Mirny Swell 96
Moldavian Platform 5
Mongolo–Okhotsk Geosynclinal Fold System 89
Mongolo–Okhotsk segment 3
Moscow Graben 39, 53
Moscow (Central Russian) Syneclise 38–40, 55, 58, 60, 61, 65
Moty suite 96
Mozambique Belt 82
Muna Swell 79, 95
Murman Block 12, 13, 29, 49
Muya Zone 85
Muyunkum (Transilian) Block, Massif 182, 190, 204

Nain Province 62
Naryan–Mar Swell 68
Naryn Basin, Depression, Trough 5, 198, 199, 216
Near Tobol ultrabasic belt 140
Nemaha–Boothia axis 38, 81
Neman Fault 38
Nepa–Botuoba Anteclise 94
Nepa Swell 95
Nesmurin series 83
Nikolaev line 232
Nikol'sko–Burluk series 178
Ninety East Ridge 81
Niyaz Anticlinorium 190
Nizhneserginsk Nappe 147
Noril'sk Depression 99
North American Craton, Platform 38, 62, 63, 81, 96
North American Platform 130
North Asian Craton 77, 84
North Baikalian (Akitkan) Volcanic Belt 85
North Caspian Depression, Syneclise 36, 39–44, 47, 48, 59–62, 64, 132, 141
North Patom Depression 97
Northern Balkhash (Kenterlau) Anticlinorium 195, 206, 209
Northern Balkhash Basin 197
Northern Bukantau Trough 230
Northern Dvina Massif 47
Northern Dzhungarian Synclinorium 196, 197, 209
Northern Ferghana Fault 234
Northern Gissar Anticlinorium 227
Northern Gissar (Major Gissar) Fault 222
Northern Karelian Trough 18, 23
Northern Kirghizian Zone 182, 183, 204, 208
Northern Ladoga (Savo–Ladoga) Trough, Zone 18, 21, 23, 25
Northern Onega Block 20
Northern Severnaya Zemlya Zone 118, 123, 123
Northern Taimyr Crystalline Massif 123

Northern Uralian (Lyapin–Isov) Anticlinorium 139, 149, 151, 153, 154
Northern Uralian (Upper Pechora) Foredeep 134, 135, 150
Northern Ust'yurt Massif 231, 244
Novaya Zemlya (Northern Novaya Zemlya) Anticlinorium 158, 159
Novouzensk Graben, Trough 45, 48, 51
Nura Synclinorium 193, 194, 209
Nuyia Downwarp 79, 97
Nyazepetrovsk Nappe 147, 148

Oboyan series 34, 49
Ob'–Zaisan (Irtysh–Zaisan) Fold System 3, 131, 174, 191, 192, 219
Odessa–Kanev Fault, Zone 32, 33
Okhotsk–Chukotka Volcano-Plutonic Belt 4, 192
Okhotsk Fault 85
Okhotsk Massif 4
Oktyabr'sk–Denisov Anticlinorium 146
Olekma Block, series 82, 83, 85, 86, 91
Olekma Pluton 87
Olenek Anteclise, Salient, Uplift 79, 82, 91–95
Olenekian litosphere Plate 91
Olsztyn Fault 40
Omolon Massif 4
Onega series 19, 50
Onot Graben 89, 93
Orekhovo–Pavlograd Fault 42
Orekhov–Pavlograd Zone 32, 34
Orsha Aulacogen 52
Orsha Trough 40
Ortatau series 202
Oslo Graben 27, 60, 64
Ovruch series, Syncline 33, 34, 43, 50

Pachelma Aulacogen 40, 41, 47, 51–54, 64
Pacific Belt 120
Pacific Mobile Belt 81, 89
Pai Khoi Anticlinorium, Ridge 134, 153, 158
Pai Khoi–Novaya Zemlya Fold System 132, 155, 157, 162, 170
Pamir–Karakorum Fold System 242
Pamir–Kuen Lun Fold System 244
Parana Syneclise 98
Parandovo series 17, 49
Patom Re-entrant, Salient 83
Pavlovsk complex 34
Pebozero series 17
Pechenga series, Trough 13, 14, 25
Pechora Basin, Syneclise 67, 69, 70, 72, 77
Pechora Fault 73
Pechora, Pechora–Kozhva Anticlinal Zone, Arch, Ridge 68, 73–77, 134
Peri-Sakmara Zone 145

Peri-Timan Trough (Timan Pericratonal Downwarp) 41, 53
Perm–Bashkirian Arch 147
Perm Swell 41
Pinsk Saddle 40
Pish Depression, Graben 21
Pistali–Bosbutau Subzone 223
Platinonosny Ultrabasic Belt 149, 150
Polar Uralian Anticlinorium 138, 154
Polar Uralian (Vorkuta, Kos'yu–Rogov) Foredeep 135, 152
Polessie Bridge, Saddle 38, 55
Polish–Lithuanian Syneclise 60
Polmos–Poros series 13
Poltava–Bredy Syncline 168
Polyudov Range (Ridge), Uplift 134, 149, 165
Popigai Astrobleme, "Graben" 95
Pripyat Aulacogen, Trough 34, 40, 42, 43
Pripyat–Dnieper–Donets, Pripyat–Donets Aulacogen, Trough system 27, 42, 43, 58–60, 69
Pskem–Sarydzhaz Subzone 223
Pyatikhatka Arch 29

Radaevo Aulacogen 52
Rebolian Epoch of Diastrophism 17
Revda Nappe 148
Riga Fault 38
Riga Pluton 37
Rioni Trough 5
Rostov Swell 27
Russian Platform 9, 21, 22, 27, 35, 37–41, 44, 47, 53, 54, 60, 62–65, 67, 71, 75, 94, 135, 136, 161–163, 166, 168, 169, 171, 173
Ryabinovka granite massif 161
Ryazan–Saratov (Pachelma) Aulacogen, Trough 40, 41

Sakmara Nappe, Zone 137, 144, 145, 148, 163, 166
Sakmara–Voznesensk Subzone 144, 145
Saksagan Dome 29
Saksagan plagiogranites 30
Salma suite 21
Sal'sk Fault Zone 27
Samarskaya Luka Flexure 41
Samgar Fault 235
Sandalash Subzone 223
Sankebai Dome 44
Sariolian series 17, 18, 25
Sarmat–Turanian (Donets–Mangyshlak–Gissar) Fault 243
Sarmat–Turanian (Central Eurasiatic) Lineament 222
Sarmatian Shield 38, 40, 51, 52, 54
Sarydzhaz Block, Massif, Salient 185, 221, 232
Sarysu Domes 181

Sarysu–Teniz block-fold zone 180–182, 202, 210–212, 214, 219
Sayak–Dzhungarian Trough 213
Sayan–Anabar Suture 91
Sayan Foredeep, Fore-Sayan Trough 80, 94, 100
Sayan Megablock 92
Scandinavian Caledonides 63
Scandinavian (Grampian) Geosyncline 9, 64, 65
Scandinavian–Severnaya Zemlya Fold System 160
Scythian Platform 5
Segozero series 19
Seletsk Epoch of Diastrophism 19, 25, 32
Selety Synclinorium 190, 210
Seljur series 26
Sernovodsk–Abdulino Aulacogen 41, 51
Serov–Mauk Ultrabasic Belt 150
Serov–Nev'yansk Gabbro-Ultrabasic Belt 138
Sharyzhalgai Salient 89, 91–93
Shchuch'ya Synclinorium 134, 155
Shiderty Synclinorium 190, 210
Shoksha suite 20
Siberian, Central Siberian Craton, Platform 1, 14, 63, 79–82, 89–91, 93, 94, 120, 121, 125, 126, 128–130
Siberian–Turanian Platform 216
Sikhota–Alin' System 4
Sin Fold Range 91
Sinian (Sino–Korean) Craton, Platform 1, 4, 200, 222, 233, 244
Smoland granite 23
Sob' Uplift 134, 152, 154
Sok–Sheshma Arch 41
Soligalich Graben, Trough 39
Sonkul' Depression 183
Sorokin (Varandei) Arch, Range, Ridge 67, 73, 74, 76, 158
Sorsele granite 23
South Barents Depression, Syneclise 67
Southern Balkhash Depression 198, 216
Southern Bukantau Zone 229
Southern Dzhungarian Anticlinorium 195, 196
Southern Dzhungarian Fault 197
Southern Ferghana Fault, Suture 174, 221, 222, 224, 238
Southern Ferghana–Dzhangdzhir Zone 223
Southern Ferghana Ophiolite Belt, Zone 224, 225
Southern Ferghana Trough 242
Southern Gissar Ophiolite Belt 224, 240, 244
Southern Karaganda Zone 194, 218, 219
Southern Karelian Depression 19
Southern Kokchetav Trough 204
Southern Onega Block 20
Southern Scandinavian Massif, Megablock 23, 26, 51
Southern Taimyr Trough, Zone 118, 124, 125, 128, 130
Southern Tien Shan Geosyncline 222

Southern Ulutau Trough 207
Southern Uralian (Belaya) Foredeep 134, 135, 141, 142, 147
Sparagmite 26
Spassk Fault, Nappe, Zone 193, 194, 213, 215
Stanovoi Belt Megablock 80, 82, 86, 89, 93
Stanovoi complex 86
Stanovoi Fault Zone 82, 85, 86, 88, 89
Stanovoi Range litosphere Plate 91
Stepnyak (Stepnyak–Betpakdala) Synclinorium 189, 208
Stoilov–Usman complex 34
Sub-Jotnian 23, 33, 50
Suisarian 19, 25, 50
Sukhan Trough 79, 95
Sukhon Arch 39
Suluterek Block, Salient 223, 228, 232
Sumian complex, series 14, 17, 18, 25
Suntar Swell, Uplift 79, 92, 101
Surmetash–Gulcha Depression 241
Sushchany–Perzhansk Fault 33
Sutam series 83
Suvanyak metamorphic complex 145
Svecofennian Megablock 11, 21
Syr Darya Depression 185

Tadzhik continental Block 238, 240
Tadzhik, Karakum–Tadzhik Median Mass 221, 228, 241, 244
Tadzhik Trough 5
Tagil–Magnitogorsk Synclinorium, Zone 137–139, 161, 164, 165, 171, 231
Tagil Nappe, Synclinorium, Zone 134, 138, 147, 149, 150, 151, 153, 154, 166–169, 172, 173
Taigonos Massif 4
Taimyr Arch 125
Taimyr Fold System 129
Taimyr Geosyncline, Trough 120, 121, 127
Taimyr–Severnaya Zemlya Fold Region, Platform Welt 118, 129, 130
Taimyr–Tunguska Protogeosynclinal System 93
Takata sandstones 136, 161
Talas–Ferghana Wrench Fault 187, 215, 216, 221, 222, 231–234, 236, 241
Tanalyk Zone 145
Taratash Swell 143, 160
Tarbagatai Anticlinorium 191
Tareya Arch 129
Tarim Median Mass 200, 221–223, 228, 232, 233, 238, 244
Tarkanda Fault 85
Taseev Megablock 93
Taseevo Syneclise 100
Tashkent–Golodnosteppian Trough 234

Tastau series 213
Tegermach Nappe 226
Tekeli Zone 195
Tekes Depression 199, 216
Tekturmas Anticlinorium 192–194, 205, 206
Tekturmas Fault 215
Teniz Basin 181, 213, 214
Teniz–Korzhunkul Syncline 210
Terek–Caspian Trough 4
Terekdavan Synform 226
Tereksai series 200
Teria Trough 95
Terskei granite pluton 208
Terskei series 183, 200
Teterev series 32
Teysseyre–Tornquist line 8, 10
Tikshezero series 17, 19
Timan–Pechora Platform 67, 69, 72, 74–77, 132, 135, 152, 161, 162, 165
Timan Range, Ridge, Uplift 67, 77, 134
Timan–Uralian Geosyncline 74
Timanide Geosynclinal System 8, 51
Timanides 70, 73
Timpton–Dzheltula series 83, 85, 86
Tisvaiz formation 153
Tokmov Swell 41
Tokrau Basin 194, 197, 210
Tokrau Batholith 213
Tonod Anticlinorium, Uplift 85
Transcontinental Arch 96
Transilian (Muyunkum) Block 182, 190, 204
Transuralian Anticlinorium, Uplift, Zone 140, 146, 160, 163, 166, 168–170, 172
Troitsk–Kengusai Anticlinorium 146
Trosnyansk–Mamonovo complex 35
Tuimaza Arch 41
Tukuringra Fault 85
Tundra series 13
Tung Fold Range 91
Tungudsk–Nadvoitsa series 19
Tungusian litosphere Plate 91
Tunguska Massif, Megablock 92, 93
Tunguska Syneclise 79, 81, 91, 94–96, 98–100, 128, 129
Turanian Platform 5, 171, 198, 222, 234, 236
Turgai Downwarp, Trough 133, 146, 179, 180, 184–186, 198
Turgai–Syr Darya Median Mass 175, 178, 187, 188, 203, 218, 221, 224, 225, 236, 238
Turgai Trough 5
Turkestan–Alai Zone 222, 225–227
Turkestan eugeosynclinal oceanic Basin 238–240
Turkestan Fault 222
Turukhansk–Noril'sk zone of dislocations 78, 94, 99–101, 121

Tyumen'–Livanov (Central Turgai) Fault 141

Ubogan Graben 216
Uchaly Zone 145
Uda–Vitim Zone 3
Udokan series 87, 88
Udzha Swell 78
Udzha Trough 78
Udzha–Zhigansk Suture 91
Ufa Amphitheater 134, 147, 148, 170
Ufalei metamorphic complex 148
Ukhta Flexure 68
Ukrainian Shield 27, 34, 35, 37, 38, 40, 50, 58, 59, 63, 65, 92, 130
Ukrainian Syneclise 44, 60–62, 64
Ukrainian–Voronezh Shield 40, 42, 51, 55
Ulkan Laccolith 88
Ulkan Volcanic Belt, series, Trough 85, 87, 88, 91
Ulsov Syncline 151
Ulutau Block, Massif, Uplift 178–180, 190, 200, 202, 204, 218
Ul'yanovsk–Saratov Trough 41, 45, 60, 61
Umov–Koiva Syncline 149
Upper Karelian complex 19
Upper Muna Trough 95
Upper Pechora Basin 68
Upper Pechora Depression, Foredeep 134, 135, 150
Ural–Novaya Zemlya Geosynclinal Fold System 67
Ural–Tien Shan Wrench Fault 231
Uralian (Cis-Uralian) Foredeep 44, 57, 60, 67, 132, 134, 135, 140, 143, 159, 170, 171
Uralian Geosyncline 38, 51, 58, 60, 64
Uralian Geosynclinal Fold System 3, 63, 132, 133, 157, 161, 168, 173
Uralo–Okhotsk (Uralo–Mongolian) Geosynclinal Belt 1, 5, 79, 89, 174, 200, 206, 217, 218, 220, 243, 244
Uralo–Siberian Belt 132, 164, 182
Uralo–Tobol Anticlinorium 134, 146, 231
Uraltau Anticlinorium 134, 136–138, 144–146, 148
Urin Anticlinorium, Fold Zone 82, 97
Urkashar Synclinorium 197
Urtyndzhal series 194, 205
Urusai Fault 243
Ushakovka series 96
Usol series 96
Uspensk Nappe, Zone 194, 212, 215
Ust–Kut Fold Range 91
Uyan series 88
Uzunkyr Zone 145
Uzunzhal granite 194

Vaigach (Kara, Southern Novaya Zemlya) Anticlinorium 158
Valer'yanov series, Volcanic Belt 141, 168, 231
Varandei Ridge 68, 73

Varegian series 26
Velmo Trough 95
Vepsian 19
Verkhnesob' Formation 153
Verkhoyansk–Chukotka Fold Region 3, 4
Verkhoyansk Fold range 91, 95
Verkhoyansk Foredeep 126, 131
Verkhoyansk Geosyncline 120, 128, 131
Verkhoyansk–Kolyma Fold System 81
Verkhoyansk Re-entrant 82
Vetreny Belt Synclinorium 18, 20, 23
Viatka–Kama Depression 61
Viluyan litosphere Plate 91
Vilyui Syneclise 79, 81, 82, 92, 97, 101
Visla–Dniester Pericratonal Downwarp 53,
    55, 60
Voikar Synclinorium 134, 154
Voikar–Syn'ia gabbro-peridotite complex,
    massif 138, 154
Volga–Kama Shield 51, 52, 54, 58
Volga–Uralian Anteclise 38–41, 58–60, 96, 166
Volyn Block 32, 33
Volyn–Kotlas Zone 51
Volyn–Orsha Aulacogen, Trough 43, 63
Vorkuta (Kos'yu–Rogov) Basin, Depression, Fore-
    deep 68, 134, 135, 152, 153, 169
Voronezh Anteclise 35, 38–41, 48, 58–60
Voronezh Massif 27, 32, 34, 49, 50, 63
Vorontsovo series 35, 50
Voronye–Kolmozero zone 13
Vyatka Arch 41

Western Ingulets Zone 30–32
Western Siberian Megasyneclise, Platform 3, 5, 76,
    119, 125, 129, 131, 132, 171, 174, 178, 198, 208,
    215
Western Taimyr Depression 125
Western Tien Shan Fault 237
Western Turkmenian Trough 5
Western Uralian Zone 136, 137, 146, 147, 150, 153,
    160, 163, 167, 169–171
White Sea (Belomorian) Megablock, series 10, 13,
    14, 18, 23, 25, 36, 47, 49
Wichita Zone 130

Yagnob Graben-Megasyncline 234, 243
Yakutsk Uplift 79
Yangoda–Gorbit Uplift 78
Yassy–Maidantag Zone 223
Yassy Zone 233
Yatulian series 17, 19, 25, 50
Yelets Zone 137, 153, 165
Yengra series 83, 86, 91
Yenisei–Khatanga Trough 78, 80, 100, 119, 121,
    125–131
Yerementau–Chuili Geosynclinal Fold System 180,
    188, 190, 191, 203, 206, 208–210, 213
Yerementau series 205
Ygyatty (Markha) Trough 79, 101
Yurmata series 137, 160–162
Yuryuzano–Sylva Depression 134, 147

Zeravshan Anticlinorium 227
Zeravshan–Eastern Alai Fold Region 223
Zeravshan Fault 222, 227
Zeravshan–Gissar (Zeravshan–Alai) Zone 222, 227,
    228, 230
Zeravshan Graben-Megasyncline 234, 243
Zeravshan–Kalmakasui Zone 223
Zeravshan (Turkestano–Zeravshan) Zone 222, 227,
    230
Zerenda granite pluton 178, 208
Zerenda series 178, 201
Zeya-Bureya Depression 5
Zhalair–Naiman Fault 215, 220
Zhalair–Naiman (Chu–Ili) Fold Zone 180,
    190, 205
Zhaman-Sarysu Uplift 194
Zhaunkar granite 202
Zhigulev Arch 41
Zhigulev–Orenburg, Zhigulev–Pugachev Swell 41,
    59
Zhiida series 179
Zhlobin Saddle 40
Zilair–In Zone 44
Zilair series 137
Zilair Synclinorium, Zone 134, 136, 137, 139, 143,
    145, 165–169
Zymnik suite 163